DESIGN OF AXIALLY LOADED PILES – EUROPEAN PRACTICE

PROCEEDINGS OF THE ERTC3 SEMINAR
BRUSSELS/BELGIUM/17-18 APRIL 1997

Design of Axially Loaded Piles European Practice

Edited by
F. DE COCK
Franki Foundations S.A., Machelen, Belgium
C. LEGRAND
Belgian Building Research Institute, Brussels, Belgium

A.A. BALKEMA/ROTTERDAM/BROOKFIELD/1997

Organized by:
The Belgian Member Society of ISSMFE

The ISSMFE European Regional Technical Committee 3 “Piles”

ERTC3

Under the auspices of International Society for Soil Mechanics and Foundation Engineering (ISSMFE)

Supported by:
Deep Foundations Institute (DFI)
and European Federation of Foundation Contractors (EFFC)

The texts of the various papers in this volume were set individually by typists under the supervision of each of the authors concerned.

Published by
A.A. Balkema, P.O. Box 1675, 3000 BR Rotterdam, Netherlands (Fax: +31.10.4135947)
A.A. Balkema Publishers, Old Post Road, Brookfield, VT 05036-9704, USA (Fax: 802.2763837)

ISBN 90 5410 873 8

Printed in the Netherlands

Design of Axially Loaded Piles – European Practice, De Cock & Legrand (eds)
© 1997 Balkema, Rotterdam, ISBN 90 5410 873 8

Table of contents

National reports on design practice for axially loaded piles

Design of Axially Loaded Piles – European Practice, De Cock & Legrand (eds)
© 1997 Balkema, Rotterdam, ISBN 90 5410 873 8

Preface

End 1994 the European Regional Technical Committee 3 'Piles' started its activities, the Belgian Geotechnical Society being appointed as the host country. F. De Cock (Franki Foundations) gladly volunteered to take up the challenge and the responsibility for presiding the committee, being assured of the secretarial assistance of C. Legrand (Belgian Building Research Institute). From the beginning, not less than 19 European countries were represented by a national expert. Later on, Dr A.A. Bartolomey (Russia) and M. Mets (Estonia) also expressed their interest to contribute in the committee work and thus joined the TC as invited experts.

By its terms of reference, the ERTC3 committed itself to establish by the ISSMFE Hamburg Conference a summarising report on the European design methods for axially loaded piles, not so much treating the large amount of scientific work on this aspect, but focusing on the present-day methods as daily used in practice, being prescribed in more or less strict national codes or recommendations.

Why going back to such a fundamental topic as the bearing resistance of a single pile under axial load? Have most aspects on this item not yet been analysed, discussed and published repeatedly over the last 50 years? Have we not come close enough to predicting the pile behaviour and is most of the available know-how (resulting from scientific work or local experience) not yet implemented in the daily practice? Why pumping energy in analysing and comparing present-day 'national' methods, while the unifying Eurocodes are being born?

The ERTC3 committee also went over these questions internally in fruitful discussions about the goals and procedures of the TC work. During a first working meeting at the European ISSMFE Conference in Copenhagen in 1995, the high scattering (up to 250%) in the predicted bearing resistance of axially loaded piles for both friction piles and end bearing piles by applying the various (standardised) national methods was strongly experienced. During a two-day workshop in Brussels – November 1995, national design regulations, philosophy, rules and formulas were presented by the participating ERTC3 members, analysed and discussed in an open and constructive atmosphere. Encouraged by the estimated added value of this workshop and the participators' enthusiasm it was generally agreed by the ERTC3 members that organising a similar meeting for a larger base of attendants would be of most extreme interest for practicians as well as researchers. That is where the idea of this International Seminar was born.

The first objective of the seminar is to give the European geotechnical engineers the opportunity to familiarise themselves with the various national design methods, methods which will most likely subsist by means of the National Application Documents for Eurocode 7. Therefore not less than 17 European countries have elaborated a detailed report on their national design methods for axially loaded piles. As far as already implemented, the application of these methods within the framework of EC7 is described as well. In order to improve the understanding of the design methods, the national papers also include aspects such as the local geology, soil investigation methods, local piling practice, limitations of the design methods and some practical examples. During the seminar, the national papers are briefly presented, alternated by the keynote lectures of Prof. W.F.Van Impe and W.F.Loftus respectively.

It is the second objective of the seminar and the strong hope of the ERTC3 and the organising committee, that better knowledge and understanding of the various national design approaches will help geotechnical engineers to interpret the results obtained, to enlarge their views on pile behaviour and to encourage to some extent a convergence of the European design rules. Therefore, the national papers are discussed during the seminar in 3 discussion sessions focused on respectively:

– Design methods based on static load tests;

– Design by calculation on the basis of field and laboratory soil tests;

– Implementation of driving formulae, dynamic tests, monitoring and quality control in the design.

Each of the sessions is further documented by a particular contribution of the session's chairmen on the considered topic.

The editors wish to thank:

– The ERTC3 members for their constructive work within the committee and the high quality of the various national papers, elaborated by their initiative;

– The invited speakers, session's chairmen and general reporters for their enthusiastic input which has certainly largely contributed to the scientific and practical interest of the seminar;

– The Belgian member society of the ISSMFE, in particular its mandataries in the organising committee who have volunteered to co-organise the seminar; without their experience, creativity and practical help, the seminar would never have taken place;

– Their respective organisations, Franki Foundations and the Belgian Building Research Institute, for the opportunities and support given to work in the ERTC3;

– The supporting organisations which have contributed to the seminar, by financial support, by promoting the seminar, by encouraging the organisers.

The Editors
F.De Cock
Franki Foundations S.A.

C.Legrand
Belgian Building Research Institute

Design of Axially Loaded Piles – European Practice, De Cock & Legrand (eds)
© 1997 Balkema, Rotterdam, ISBN 90 5410 873 8

Welcome address

United Europe is not longer a fiction, it becomes more and more a reality. In the next millennium the European Community will no longer pay with Gulden, Deutsche Mark or Francs, but with 'Euro'. More countries will join the Community and I hope that all countries of geographical Europe will become the U.S.E. (United States of Europe) in the next decades.

Not only political, economical and social aspects will melt together, but also technical approaches will be standardized in the next years. But there is no future without a past..., and that is the reason why the ERTC3 proposed to organize a seminar on '*Design of axially loaded piles: European Practice*'.

The Belgian Society for Soil Mechanics and Foundation Engineering is very proud to have the honour to host this important seminar in Brussels, the heart of Europe. We attach a great significance to the fact that this seminar combines all aspects of European practice in axially loaded pile design: design based on field and laboratory tests as well as design based on static and dynamic load tests. It is the first time that all European knowledge and practice in pile design is bound together in one volume.

In these times of rapid technological progress engineers still face their most important challenge: how to translate the needs of humanity to technical solutions which meet with safety, durability and economy. I sincerely hope that this seminar will largely contribute to this noble objective, and that this volume will serve as a constant reminder of technical achievements.

I am convinced that the Belgian Geotechnical Society will offer you a stay that you will remember. Not only due to different approaches for pile design you were faced with, or due to the astonishing design approach of other colleagues, but also due to the good climate in the seminar, the appreciation you build up for each other and the warm friendship you never will forget.

L. Maertens
Chairman of the Belgian Member Society of ISSMFE

Design of Axially Loaded Piles – European Practice, De Cock & Legrand (eds)
© 1997 Balkema, Rotterdam, ISBN 90 5410 873 8

Foreword

For several decades, the vivid enthusiasm supporting the research linked to pile foundation design, resulted mostly in very interesting, but rather esoteric, academic uplevelling of the knowledge without actual great benefit to the daily contractors' practice.

Several explanations can be found for such discrepancy between scientific progress and on site work:

– There first of all, till the late seventies, was the favourable economic afterwar booming of the construction activities; over conservative design was not severely punished by the clients;

– the pile type variety remained negligibly small until two decades ago; the large experience with the wellknown existing pile type could cover the daily design needs;

– most of the existing codes and standards were restrictive to the extent of even slowing down new initiatives and engineering spirit.

The world economics, since the late seventies, pushed contractors for opening new horizons, creating new pile type and pile installation techniques. The overwhelming lack of clear understanding of pile installation parameters and of design impact of quality of pile work on site, became scaringly obvious.

The ISSMFE interacted promptly and installed many technical committees on deep foundations. But even at the end of this millennium, some fundamental questions related to scientifically and economically justified design of piles remain to be answered.

This is all the more important since Eurocodes are gradually standing up. More particularly for pile design, the philosophy and the principles to be followed are gathered in the EC7. It fortunately only gives a framework for design methods without too specific rules. Within such framework, completed by some prescriptions of the various National Application Documents, the geotechnical consultants and foundations contractors could still interact and implement their engineering skills into the final concept and design of the pile foundation.

The European Technical Committee ERTC3, hosted by the Belgium Geotechnical Society, has taken up the challenge to provide us, at the occasion of the ISSMFE Hamburg Conference event – September 1997 – with both a comprehensive report on the single axially loaded pile design practice, relevant for the European academic and on site approach, and the possible implementation of Eurocode 7 in such design and pile performance philosophy. The Brussels' seminar in April 1997 organized by the ERTC3 group of experts therefore can be considered as a main contribution to our state-of-the-art knowledge and understanding of advanced modern single pile design.

The ISSMFE, and the European Member Societies more specifically, are grateful to its chairman and all ERTC3 members for the strong involvement and important scientific support required to comply with goals specified in the terms of reference of this committee work. I may wish the committee a very fruitful seminar and all the success in preparing the report of their work for presentation during the coming ISSMFE conference in Hamburg (September 1997).

Prof. Dr Ir W.F. Van Impe
ISSMFE Vice-President Europe

Design of Axially Loaded Piles – European Practice, De Cock & Legrand (eds)
© 1997 Balkema, Rotterdam, ISBN 90 5410 873 8

Organization

ERTC3 MEMBERS

Chairman:
F. De Cock, Franki Foundations S.A., Belgium
Secretary:
C. Legrand, Belgian Building Research Institute, Geotechnical Division, Belgium

Members:
Prof. E. Toshkov, Building Research Institute, Bulgaria
Dr J. Feda, Laboratory of Soil Mechanics, Institute of Theoretical & Applied Mechanics, Czech Republic
R. Skov, cp test a/s, Denmark
Prof. J. Hartikainen, Tampere University of Technology, Finland
Dr M. Bustamante, Laboratoire Central des Ponts et Chaussées, France
Prof. R. Katzenbach, TH Darmstadt, Germany
Dr G. Petrasovits, TU Budapest, Geotechnical Department, Hungary
Dr B. Lehane, Department of Civil, Structural & Environmental Engineering, Ireland
A. Mandolini, University of Naples 'Federico II', Inst. di Tech. delle fond. e constr. in terra (Geotecnica), Italy
H. J. Everts, Delft Geotechnics, Netherlands
A. Simonsen, Noteby, Norway
Dr K. Gwizdała, Technical University of Gdansk, Geotechnical Department, Poland
Prof. I. Manoliu, Civil Engineering Institute, Romania
Prof. B.V. Bakholdin, NIIOSP, Russia
T. Virdia, Terratest – I.C.O.S., Spain
C.-J. Grävare, Palanalys AB, Sweden
Dr F. Bucher, Swiss Federal Institute of Technology, Switzerland
J.D. Findlay, Stent Foundations Ltd, United Kingdom

Invited experts:
Dr A.A. Bartolomey, Perm State Technical University, Russia
M. Mets, AS Geotehnika Inseneribüroo GIB, Estonia

SEMINAR ORGANIZING COMMITTEE

Chairman:
F. De Cock, Franki Foundations S.A.
Secretary:
C. Legrand, Belgian Building Research Institute

Members:
E. Dupont, Fundex
Prof. A. Holeyman, Catholic University of Louvain
Prof. P. Mengé, Department LIN/Ghent University
Prof. L. Maertens, Besix
Prof. J. Nuyens, Free University of Brussels
P. Welter, MET

Keynote lectures

Design of Axially Loaded Piles – European Practice, De Cock & Legrand (eds)
 ISBN 90 5410 873 8

Influence of screw pile installation on the stress state in the soil

W.F. Van Impe & H. Peiffer
Laboratory of Soil Mechanics, Ghent University, Belgium

ABSTRACT : In the doctoral research of the second author, one sub-topic was focussing on the experimental evaluation of change in stress state due to pile installation. Especially measurements before, during and after pile installation were carried out. The experimental results have put the problem of prediction of the shaft capacity in a new context. The experimental results are in contradiction with simplified prediction models for evaluating the influenced zone and the magnitude of the soil stress changes.

1. INTRODUCTION

Auger cast in place piles became increasingly popular over the last 15 years. There can be listed several reasons for preferring screwed or auger piles, such as the absence of vibrations in the surrounding soil or important reduction of noise during pile installation. Moreover, some of the auger pile systems are called to be of the soil displacement type, which should be reflected in a favourable load-settlement behaviour. However, the influence of pile installation on the real behaviour of the pile is very important, since the pile behaviour quality seems to be extremely sensitive to it.

To evaluate the pile installation one can use a monitoring system registrating the characteristics of pile execution or execute comparable in situ tests before and after pile installation. In this contribution the dilatometer is discussed as a tool for monitoring the pile installation and to evaluate the change in stress state close to the pile shaft due to such pile installation.

2. DILATOMETER TEST (DMT)

The initial scope of the dilatometer was the evaluation of the horizontal stress and stiffness of the soil close to a driven steel pile, in order to make a reliable estimation of the behaviour of a horizontally loaded pile. Extensive research programs in the years lateron resulted in an important number of correlations for the determination of a wide spectrum of soil characteristics.

Regarding the suitability of the DMT test for the evaluation of change in stress state in the soil during pile installation, it is important to focus on two important features :

1. The soil distortion due to the DMT blade insertion remains small, as compared to other in situ tests, mainly because of the shape of the DMT-blade (fig. 1).
2. Based on a theoretical analysis (theory of Durgunoglu and Mitchell (1975)) it can be

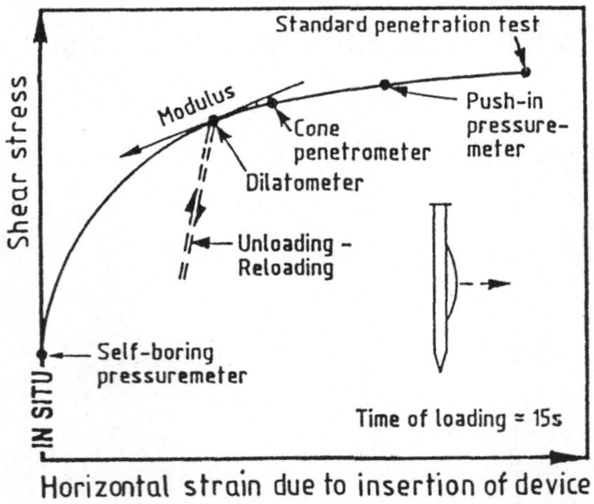

Figure 1. Qualitative estimate of insertion of in situ testing devices

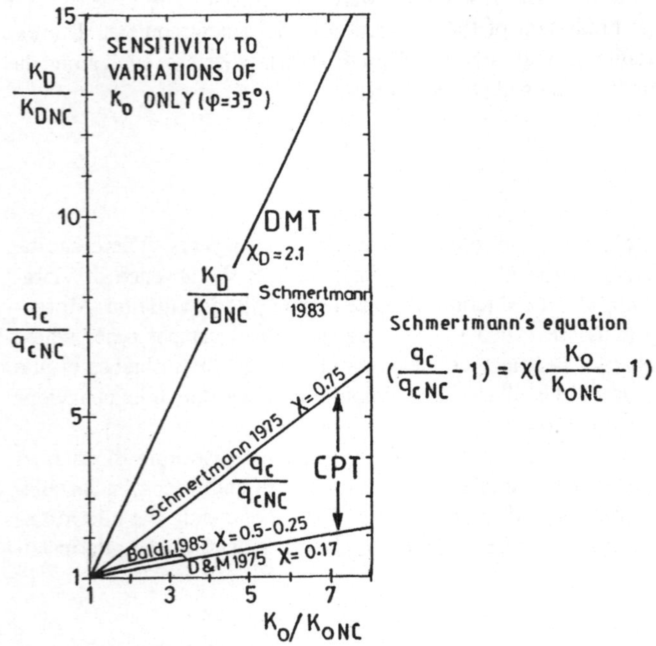

Figure 2. Sensitivity of q_c and K_D to K_0

shown that the basic DMT parameter K_D is more sensitive to changes in K_0 as compared to the CPT test results (fig. 2).

Based on this we could reasonably well expect more pronounced results related to the DMT-quality of pile installation evaluation, as discussed by e.g. Van Impe (1989). One has however to take in account that the disturbed zone around the blade will influence those measurements. In the doctoral research of the second author, the influence of the soil disturbance around the DMT blade due to insertion was investigated theoretically and experimentally.

First of all one has to make a distinction between drained and undrained penetration. Fully drained penetration occurs in soils with I_D (material index) greater than 1.8. Fully undrained penetration occurs in soils with $I_D \leq 1.2$.

A theoretical analysis shows that for cohesive soils, the DMT-lift-off pressure P_0 is dominantly determined by the pore water pressure (80 to 100 % of the total stress immediately after penetration, decreasing with increasing overconsolidation ratio). These results are in good agreement with the experimental results for normally consolidated to slightly overconsolidated clays (OCR < 4). For the evaluation of change in stress state due to pile installation, using the DMT test, one has to deal in such cases with a total stress approach (for soils with $I_D \leq 1.2$). An effective stress approach only comes in for soils with $I_D \geq 1.8$. Although only total stresses are known, one can use these stresses to judge (qualitatively) for the change in stress state. Indeed the excessive pore water pressures are determined by the undrained shear strength, stiffness of the soil, overconsolidation ratio and soil structure.

3. GLOBAL CLASSIFICATION OF SCREW PILES

Nowadays one can make a global distinction between the different types of screwed or auger piles, (fig. 3), referring to the in time rather well defined generations of screw pile developments :

1. auger piles allowing for excavating the soil during pile installation : type CFA (CFA = continuous flight auger), being the first (and oldest) generation (early sixties).
2. auger piles with unique lateral displacement of the soil, second generation (seventies). Some types of pile, like the PCS-pile (PCS = "Pressured Concrete Screw") belong to both categories. Also there is a big variety in type of soil displacement. The displacement is dominantly unique (in the stage of screwing down the casing) as in the Fundex-auger pile type.
3. Auger piles with double soil displacement action (third generation : starting from the eighties). The basic feature is the twofold soil displacement (in both the stages of screwing in and out of the casing) as in the case of the Franki-Atlas auger pile and the latest development such as the Omega screw pile.

Such type of classification based on changes of the soil stresses due to installation can be extended towards other types of pile. The distinction is determined by what happens at a considered depth between the moment the pile tip or casing passes the first time (downward movement) and the moment the concrete has built up sufficient resistance to resist the horizontal possible inward soil displacements.

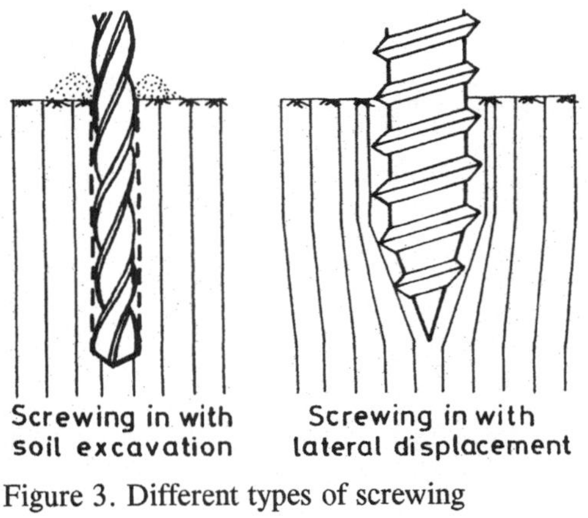

Figure 3. Different types of screwing

4. NUMERICAL AND ANALYTICAL APPROACH FOR THE DETERMINATION OF THE STRESS STATE IN THE SOIL DURING PILE INSTALLATION

The expansion of a soil cavity has been studied by an important number of authors. Only in cases of fully undrained expansion, reliable theoretical approaches are available for continuum materials. For the expansion in drained conditions, when no excessive pore water pressures are generated, or even in partially drained conditions, however, the number of available theoretical models is still limited. Experimental results only refer to experiences with the quasi static penetration of the piezoblad, where the membrane of the dilatometerblade is replaced by a porous stone (Davidson and Boghrat, 1983). The dissipation of the excessive pore water pressure due to penetration after 1 minute was ± 15 % for silty soils, ± 25 % for sandy silts, ± 80 % for silty sands and 100 % for sands.

4.1 *Partially drained expansion*

Datta (1982) derived from his experiments with driven piles an excess pore pressure dissipation time in coarse sand of less than 5 minutes, and in fine silty sand of less than 45 minutes. He also measured a maximum excess pore water pressure in such soils of about 20 % of the vertical overburden.

Möller and Bergdahl (1981) found for fine sand (D'_r = 80 to 90 %), based on their experiments, changes in pore pressure to a distances of 5 to 7 times pile diameter. These results have to be interpreted cautiously. First of all neither the influence of the granulometry, nor of the stress state around the pile was investigated. Moreover the tests were carried out in a small sand box, height 40 cm and diameter 23 cm, for piles with diameter 2 cm. The scaling of the soil-pile stiffnesses, soil stresses and boundary conditions was not properly taken care for.

4.2 *Undrained expansion*

When excess pore water pressures are generated, an additional parameter comes forward. For the undrained expansion of a spherical or cylindrical cavity one can refer to Butterfield and Banerjee (1970), Vésic (1972), Randolph and Wroth (1979), Carter et al. (1979). For our research, the results of the measurements close to piles during installation are compared with the results from theory of cylindrical expansion. Therefore we went out from cylindrical expansion, using the elastic-perfectly plastic model of Tresca for the soil. This criterium coincides with the criterium of Mohr-Coulomb for a shear angle $\varphi = 0°$, so for cohesive and fully saturated soils under a quick (undrained) loading.

Considering :

- the soil to be homogeneous and isotropic
- the deformations occur only in a horizontal plan ($\varepsilon_z = 0$)
- the pore water being incompressible
- undrained condition : a fast enough expansion in order not to have dissipation during the expansion stage.

The plastic radius R_p around the pile is given by (fig. 4)

$$R_p = \sqrt{\frac{G}{c_u}} \cdot r_0 \qquad (1)$$

with G = shear modulus; c_u = undrained shear strength ; r_0 = radius at the end of expansion.

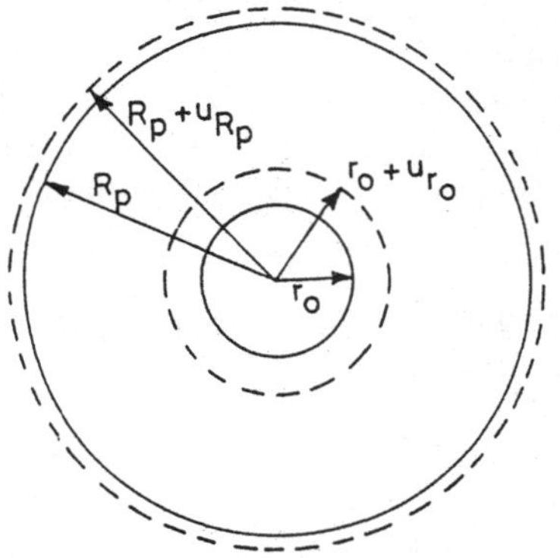

Figure 4. Cavity-expansion model

The stresses close to the piles, in the plastic zone, are given by (r = distance at the end of expansion) :

$$\Delta\sigma_{rr} = 2 \,.\, c_u \,.\, \ln\frac{R_p}{r} + c_u \qquad (2)$$

$$\Delta\sigma_{\theta\theta} = 2 \,.\, c_u \,.\, \ln\frac{R_p}{r} - c_u \qquad (3)$$

$$\Delta\sigma_{zz} = 2c_u \,.\, \ln\frac{R_p}{r} \qquad (4)$$

$$\Delta u = 2c_u \ln\frac{R_p}{r} \qquad (5)$$

4.3 *Consolidation around the cavity*

One can evaluate the consolidation around a cavity using one of following consolidation theories :

- the coupled consolidation theory (Biot (1941)), assuming that the mean total stress varies with time and the variation is determined by the variation of pore water pressures

$$\frac{E.k}{3\delta_w(1-2\nu)} \,.\, \nabla^2 u = \frac{\partial u}{\partial t} - \frac{\partial p}{\partial t} \qquad (6)$$

- the uncoupled consolidation theory (Terzaghi (1923) and Rendulic (1936) assuming a constant mean total stress :

$$\frac{\partial u}{\partial t} = c.\nabla^2 u \qquad (7)$$

5. OVERVIEW OF PILE TYPES EXAMINED

The research was mainly focused on the experimental determination of the stress state close to the pile shaft of screw piles of the displacement type. Besides of the torque available, also the design and the shape of the drilling tip is of big importance, not only to easily reach the bearing layer but also for the assurance of a good pile shape.

The auger tip of a screw pile of the soil displacement type has a specific shape, designed to penetrate with displacement as quickly as possible, depending on the mechanical soil properties (fig. 5). Following piles were examined (tests executed indicated in the table 1) .

6. PROCEDURE FOR THE EXPERIMENTAL DETERMINATION OF THE INFLUENCE OF PILE INSTALLATION ON THE STRESS STATE IN THE SOIL

6.1 *General considerations*

Generally, foundation piles are installed to transfer the load from the surface to a deep bearing layer. The pile shaft is mainly surrounded by weak layers (weak clays, loose sandy layers). In order to obtain reliable data of the influence of pile installation on the stress state in the soil, one has to make a judicious choice for the interdistance pileshaft to DMT-blade.

Table 1. Overview of pile types examined

Type of pile	DMT before	DMT during	DMT after
	pile installation		
Omegapile	x	x	x
Atlaspile	x	x	x
Tubelar screw pile	x	x	x
PCS-pile	x	x	x
gravel column	x	x	
driven cast in situ pile	x	x	x

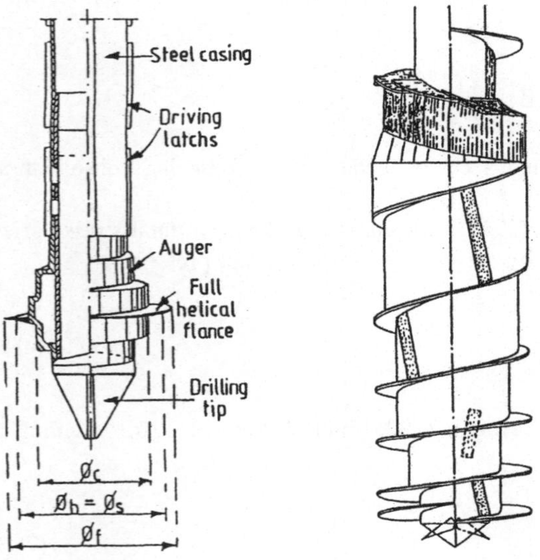

a) Atlaspile b) Omegapile

Figure 5. Different types of auger tip

This distance has to be as small as possible in order to measure important variations in horizontal stress. Preferably one measures in the zone where plastic soil deformations have occurred. This zone is determined by

- the diameter of the pile
- the soil characteristics, (especially for cohesive soils the deformation and shear resistance characteristics and for non cohesive soils also the relative density).

Taking into account :

- the distance pile shaft to DMT-blade has to be as small as possible in order to perform reliable measurements in the plastic zone around the pile
- the distance to be sufficiently large on the other hand in order to avoid damage of the membrane and the blade on one hand, and to reduce the influence of non-verticality on the other hand
- the extent of the plastic zone varies with depth
- preferring only one procedure for all tests, the DMT-blade is placed at a distance smaller than the theoretically predicted plastic radius R_p (Cavity Expansion theory).

For the model characteristics, the characteristics for the Ypresian clay were taken, because the research was mainly dealing with pile installation in this type of soil.

6.2 *Theoretical prediction for the distance DMT-blade - pile shaft*

The expression for R_p is given by (1). The Ypresian clay is a tertiary overconsolidated fissured clay. For the relation between cone resistance and undrained shear resistance, the extensive research of the Ghent University lab in this type of soil, allows for the evaluation of the Ypresian clay parameters as :

$$q_c = N_k \cdot c_u + \sigma_{v,0} \qquad (8)$$

where $\sigma_{v,0}$ = vertical total stress at the depth considered and $N_k = 27$ for stiff fissured clay having a marine origin

For the test site in Koekelare where the Ypresian clay was situated beneath a depth of five meters and taking in account a pile length of 13 m one can estimate c_u as follows : $\sigma_{v,0} \approx 180$ kPa and $q_c \approx 2.7$ MPa. This results in $c_u = 93$ kPa. This is in good agreement with De Beer (1979) : $c_u = 64 + 16.5 \cdot z$ (z = depth in m, c_u in kPa). For slightly overconsolidated soils, $M \approx (1.3 \text{ to } 3.3) \cdot q_c$ or $M \approx 6.25$ MPa for the test site in Koekelare. Out of the theory of elasticity one can derive $G_0 = 1.04$ MPa ($\nu = 0.40$); eq. (1) results finally in $R_p \approx 3.3 \cdot r_0$. Based on this result the interdistance between pile shaft and dilatometerblade is taken equal to the pile diameter. The same interdistance is taken for non-cohesive soils, because of the lack of a reliable theoretical model on the one hand and the required uniformity for the test procedure on the other hand.

7. CHANGES IN TOTAL HORIZONTAL STRESSES DURING PILE EXECUTION

7.1 *Omegapile at Vorst test site*

In a research program on Omega piles (Socofonda) the test site of Vorst was selected. The pile diameter was 51 cm. The dilatometer blade was placed at a distance of 77 cm from the pile axis and a depth of 10.60 m. The total pile length was 25 m. Out of the available geotechnical maps the Ypresian clay appeared to be situated between 9.60 m and 23.60 m.

On figure 6 one can see that the pressure P_0 reaches a first peak value the moment the auger tip passes the installation depth of the DMT-blade (downward movement). A second peak value is reached when the auger tip passes again this depth (upward movement). Immediately after reaching these peak values there is a considerable drop down of the total horizontal pressure. This can be explained by exceeding the tensile strength in the soil leading to the formation of micro fissures. This results in temporary faster pre pressure dissipation until the fissures again are closed by the inwand soil deformation. This theoretical explanation is discussed under 8. Besides this, the decrease of P_0 is explained by decompression of the soil after the passing by of the auger, the dissipation of the excessive pore water pressure and the changes in total stress state around the dilatometerblade.

7.2 *Franki-Atlaspile at Koekelare test site*

For this program the test setup is presented in figure 7. Two dilatometerblades and 1 pore water pressure transducer (CPTU-cone) were installed close to the pile, all at the same distance from the pile axis. The results are presented in figure 8 for the four piles examined. One can detect only a small change in total horizontal stresses at a depth of 5 and 7.5 m. At a depth of 10 m and 13 m the stress change is more pronounced. This probably can be explained by an inappropriate the installation of the dilatometerblade (outside the plastic zone around the pile).

7.3 *Tubelar screw pile at Koekelare test site*

The same test setup as for the Atlaspiles was taken. The results are presented in figure 9.

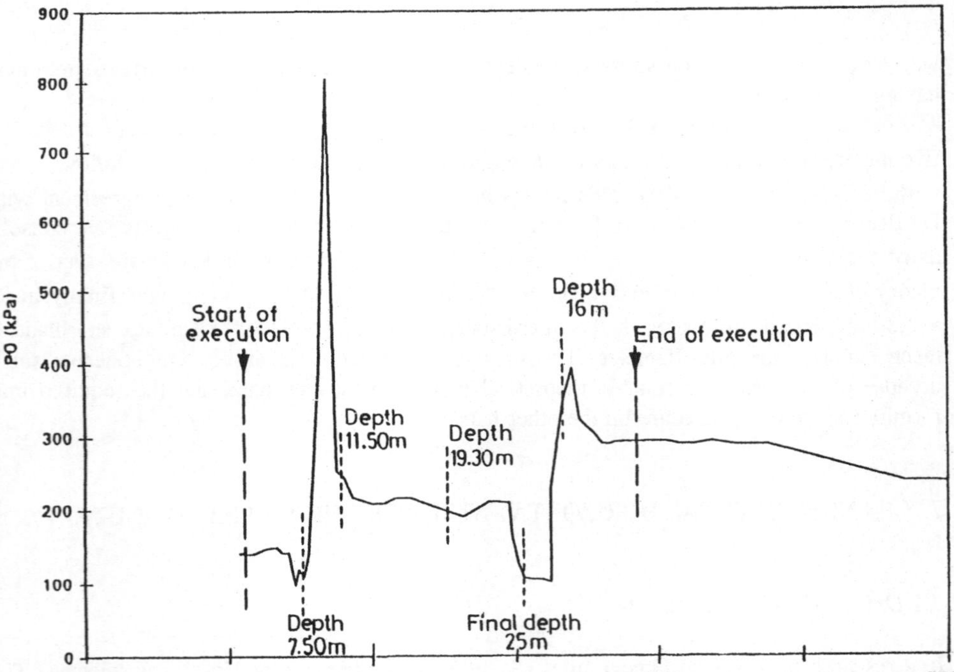

Figure 6. DMT-test during installation Omegapile at Vorst test site

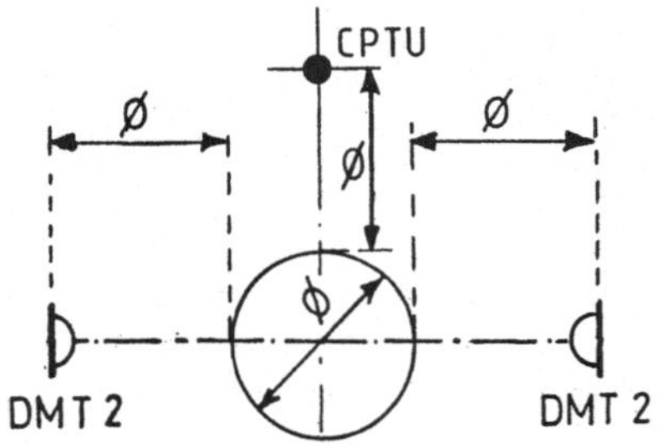

Figure 7. Test setup for pile evaluation

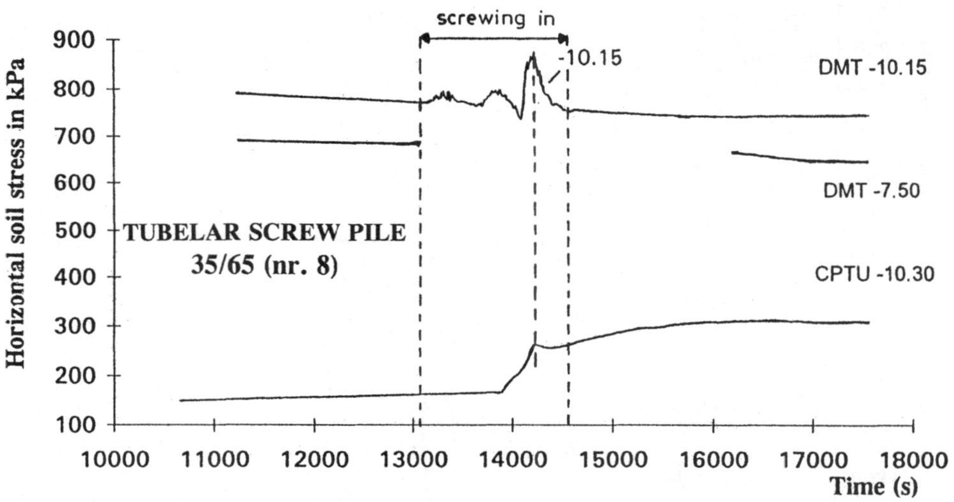

Figure 9. DMT test during installation of Tubelar screw pile

7.4 *PCS-piles in sand*

Two PCS-piles of Socofonda company were examined. The details are gathered on the following table

Table 2.

Site	Pile length	Pile diameter	Installation depth DMT	Material index I_D
Doel	19.07 m	0.40 m	8.5 m	2.9
Dendermonde	16.52 m	0.45 m	7.5 m	2.0

Both test programs were performed this time in a sandy soil. The DMT test measures therefor predominantly during most of the pile installation period changes in effective stress around the pile. At Doel test site, the dilatometerblade was directed tangentially (deformation of the membrane perpendicular to the radius from pile axis)

Doel (fig.10)

When the installation depth of the DMT-blade is reached, the total stress shows a sudden

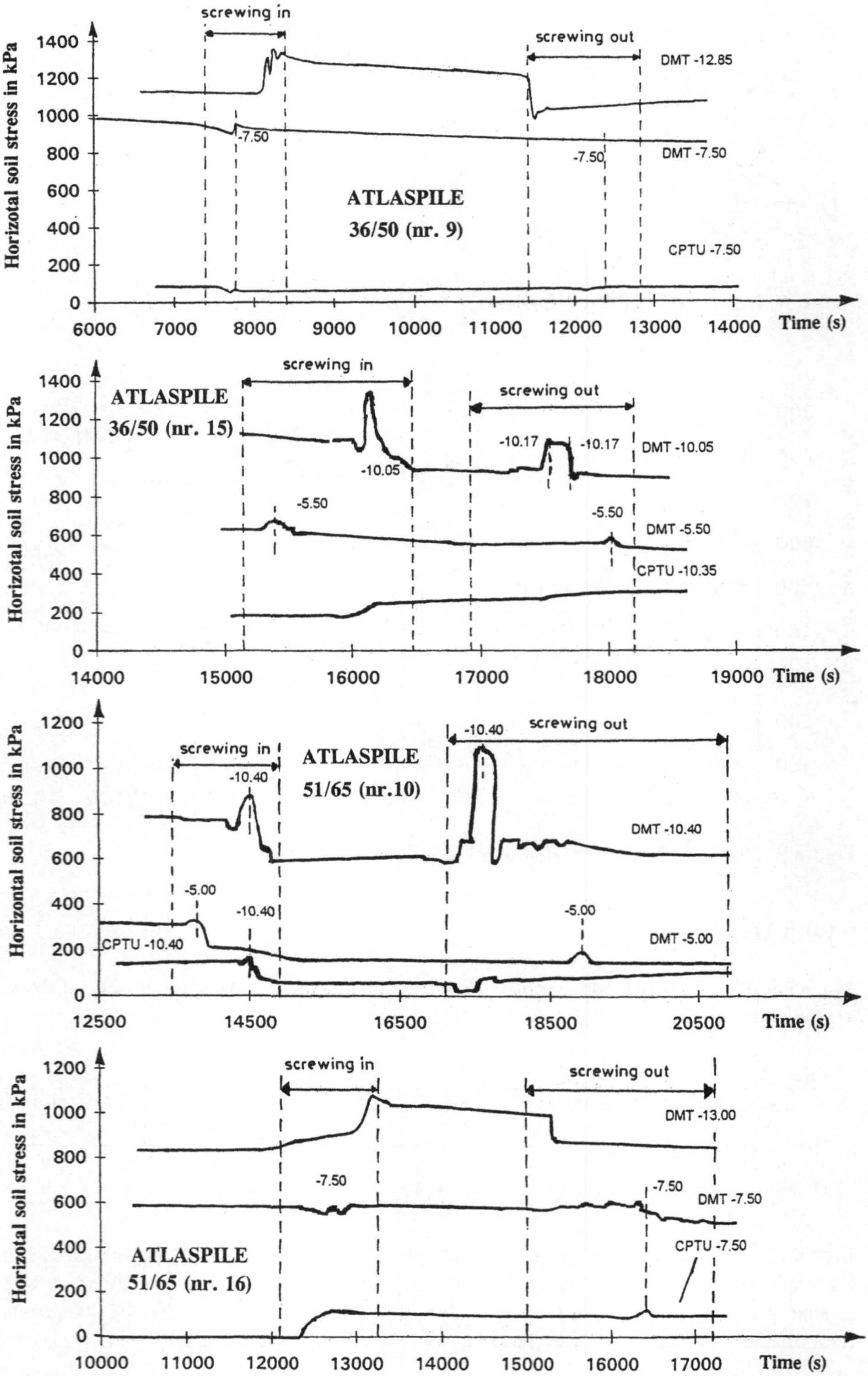

Figure 8. DMT tests during installation of Atlaspiles

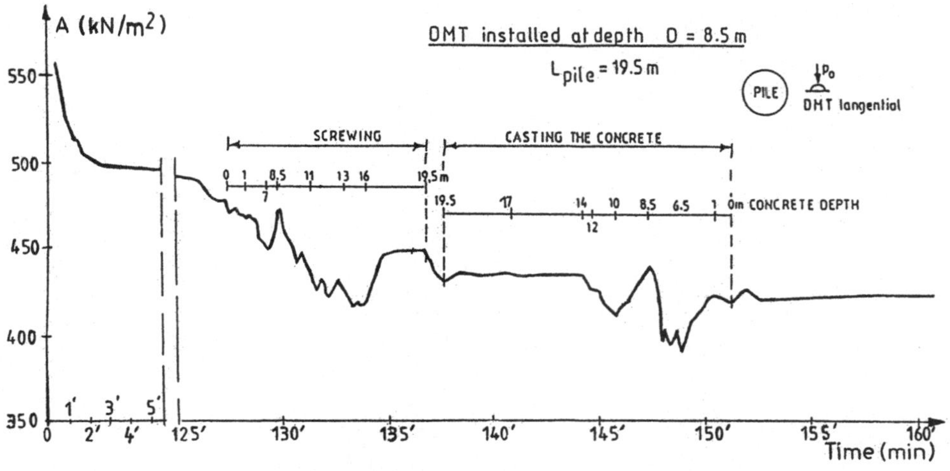

Figure 10. DMT test during pile installation of PCS-pile at Doel test site

peak value, what can be explained by a sudden pore water pressure peak followed by an immediate dissipation.

Afterwards, due to arching, the tangential stress will increase because of soil decompression. During casting the concrete a similar phenomenon can be seen. At the moment the augertip passes, pore water pressures are induced but they immediately are dissipated. At the end, when the concrete overpressure is taken away, some decrease in tangential stress is induced leading to an increase in tangential stress. One can notice that the difference in tangential stress for the upward movement is much less than during the downward movement. This can be explained by higher decompression during the boring stage of pile execution, what could be expected, since the PCS screw pile is basically only a CFA type, with some precautions on high enough vertical downward speed and pressurized concrete casting.

Dendermonde (fig. 11)

Here one can see a quite similar picture as at the Doel test site. The dissipation of pore water pressure goes on however more slowly. This can be explained by the more silty sand character of the soil. A second remarkable difference comparing with the Doel test site is only a partial reduction of this radial horizontal stress after taking away the concrete overpressure. Also this can be explained by a somewhat delayed dissipation in this soil type.

8. MICRO FISSURES WHEN EXCEEDING THE TENSILE STRENGTH IN THE SOIL ('FRACTURING')

Excess pore water pressures arise when a saturated soil with low permeability is suddenly loaded.

Assuming that
- the soil is homogeneous, isotropic and elastic
- the soil is saturated
- the water and the grains are incompressible
- no water can drain from the soil volume considered (undrained behaviour)

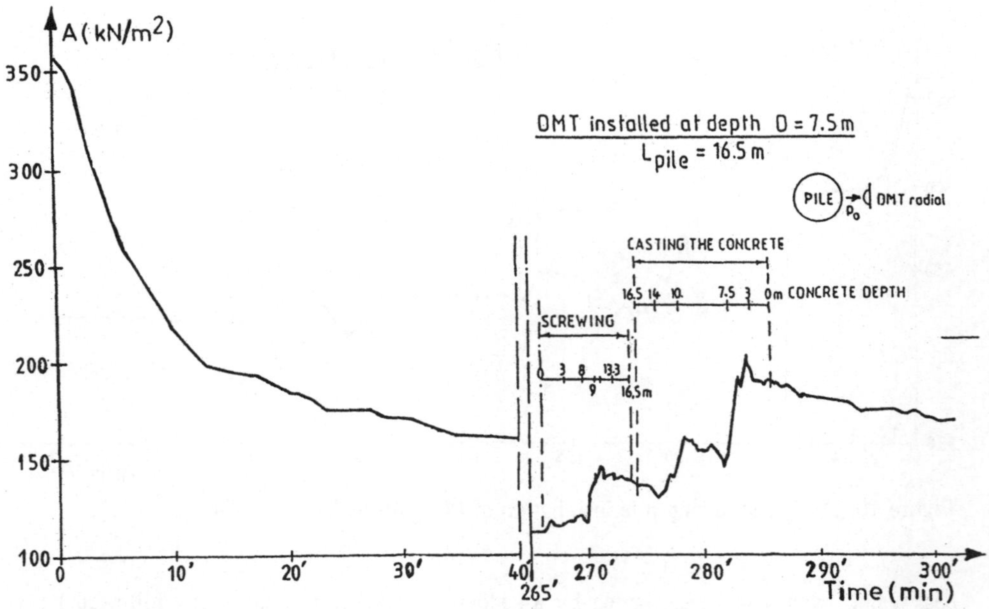

Figure 11. DMT test during pile installation of PCS-pile at Dendermonde test site

So, the relative volume change $\left(\frac{\Delta V}{V}\right) = 0$

One can prove that for triaxial conditions the α-coefficient from Henkel and the A-coefficient from Skempton are interrelated as :

$$\alpha = \frac{\sqrt{2}}{2}(3A\text{-}1) \tag{9}$$

This leads to the following effective stress ratios (Peiffer, 1997) :

$$\frac{\Delta\sigma'_{rr}}{c_u} = 1 - \frac{1}{\sqrt{3}}(3A\text{-}1) \tag{10}$$

$$\frac{\Delta\sigma'_{\theta\theta}}{c_u} = -\left(1 + \frac{1}{\sqrt{3}}(3A - 1)\right) \tag{11}$$

$$\frac{\Delta\sigma'_{zz}}{c_u} = -\frac{1}{\sqrt{3}}(3A - 1) \tag{12}$$

The tangential effective stress becomes zero (fracturing could start) when :

$K_0 \,.\, \sigma'_v + \Delta\sigma'_{\theta\theta} = 0$ (assuming $\sigma'_h = \sigma'_{rr} = \sigma'_{\theta\theta}$: isotropic soil) (13)

Out of (11), (13) leads to

$$\frac{K_0 . \sigma'_v}{c_u} = 1.73\ A + 0.42 \qquad (14)$$

In table 3 the data for the site in Koekelare are gathered

Table 3.

Depth	σ'_v (kPa)	K_0	$(\Delta\sigma_1 - \Delta\sigma_{III\ failure}$ (kPa)	$\Delta u_{failure}$ (kPa)	$A = \frac{\Delta u_{failure}}{\Delta\sigma_I - \Delta\sigma_{III}}$	$\frac{K_0 . \sigma'_v}{c_u}$	$\frac{K_0 . \sigma'_v}{c_u}$ - 0.42-1.73A
5m	53	1.8	116	40	0.34	0.65	-0.36
7.5m	78	1.7	206	38	0.18	0.71	-0.02
10m	103	2.0	211	73	0.35	0.90	-0.12
13m	133	1.5	259	59	0.23	0.91	+0.09

Out of these theoretically predicted results, one can see that, with the exception of the depth of 13 m, the effective stresses became negative, what the probable explanation can be for the microfissuring in the plastic zone, the (temporarary) increased dissipation and shorter reconsolidation time.

9. COMPARISON DMT MEASUREMENTS BEFORE AND AFTER PILE INSTALLATION

The scope of these measurements was to evaluate the stress state, as finally changed by the pile installation. In fig. 12 the results of this comparative analysis are compiled. In order to evaluate these measurements, P_0 was normalised with respect to an approach of the mean total stress around the dilatometerblade. The normalization procedure is the same as proposed by Fretti et al. (1989) : $P_0 = \frac{P_0}{\sigma_p} = \frac{P_0}{\frac{(P_0 + \sigma_v)}{2}}$. In this way the horizontal stress in the direction parallel to the largest DMT cross section is not take into account. This can be accepted by the fact that the stress in this direction is much smaller than P_0.

9.1 *Discussion of the general pattern of the results'graph*

The normalized curve $\frac{P_{0,f,N}}{P_{0,i,N}}$ as a function of $P_{0,i,N}$ shows some general trends. The shape is determined by :

- the stress state in the soil around the pile shaft before pile installation
- the distance dilatometerblade to pile shaft
- type of pile
- the installation procedure and execution parameters of the pile
- the time lag between pile execution and the DMT-measurements (influence of the consolidation around the pile)
- the time lag between the moment the DMT blade reaches a certain depth and the moment the A-pressure is measured (influence of the consolidation around the DMT-blade).

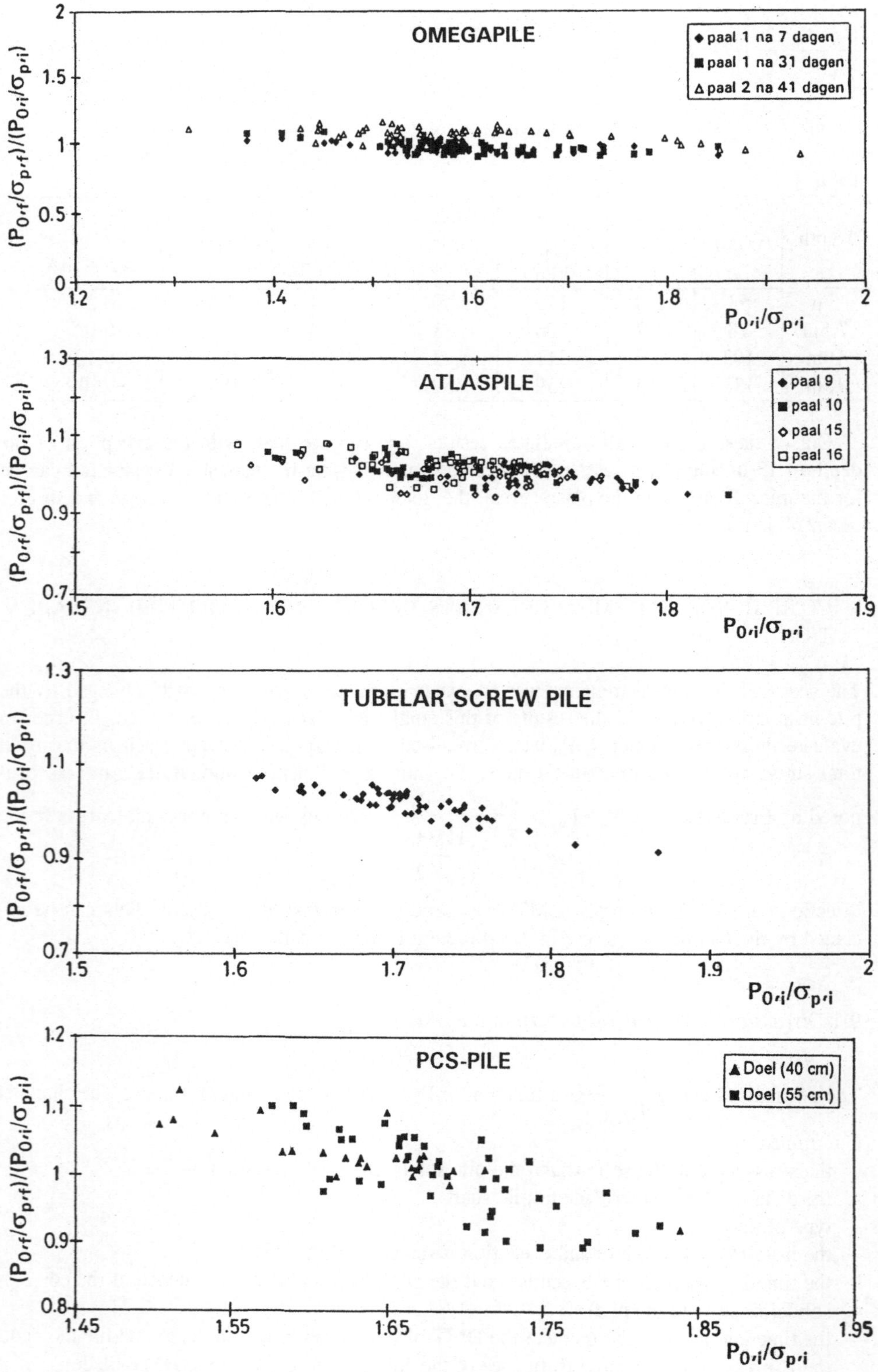

Figure 12. Changes in normalized total horizontal stress ($I_D \leq 1.2$)

I_D ≤ 1.2 (undrained penetration of DMT)

When we consider the mechanical overconsolidation as a measure for eventual soil improvement due to pile installation, then the degree of soil improvement decreases for increasing stiffness of the cohesive saturated soil.

In our opinion, logically expected is that in such soil the maximum ($P_{0,f}/P_{0,i}$)-ratio was about 10 % for all type of displacement pile, independent the degree of soil displacement they forced. A stronger piled soil displacement so not necessarily implements (more pronounced cohesive soil improvement). In fig. 12 the results for the different types of piles are presented.

I_D ≥ 1.8 (fully drained penetration)

For these soils the variations of the ratio ($P_{0,f}/P_{0,i}$)- of horizontal normalised stress were more pronounced : 75 % for the Omegapile, < 20 % for the PCS-pile (maximum). The difference (compared to soils with $I_D \leq 1.2$) can be explained by the fact that here directly the effective stress state is influenced. Soil decompression appeared for the Omegapile from an initial normalised horizontal stress of about 1.6 on, (for the PCS-pile already from 1.2 on). This is a clear boundary for the applicability of this pile as a displacement pile. For the other pile type this as far is not get investigated ; which is a serious limitation for the determination of a reliable field of application. In fig. 13 the results for the different types of piles are presented.

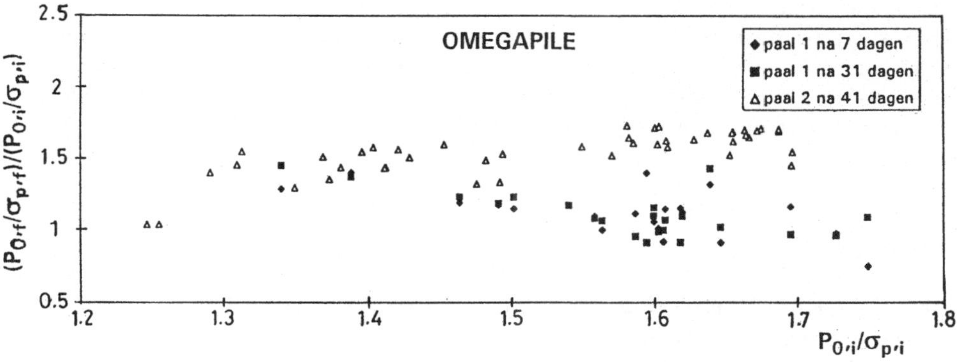

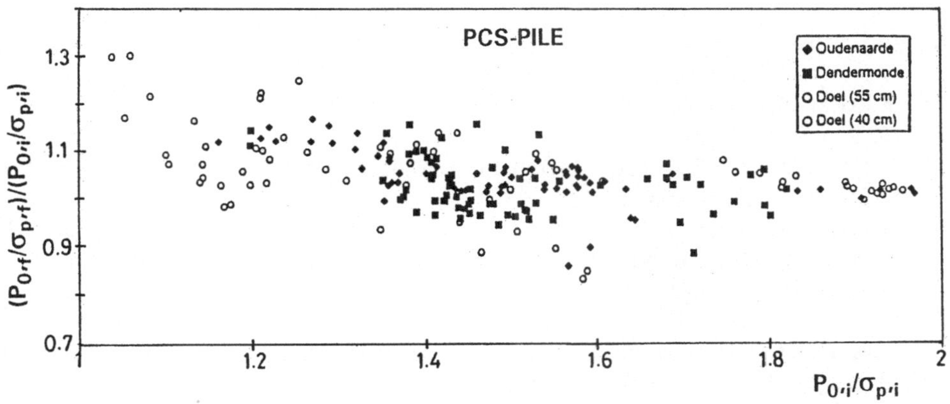

Figure 13. Changes in normalized total horizontal stress ($I_D \geq 1.8$)

10. EXTENSION TOWARDS SHAFT CAPACITY

An attempt was made towards the method of direct evaluation of the unit shaft capacity of piles by DMT. For the Atlaspiles, at the Koekelare test site, evaluation of the shaft bearing capacity was available out of pile instrumentation.

This resulted in following successful preliminar, semi-empirical, relationship for the type of piles performed overthere :

$$q_{u,s} = 0.0035 \cdot P_0 \cdot K_D^{1.25} \qquad (15)$$

One can expect this method to be promissing also for the prediction of the unit shaft bearing capacity for other soil displacement type of pile; because for each of these types a relationship between the horizontal effective stress before and after pile execution, depending on the pile execution parameters should anyhow be quite clear (see 9). More research in this respect for the various pile types mentioned is necessary for the moment.

11. CONCLUSIONS

The execution of DMT tests close to piles leads to a better understanding of the changes in stress state in the soil due to pile installation. The DMT test at this stage offers a qualitative tool for the judgement of the quality level of execution of the piles. Extension of the procedure towards the direct evaluation of the shaft capacity becomes plausible and is highly recommended. It is not a real (type A) design method because such estimates are obtained after the pile is executed, nor is it a real type (load test) because the bearing capacity is estimated from soil properties and not by loading the pile. It is widely recognized that pile capacity largely depends on execution next to the soil type. Hence one can not pretend to estimate the pile bearing capacity based only on measurements on the original soil ; one should more rationally base such estimates on measurements on the soil parameter before and after pile installation.

12. ACKNOWLEDGEMENT

The research was funded by the Fund for Scientifique research (Flanders) and the piling contractors Atlas N.V., Franki N.V. and Socofonda N.V.

REFERENCES

Baligh, M.M. & Levadoux, J.N. 1980. Pore pressure dissipation after cone penetration, MIT Report, N° MITSG 80-13, 368 pp.

Biot, M.A. 1941. General theory of three dimensional consolidation. Journal of Applied Physics, Vol. 12, pp. 155-164.

Butterfield, R. & Banerjee, P.K. 1970. The effect of pore water pressures on the ultimate bearing capacity of driven piles. Proceedings of the 2nd South-East Asian Conference on Soil Engineering, Singapore, pp. 385-394.

Carter, J.P., Randolph, M.F. & Wroth, C.P. 1979. Stress and pore pressure changes in clay during and after the expansion of a cylindrical cavity. International Journal for Numerical and Analytical Methods in Geomechanics, Vol. 3, N° 4, pp. 305-322.

Datta, M. 1982. Pore water pressure development during pile driving and its influence on driving resistance. Proceedings Conference on Behaviour of Offshore Structures, BOSS, Houston, Vol. 2, pp. 295-304.

De Beer, 1979, Historiek van het kanaal Leie-Ieper - Eigenschappen en gedragingen van de Ieperiaanse klei. Tijdschrift voor Openbare Werken nr. 4, 5 en 6.

Fretti, C., Ghiona, V.N., Jamiolkowski, M., Lo Presti, D. and Tortella, M.L. 1984. Dilatometer stiffness of Ticino sand. ASCE Journal.

Henkel D.J., 1961. The shear strength of a saturated remoulded clay. Proceeding Americ Soc. of Civil Engineering., Research Conference on shear strength of cohesive soils.

Ladanyi, B., Expansion of a cavity in a saturated clay medium, Proceeding Am. Soc. Civ. Eng., July, 89 SM 4, pp. 127-161.

Massarsch, K.R. en Broms, B.B. 1977. Fracturing of soil caused by pile driving in clay, IX International Conference on Soil Mechanics and Foundation Engineering, Tokyo, pp. 197-199.

Meigh, A.C. 1987. Cone penetration testing, methods and interpretation, CIRCIA Ground Engineering Report : In-situ Testing, pp. 141.

Möller, B. & Bergdahl, U. 1981. Dynamic pore pressure during pile driving in fine sand. Proceedings 10th International Conference on Soil Mechanics and Foundation Engineering, Stockholm, Vol. 2, pp.791-794.

Peiffer, H.. 1997, Interpretatie en aanpassing van de dilatometerproef, uitgebreid tot de beoordeling van de spanningstoestand naast schroefpalen, PhD thesis, RUG.

Randolph, M.F. & Wroth C.P. 1979. An analytical solution for the consolidation around a driven pile. International Journal of Numerical and Analytical Methods in Geomechanics, pp. 361-383.

Torstensson, B.A. 1977. The pore pressure probe. Nordiske Geotekniske MØTE, Oslo, Paper n° 34, pp. 34.1-34.15.

Van Impe, W.F. 1989. Auger piles, planning and use. Conference Association of Finnish Civil Engineers (RIL), Helsinki, Finland, pp. 1-39.

Vesic, A.S., 1972. Expansion of cavities in infinite soil mass. Journal of the Soil Mechanics and Foundation Division, Proceedings of the ASCE, Vol. 98, SM3, pp. 265-290.

Design of Axially Loaded Piles – European Practice, De Cock & Legrand (eds)
© 1997 Balkema, Rotterdam, ISBN 90 5410 873 8

Some comments about United States practice of design of axially loaded piles

William F. Loftus
William F. Loftus Associates & Deep Foundations Institute, Englewood, N.J., USA

Due to the vast difference in subsurface conditions North to South and East to West in the United States, various design practices have evolved which, when analyzed, have at times defied uniformity. Each of the states, particularly those which have major cities, developed procedures which fit one state forgetting that some day a bridge may span the river, separating that state from its neighbor. The abutments of that bridge may have to have different design, subsurface investigation and construction standards.

This paper will highlight some of the examples of the above and present the strides that were made toward unifying these variations.

To give some examples of some design peculiarities, we must first have a look into the construction process in the U.S. and hope that the European process may be or may be made simpler.

Aside from the major city building codes, there are state codes which generally are cited by each of the smaller cities. At this writing, variations exist among all of these municipalities. Next, there are other agencies which operate outside the jurisdiction of the city and state, among them would be the federal agencies - Federal Highway; Federal Aviation; Department of Parks, U.S. Army Corps of Engineers, both civilian and military; Navy, civilian and military; NASA; Voice of America; various Port Authorities (New York, Houston, Baltimore, Philadelphia, San Francisco, Oakland, San Diego, etc.); major River City authorities (St. Louis, New Orleans, Pittsburgh, Cincinnati, Memphis, etc.); major Lake City authorities (Great Lake cities); and so forth. Any one of these (or more) may claim jurisdiction over a project and each will try to impose its own set of rules, if only where the granting of permits is concerned.

A) With respect to "driven piles", there is general agreement that allowable stresses in the three types of pile materials - steel, concrete, and timber - be addressed and quantified. Since these are given in most codes, let us first examine them.

1) Steel - Allowable axial stresses given in the texts vary from state to state, from 30% to 35% of yield (yield always given as 36 ksi. Only by special amendment or over-riding specifications do the codes acknowledge that higher yield steels exist.)

2) Concrete - Allowable axial stresses given in the texts vary from 25% to 30% of design-mix yield. Most often used is 4 ksi concrete. Higher strength mixes aren't discussed but can be used providing a good discussion of the use of admixtures to deal with the higher temperatures of the mix is presented to the examiners.

3) Timber - Allowable axial stresses will vary with wood species but generally are listed as 1000 to 1200 psi.

Applying these ranges of allowables to some of the cities, a 254 mm end-bearing H-pile can support 74 tons in New York, Los Angeles, and Chicago, but only 57 tons in New Orleans and Denver; a cast-in-situ light corrugated shell pile, 254 mm diameter can support 35 tons in San Francisco, Los Angeles, Denver, Atlanta. and New Orleans, but in Boston, 43 tons, and in Houston, 62 tons; a timber pile can support 50 tons in Cleveland and only 35 tons in Atlanta.

B) Most codes do not address the soil considerations of the project within the pile section except that pile tips may not be founded in loosely placed fills nor in soft, highly sensitive and weak soils. Some will allude to the local soil conditions but only obliquely. For example:

1) Along parts of the Gulf Coast, bedrock is very deep. Alabama, Northern Florida, Louisiana, and Southeastern Texas, all have rock elevations up to 1000 meters below the surface. Yet, the codes often speak of End Bearing! In areas of varved clays, very high gains of pile capacity are acknowledged not necessarily by code but more by practice. So, too, pipe pile collapses in these clays may cause heavier wall piles to be used, this by practice rather than codes.

2) In those same areas, soil subsidence is prevalent but little or no discussion of "negative-friction" (down drag) can be found in the codes of some of these cities.

3) Where coral and sinkholes are found cautions to those attendant problems do exist. Both codes and practice address them very well.

4) All codes require that an adequate and sometimes very extensive soil investigation program for any project be conducted . SPT's and/or CPT's will be used at the discretion of the Geotechnical Engineer. He will instruct the driller as to number of borings (generally set by codes as one hole per unit of building area.) As to depth, the Engineer will determine that with strong aid from local practice and his own experience in the area.

5) In the design, the engineer will consider all loads, i.e. dead. Live, roof (snow), wind, earthquake, buoyancy, scour (in streams and rivers), overturning, sliding (in dams), and negative friction (downdrag). These will translate into pile choice and capacities to be reached for (and tested for.)

C) Job Plans and Specifications are set with the joint efforts of the Structural and Geotechnical Engineers, each being guided by the code and local practice. 1) In the U.S., it is prevalent that once the pile and its capacity are set, cost proposals from General Contractors (GC) are received. The pile drivers in turn will present their cost proposals to the GC. If the pile capacity is to be over 40 tons, most codes require a number of test piles to be statically tested but only to prove that the axial capacity of the pile is at least enough to satisfy the design. More capacity is not sought after since the design is already made. Little or no attempt is made to change the design by either pile choice or higher capacity.

1) Dynamic testing of piles although being used minimally is not wide spread and still requires that static testing be done.

2) Virtually no attempt is made toward using newer and innovative pile types and/or equipment.

D) Most codes do not discuss equipment. Energies of hammers are usually considered such that the hammer and pile are at least compatible. Wave equation analysis is becoming prevalent as a requirement of the Engineers, not the codes. Little discussion of hammer types is given, however. In the major cities of the northeast - New York, Boston, Philadelphia, etc. - air-stream hammers prevail, while diesels are predominant elsewhere. Air pollution problems, however, are causing a move to hydraulic hammers and now vibrators are being used even for piling in structures. For the latter, some proof testing with impact hammers is generally required, however.

In the very "early" times of construction, say mid 1800's, pile installation was used only at marine structures and then only timber piles were used. Treatment of the wood came into play enabling structures to last considerably longer. Very little thinking in the technical sense was incorporated in the art of pile driving but when it was, the formula

$$P = \frac{2E}{S+0.1}$$

where P = Design Load

E = Energy delivered by the hammer system

And S = Unit of length (inches) per blow

came to be relied on. To this day, that very same formula is found in many of the codes even though the 50 year old Michigan Test Study found this formula to have Factors of Safety of as much as eight built in to it.

E) As to specifics, the types of piles prevalent in the U.S. are as follows:

1) H-Piles. These are usually designed as end-bearing piles and an example of disparity would be the splicing detail. To support an elevated highway designed by the New Jersey Department of Transportation (NJDOT), 12 HP 53 piles were used for which splices called for "full-butt-weld" with plates on webs, and flanges. The highway passed over an older bridge which had been founded on H piles too. The splices of these piles utilized the "Channel Splice" - no plates at all. When asking if the "Channel Splice" could be used on the new structure, the contractor was told "--it's not approved!"

Near St. Louis on the Mississippi River, a Lock and Dam was being constructed between Missouri and Illinois. The contractor wishing to use a vibrator to install the H-piles petitioned the U.S. Corps of Engineers to do so. The answer was delayed only because Illinois had never used the vibrator before even though the request stated that each pile would be seated with an impact hammer.

2) Cast-in-place piles (CIP). Perhaps the most discussed and variable type of pile, it is defined as a shell driven either with or without an internal mandrel (top driven or bottom driven.) For years (1930-1980), the bottom driven pile, either tapered or cylindrical, was protected by patents in the U.S. As such it was accepted minimally throughout the country since it was considered a contractor's tool and thus mistrusted by the engineering community. The loading and reinforcing was thus over designed due to this mistrust. At times, no reinforcing was required, in others full=length cages were required. Rules were temporarily instituted, however, which limited the unreinforced capacity to 40 tons, regardless of the concrete design or diameter of the pile. If greater than 40 tons was needed, full-length internal steel was designed sometimes using square bars (50 mm x 50 mm) inserted into the pile. Centering of this bar was a problem but worse the cost of this pile almost put it out of reach. Finally, cooler heads prevailed and proper designs were allowed and capacity became a function of concrete cross section only.

The bidders who were not protected by patents started to use heavier wall pipe piles; combinations of various diametered piles with still varied wall thicknesses; tapered fluted piles; newer splicing methods; and newer end-closure plates or cutting shoes; virtually any shape or form that would fit the loosely written, somewhat confusing and "non-standard" code specifications. Often, the state highway projects would have abutments and piers of one span of a bridge having one type of CIP piles while a second span of the same bridge would have another.

Timber and precast piles. In an effort to allow competition to exist, these two piles were by code given design parameters such that they were equal. Having warm climates in the south, precast (and later, prestressed) piles are prevalent since outdoor casting and storage is readily available. Wood on the other hand, though the growth and treatment facilities are also in the south, can be trucked north on easily unloaded flatbed trailers. With 20 or 30 pieces per truck. So precast remained prevalent in the south, wood in the north.

Treatment of the timber, originally all creosote, now is changing to the environmentally friendly salt treatment. The piles are more brittle with the salt however. Being tapered, the question of "What and at what elevation" is the proper pile diameter to figure the piles' capacity. This varies throughout the codes at anywhere from the tip to the top.

Precast/prestressed piles are widely used in the southerly coasts both east and west mostly in the sandy areas. Handling is a big cost and often precludes its use elsewhere.

CONCLUSION

The problem as I see it is that the code writers and enforcers are not experts in foundation design and performance. They must know about roof construction, plumbing, heating, windows, flooring bathrooms, pavement, recreational facilities and pile driving. Who is so talented? If they could be convinced, they would allow us - the members and experts of our industry - to write the codes, to review the new innovations and bring the state-of-the-art to the level it belongs.

In the U.S. it is slowly happening but our impatience is wearing thin. Either turn us loose or get them to attend the conferences such as this one so that they will understand that: 1) Aggregate separation does not occur while pouring concrete into a dry pile; 2) Vibrations from pile driving do not effect freshly poured concrete in a pile after its initial set; 3) Poured concrete in a pile need not be vibrated; 4) Up to 50% of yield stress should be considered in foundation design; 5) Driveability location and alignment tolerances should be analyzed on a job basis; 6) Static testing when implemented should always be carried to higher than twice the design value to perhaps give more capacity to the pile; 7) Quick testing should be considered; 8) Dynamic Testing with restrike should be used 9) Wave Equation with better pile/soil modeling capability should be developed. 10) Group testing should be implemented 11) Understanding of hammer systems should be learned, including types, energies, differences, cushion materials, leads/guides/cranes/power, etc. 12) Air and Noise Pollution should be addressed. These are only some of the items that are being addressed in the U.S.A. when piling is necessary for a project.

Also, as if in cooperation with the intent of this paper and this conference, the Deep Foundations Institute in this month's issue of Fulcrum reports that although early, the International Building Code (IBC) is to be produced by the year 2000. It is hoped that all of the codes will then be unified into one in the U.S.. More discussion on this will follow in future issues.

Special lectures

Design of Axially Loaded Piles – European Practice, De Cock & Legrand (eds)
© 1997 Balkema, Rotterdam, ISBN 90 5410 873 8

Improved pile design methods from field testing research

R.J.Jardine
Imperial College of Science Technology and Medicine, London, UK
F.C.Chow
Geotechnical Consulting Group (Formerly: Imperial College, London, UK)

ABSTRACT: Existing methods for predicting pile axial capacity are relatively unreliable. This paper illustrates the vital role that research involving static load testing can play in improving deep foundation design, describing an extended study undertaken by a group from Imperial College into the axial capacity of driven pipe piles. A precis is given of the main features of the new design procedures which have been developed from the work. The latter are shown to offer considerable theoretical and practical advantages over existing calculation procedures.

INTRODUCTION

Statistical assessments of the reliability of existing procedures for calculating axial pile capacity show that none of the currently popular methods is particularly reliable. As an example, Figure 1 shows the probability distributions generated by Briaud and Tucker (1988) on the basis of 98 pile load tests performed in the southern states of the USA. In such a plot an ideal method would present a Gaussian type curve centred around [predicted/measured] = 1.00, with a narrow spread of data falling to either side. None of the thirteen methods considered gave highly reliable results; several showed severe skews or very wide scatter bands.

Briaud and Tucker's data base mainly involved relatively short piles, of various types, installed for highway structures. In such circumstances static load tests can be used to check performance, and modify designs when necessary, so reducing the risk of failure or poor performance.

Most offshore (usually hydrocarbon producing) structures rely on large driven pipe piles to carry their foundation loads safely. The design of their piles merits special attention, as site specific load tests are rarely feasible and the consequences of failure to human life and the environment are enormous, as are the potential economic disbenefits. To meet this need, the international working groups of the American Petroleum Institute (API) have sought for many years to improve and update offshore pile design, and to optimise their foundation design recommendations. The work of the API has been recognised internationally and has been adopted as the basis for the new ISO recommendations for offshore structures. In addition, many Civil Engineers follow the API/ISO procedures when designing large driven piles for major structures such as bridges or harbours.

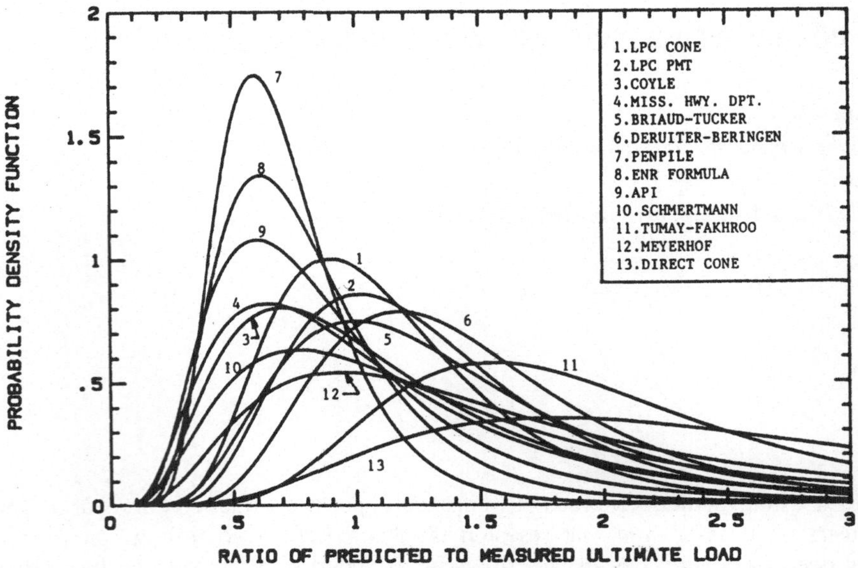

Figure 1. Probability distributions assessed by Briaud and Tucker (1988) for thirteen methods of predicting pile axial capacity. Based on 98 pile load tests.

The statistical reliability of the API RP2A 20th edition (1993) recommendations for assessing the axial capacities, Q, is generally considered to be at least as good for large driven pipe piles as that of any other conventional method. As detailed later, when distributions of $Q_{calculated}/Q_{measured}$ (or Q_c/Q_m) are assessed for test piles analysed according to the API procedures, the Coefficients of Variation (COV defined as the standard deviation, s, divided by the mean value, μ) fall between 0.5 and 0.7. While these COV values are better than those of most methods, they are not compatible with the relatively low safety margins (typically 1.5) adopted for most offshore pile designs.

At the same time, many offshore engineers consider the existing API methods to be over-conservative in certain circumstances. For example, it is recognised that in sand the approach leads to strong and systematic skewing of Q_c/Q_m with respect to pile slenderness and sand relative density, leading to underpredicted capacities for short piles in dense strata.

Foundation problems are relatively rare among the existing population of piled offshore structures. However, a clear need exists to improve predictive methods to obtain economies in some cases and enhance performance, and safety, in others. The development of improved methods needs to be co-ordinated with parallel developments in reliability theory and the characterisation of environmental loading.

A group from Imperial College, London has undertaken a research programme to improve the reliability of driven pile design. A brief resume of the work is given in the following paragraphs. More detailed descriptions of the experimental work, background theory and validation procedures are given in the references listed in the bibliography given at the end of the paper, while Appendix I defines the notation and symbols used to describe the results.

RESEARCH PROGRAMME

The Imperial College research programme has been aimed towards achieving: (i) a more fundamental and thorough understanding of pile behaviour and (ii) practical simple design methods that capture the basic mechanics of driven piles. The main tasks were to identify:

- How piles behave in different soils and layering sequences
- The scaling laws that relate the behaviour of models to that of full-scale piles
- The effects on capacity of pile properties (dimensions, wall thickness, end conditions, surface roughness, material hardness etc) and installation methods
- Any changes in capacity and stiffness associated with time after pile installation
- The response to different loading types, including group effects and cyclic loading
- The controlling soil parameters that should be measured in site investigations

Field tests with instrumented piles

Until recently, the stress conditions surrounding driven piles have been open to conjecture. A central feature of the recent IC research has been the development of accurate and reliable on-pile instrumentation to study the pore pressures, radial total stresses, local shear stresses and temperatures developed on pile shafts. The gauges were mounted on 6m to 20m long, 102mm diameter, closed-ended, steel pipe piles (termed ICPs) and used in intensive test programmes involving a wide range of geomaterials between 1986 to 1994 at the six sites identified in Table 1 and Figure 2.

Table 1. Summary of recent Imperial College pile research sites.

Site		Soil conditions
1.	Canons Park	London clay: stiff to very stiff, high plasticity Eocene marine clay; high YSR
2.	Cowden	Cowden till: stiff to very stiff lean glacial lodgement till; high YSR
3.	Bothkennar	Carse clay: soft, high plasticity, moderately organic Holocene marine-estuarine clay-silt, lightly cemented; moderate YSR
4.	Labenne	Dune sand: loose to medium dense, medium sized, Holocene; low YSR
5.	Pentre	Glacio-lacustrine clay-silt and laminated clays: very soft to firm, low plasticity and low YSR
6.	Dunkirk	Marine sand: dense to very dense shelly medium sized sand, Flandrian; low YSR

Note: Yield Stress Ratio (YSR) is the apparent OCR, as defined in Appendix I

The ICPs were installed by fast jacking, allowing comprehensive measurements of the effective stress conditions developed close to the shafts to be made at multiple levels during installation, long-term equalisation and load testing to failure. Detailed site investigations were also performed, involving in-situ tests and advanced laboratory experiments. 'Strain Path Method' numerical simulations of the ICP tests performed at Canons Park and Bothkennar were also carried out in conjunction with Professor A. Whittle from MIT as described by Bond (1988) and Lehane (1992).

The Pentre piles (Figure 2) were installed close to the large-scale (termed 'LDP') driven piles described by Clarke (1993); tests on less intensively instrumented open-ended driven piles were conducted at Canons Park, Cowden and Dunkirk.

Results from Phases 1, 2 and 3

The research has taken place in three Phases. The first involved developing the ICP instruments and experimental procedures, and performing multiple ICP tests and other experiments at the Building Research Establishment (BRE)'s Canons Park test site. The research was summarised by Bond (1988) and Bond and Jardine (1990).

The scope was broadened in Phase 2 to cover tests in sand at the French Ponts et Chausseés test site at Labenne, the BRE's stiff till site at Cowden and the EPSRC's national soft clay test bed-site at Bothkennar. At each location an advanced site investigation was performed, a field pile testing facility established and a programme of multiple (closed-ended) ICP tests carried out. Clear and striking results emerged from the experiments which allowed new design approaches to be proposed for closed-ended piles. The Phase 2 work was reported by Lehane (1992) and Jardine and Lehane (1994).

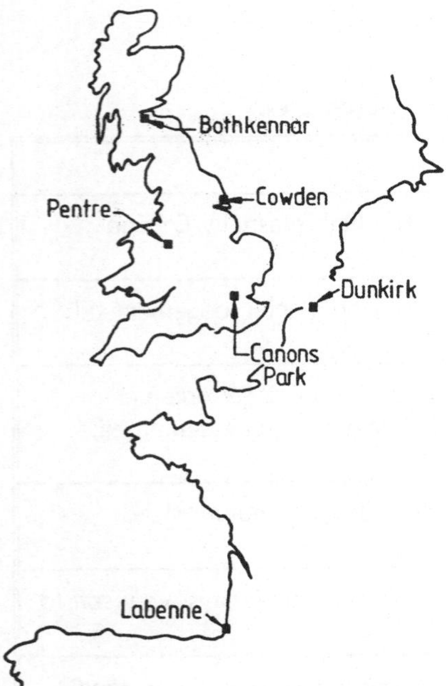

Figure 2. **Location of Imperial College test sites**

The third Phase, which has recently been completed, involved:

1. Establishing test facilities and performing advanced site investigations and multiple ICP tests at the Pentre (clay-silts/laminated clays) 'LDP' research site and at the Dunkirk 'CLAROM' dense sand research site.

2. Interpreting and performing tests on full-scale driven open-ended piles (with diameters up to 760mm) at the ICP sites to assess the effects of scale, installation methods and pile end conditions.

3. Experiments to assess pile group and ageing effects in dense sand.

4. Using the above to refine the new approaches for closed-ended piles and extend the design methods to cover open-ended driven piles.

5. Collating an up to date and critically approved data base of full-scale pile tests which met rigorous quality criteria.

6. Using the above to calibrate and validate the new methods for a wide range of practical applications.

The work is reported by Chow (1997) and Jardine and Chow (1996).

IMPROVED DESIGN PROCEDURES

New effective stress design procedures have been derived from the research for tubular piles installed in sands and clays. As summarised by Jardine and Chow (1996), completely new approaches have been developed for predicting the local distributions of radial effective stress developed on the pile shaft (i) once equilibrium has been reached after installation and (ii) during loading to failure (in either tension or compression). Listed in order of importance, the equilibrium local radial effective stresses, σ'_{rc}, developed in granular soils were found to be controlled by:

- The initial state of the sand, as expressed by the local CPT tip resistance q_c

- The relative position of the pile tip, characterised as the distance h between the pile tip and the layer in question, as identified in Figure 3, divided by R*, where:

$$R^* = (R^2_{outer} - R^2_{inner})^{1/2}$$

- The free field vertical effective stress σ'_{v0}

The most appropriate equation for local σ'_{rc} in granular soils is:

$$\sigma'_{rc} = 0.029\ q_c\ (\sigma'_{v0}/P_a)^{0.13}\ (h/R^*)^{-0.38}$$

With clays the values of local σ'_{rc} depend on the following factors, given again in the order of their importance:

- The free field vertical effective stress σ'_{v0}

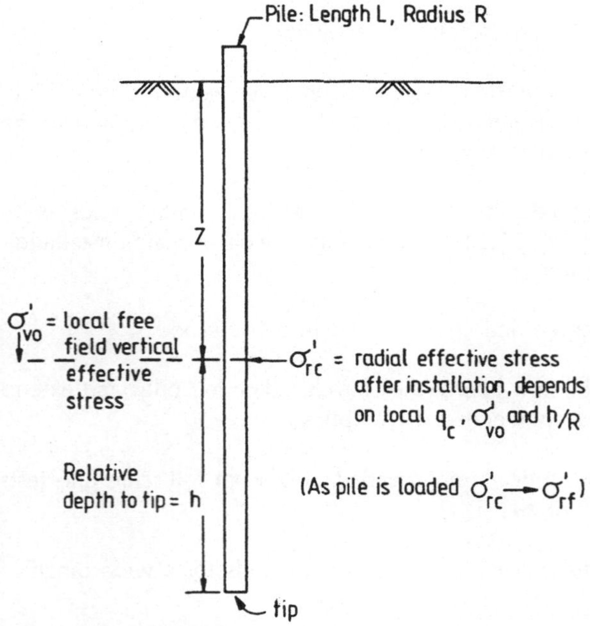

Figure 3. Definitions of terms for radial effective stresses

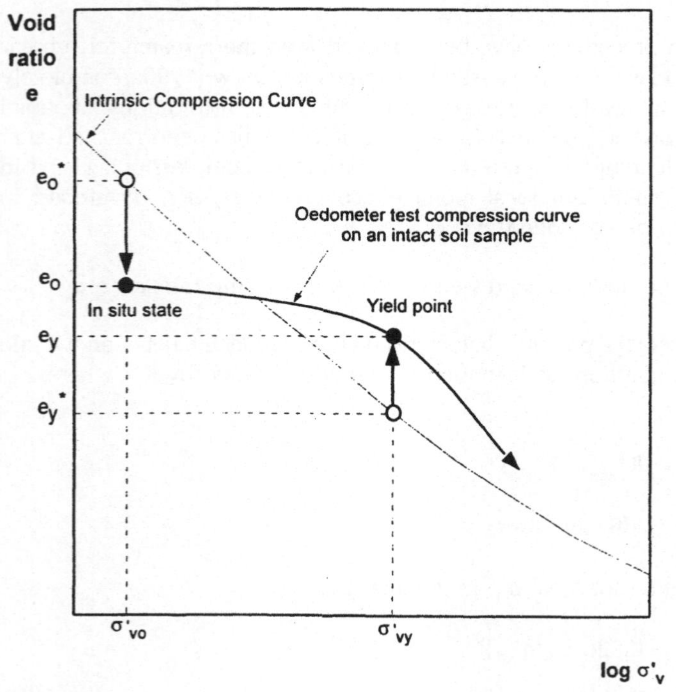

Figure 4. **Terms used to define clay sensitivity from oedometer tests. $I_{rr} = (e_0 - e_0^*)/C_c^*$; $\Delta I_{vy} = (e_y - e_y^*)/C_c^*$; YSR $= \sigma'_{vy}/\sigma'_{v0}$.**

- The initial state of the clay, as expressed by the local value of apparent overconsolidation ratio, or Yield Stress Ratio (YSR)

- As with sands, the relative depth to the pile tip, h/R*

- The initial sensitivity of the clay, expressed by either the conventional measure S_t or two new parameters I_{vr} or ΔI_{vy}, which are obtained by comparing the oedometer curves of intact and reconstituted samples, as shown in Figure 4.

The corresponding equations for local σ'_{rc} in clay soils are:

$$\sigma'_{rc} = \sigma'_{v0} \, [2 - 0.625 I_{vr}] \, YSR^{0.42} \, (h/R)^{-0.20} \quad \text{or}$$

$$\sigma'_{rc} = \sigma'_{v0} \, [2.2 + 0.016 YSR - 0.870 \Delta I_{vy}] YSR^{0.42} (h/R)^{-0.20} \quad \text{where}$$

$$\Delta I_{vy} = \log S_t$$

Note that for points where $h/R^* > 8$ the granular and clay equations should be evaluated with $h/R^* = 8$.

The new methods also specify the changes that should be expected in σ'_r as piles are loaded to failure in either clays or sands, with the changes in sands being generally more significant due to interface dilation and principal stress rotation effects. In clays:

$$\sigma'_{rf} = 0.8 \, \sigma'_{rc}$$

While in sands

$$\sigma'_{rf} = \sigma'_{rc} + \Delta\sigma'_{rd} \quad \text{in compression and}$$

$$\sigma'_{rf} = 0.9 \, (0.8 \, \sigma'_{rc} + \Delta\sigma'_{rd}) \quad \text{in tension}$$

Where:

$$\Delta\sigma'_{rd} = [4 \, G \, R_{cla}] / R$$

With R being the pile radius, R_{cla} the pile's centre-line average roughness (typically 10μm with steel piles) and G is the operational shear modulus. Jardine and Chow (1996) relate G to CPT resistance via a simple equation; the $\Delta\sigma'_{rd}$ term is relatively small for large diameter piles.

The local shaft resistances τ_{rf} are then evaluated as:

$$\tau_{rf} = \sigma'_{rf} \tan \delta$$

Where σ'_{rf} is the local σ'_r value at failure and δ is the ultimate interface shear friction angle, which should be measured in appropriate laboratory tests or evaluated from the charts given by Jardine and Chow (1996). There are no upper limits to τ_{rf} in granular soils and δ is independent of relative density and tends to decline with particle size. With clays, δ may be controlled by residual strength phenomena and often tends to decline with plasticity index. Jardine and Chow (1996) present strong evidence showing that shaft capacity grows with time in sands; the above equations predict the medium term capacity available at a nominal 50 days after driving.

In many respects the experimentally proven trends for σ'_{rf}, τ_{rf} and δ completely contradict those assumed in the API method of pile design. One vital feature of the new methods is the scale effects that they imply. In both sands and clays σ'_r decreases systematically as h/R* increases so that, when all other factors are constant, the average unit shaft resistance decreases markedly as the pile slenderness ratio, L/R, increases. With sands the local shaft friction capacity decreases with pile radius, because the component of capacity associated with $\Delta\sigma'_{rd}$ is inversely proportional to R.

The new design methods incorporate different rules for evaluating base resistance, and the potential for plugging with open ended piles. These rely principally on the local CPT resistance and incorporate a strong scale effect for piles in sand. Generally they predict lower resistances with large diameter piles than the API procedures. For the two case histories considered here the piles did not plug and following base capacity equation applies:

$$Q_b = q_c \, \pi \, (R^2_{outer} - R^2_{inner})$$

VALIDATION BY FULL SCALE STATIC TESTS

Jardine and Chow (1996) report how a large data base of full scale, high-quality, pile tests was assembled and used to validate the new procedures for practical use by Industry. As detailed below, the new methods are more than twice as reliable as the existing API procedures.

Piles in sand

Table 2 summarises the ranges of $Q_{calculated}/Q_{measured} = Q_c/Q_m$ found for predicted and measured shaft capacities when the new method and the API RP2A recommendations were tested against a new data base of 65 high quality pile tests. the three methods. We recall that Coefficient of Variation (COV) is defined as standard deviation (s) divided by the mean (μ). The COV and s should all be as low as possible.

The new IC method is clearly far more reliable than the API RP2A recommendations: it leads to a mean Q_c/Q_m close to unity, and a 50% lower COV. Jardine and Chow (1996) show that the new IC method eliminates the strong skewing produced by the API method with respect to relative density, D_r and normalised pile length, L/R. The new IC methods gave far better results and were equally reliable for open-ended and closed-ended piles and can be applied confidently to a wide range of conditions.

Similar results were obtained when the new procedures for estimating base resistance were tested against another data base of 42 pile tests, with the API predictions being strongly skewed with respect to pile diameter.

Piles in clay

The reliability of the new method for predicting medium term, single pile shaft capacity was assessed through comparisons with 55 test cases which met the stringent quality criteria set by Chow (1997). Table 3 summarises the ranges of Q_c/Q_m found for the new IC method and the API RP2A procedure.

Table 2 Assessment of shaft capacity predictions for 65 pile tests in sand

Method	Mean, μ	Standard deviation	COV
API RP2A(1993)	0.86	0.56	0.65
New IC method	0.97	0.28	0.30

Table 3. Assessment of peak clay shaft capacity predictions for 55 pile tests

Method	Mean, μ	Standard deviation, s	COV
API RP2A (1993)	0.98	0.33	0.34
New IC approach	1.01	0.18	0.18

The assessment confirms that the new Imperial College approach gives good predictions for axial capacity under a wide range of circumstances. In comparison with the API approach the method was marginally non-conservative, but produced less scatter.

Jardine and Chow (1996) show that the API procedure leads to systematic skewing with respect to the length ratio (L/R) and apparent overconsolidation ratio, or YSR. The new IC method works well with both open and closed ended piles and offers alternative ways of characterising clay sensitivity. These changes lead to significant improvements in reliability.

A similar exercise demonstrated that the new IC methods for base resistance in clay are also far more reliable than the existing API RP2A procedure.

SUMMARY

1. Existing procedures for calculating pile axial capacity have a low reliability. Site specific pile tests and research involving field experiments with piles are the key activities required to improve deep foundation design.

2. The value of field research has been illustrated by showing how new methods for designing piles driven in clays and sands have been developed through a long term research programme carried by a group from Imperial College, London.

3. No mention has been made of predicting load-displacement behaviour, but previous research on this area, including field measurements, has been reported by Jardine and Potts (1988), (1992).

4. The new capacity calculation procedures are relatively simple and can easily be applied in practice.

5. Some of the input soil parameters (particularly interface shear tests) are not normally

measured in routine site investigations and a new emphasis is placed on the importance of CPT testing (particularly with sands). Site survey practice will need to be revised to deal with this. Some parameters, such as clay sensitivity and YSR, may be derived by more than one type of procedure.

6. The new methods offer major advantages over the existing API approaches. When tested against a newly assembled data base of high quality field tests, they lead to much more reliable predictions for the medium-term shaft and base load capacities of single piles in both sands and clays.

7. The research work has also identified important effects of time and group action for piles in sand.

ACKNOWLEDGEMENTS

The most recent Imperial College research programme on "Offshore Piling" was funded by the UK Engineering and Physical Science Research Council (through MTD Ltd.), Amoco (UK) Exploration Company, Building Research Establishment, Health & Safety Executive and Shell Exploration and Production, with further support from the Institut Francais du Petrole (IFP). Additional participants in earlier JIP projects include British Petroleum, Conoco (UK) Ltd., Exxon Production Research Co., Mobil Research and Development Corporation and Saga Petroleum.

The support of the sponsors and the efforts of current and former colleagues at Imperial College, particularly Dr A Bond and Dr B Lehane, are also gratefully acknowledged.

APPENDIX I: LIST OF NOTATION

COV	=	Coefficient of variation: s/μ
C_u	=	Undrained shear strength
C_c	=	Coefficient of compressibility = $-\Delta e/\Delta \log \sigma'_v$ for natural soil
C_c^*	=	Coefficient of compressibility = $-\Delta e/\Delta \log \sigma'_v$ for reconstituted soil (on ICL)
D_r	=	Relative density (%)
e	=	Void ratio of natural soil (see Figure 4 for e_0, e_y etc)
e*	=	Void ratio of reconstituted soil on ICL (see Figure 4 for e^*_0, e^*_y etc)
G	=	Operational shear modulus
h/R	=	Normalised height above pile tip
I_{vr}	=	Relative void index, see Figure 4
ΔI_{vy}	=	Relative void index at yield, see Figure 4
K	=	σ'_r/σ'_{v0}
P_a	=	Atmospheric pressure
q_c	=	CPT end resistance
q_b	=	pile end bearing stress
Q_b	=	Base capacity
Q_s	=	Shaft capacity
R	=	Pile radius
R_{cla}	=	Centre-line average roughness
R*	=	Equivalent radius for open-ended piles
s	=	Standard deviation
S_t	=	Clay Sensitivity

YSR	=	Yield Stress Ratio, or apparent OCR, = $\sigma'_{vy}/\sigma'_{v0}$
δ	=	Interface angle of friction
δ_h	=	Radial normal movement (also δ_r)
σ'_{h0}	=	Free-field horizontal effective stress
σ'_r	=	Radial effective stress
σ'_{rc}	=	Equalised radial effective stress
σ'_{rf}	=	Radial effective stress at point of shaft failure
$\Delta\sigma'_r$	=	Change in σ'_r during loading [also $\Delta\sigma'_{rd}$ in sands]
σ'_{v0}	=	Free-field vertical effective stress
σ'_{vy}	=	Vertical effective yield stress
τ_f	=	Peak local shear stress
μ	=	Mean value

General meaning of subscripts

c	=	at equilibrium (after consolidation)
f	=	failure
0	=	free-field, before pile installation

BIBLIOGRAPHY AND REFERENCES

American Petroleum Institute (1993), RP2A-WSD: Recommended practice of planning, designing and constructing fixed offshore platforms - Working stress design, 20th edition, Washington, pp 59-61.

Baldi, G., Belottini, R., Ghionna, V.N., Jamiolkowski, M. & Lo Presti, D.L.F. (1989), "Modulus of sands from CPT's and DMT's", Proc. 12th Int. Conf. Soil Mech. & Fndn Engng, Rio de Janeiro, 1, pp 165-170.

Bond, A.J. (1989), "Behaviour of displacement piles in overconsolidated clays", Ph.D. Thesis, Univ. of London (Imperial College).

Bond, A.J. & Jardine R.J. (1990), "Research on the behaviour of displacement piles in an overconsolidated clay", UK Dept. of Energy OTH Report, OTH89 296, HMSO, London.

Bond, A.J., Jardine, R.J. & Dalton, J.C.P. (1991), "The design and performance of the Imperial College instrumented pile", Am. Soc. for Testing Materials, Geotech. Testing J., 14(4), pp 413-424.

Bond, A.J. & Jardine, R.J. (1991), "Effects of installing displacement piles in high OCR clay", Geotechnique 41(3), pp 341-363.

Bond, A.J. & Jardine, R.J. (1995) "Shaft capacity of displacement piles in a high OCR clay", Geotechnique, **45**(1), pp 3-23.

Briaud, J-L. & Tucker, L.M. (1988), "Measured and predicted axial load response of 98 piles", J. Geotech. Engng, ASCE, **114**(9), pp 984-1001.

Clarke, J. (1993), Editor of Conf. Large-Scale Pile Tests in Clay, London, 23-24 June 1992, Thomas Telford, London.

Chow, F.C. (1995), "Field measurements of stress interactions between displacement piles in sand", Ground Engng, **28**(6), pp 36-40.

Chow, F.C. (1997). "Investigations into displacement pile behaviour for offshore foundations". Ph.D Thesis, Univ. London (Imperial College)

Chow, F.C., Jardine, R.J., Nauroy, J.F. & Brucy, F. (1997), "Time-related increases in the shaft capacities of driven piles in sand", Géotechnique, (in press).

Jardine, R. J. and Potts, D. M. (1988) "Hutton Tension Leg Platform Foundations: an approach to the prediction of pile behaviour" Geotechnique, Vol.38, No.2, pp 231-252.

Jardine, R.J. & Christoulas, S. (1991), "Recent developments in defining and measuring static piling parameters", Proc. Conf. Fondations Profondes, Paris, Presse de l'ENPC, pp 713-745.

Jardine R J and Potts D M (1992) "Magnus foundations: Soil properties and predictions of field behaviour". Proc. Conf. Large scale pile tests in clay. Thomas Telford, London, pp 69-83.

Jardine, R.J., Lehane, B.M. & Everton, S.J. (1992), "Friction coefficients for piles in sands and silts", Proc. Conf. on Offshore Site Investigation and Fndn Behaviour, Society for Underwater Technol., London, Kluwer (Dordrecht), pp 661-677

Jardine, R.J. & Lehane, B.M. (1994), "Research into the behaviour of offshore piles: Field experiments in sand and clay", UK Health and Safety Executive, OTH Report, OTH93 401, HSE Books, London.

Jardine, R. J. and Chow, F.C. (1996), "New design methods for offshore piles". MTD Publication 96/103, Marine Technology Directorate, London.

Lehane, B.M. (1992), "Experimental investigations of pile behaviour using instrumented field piles", Ph.D. Thesis, Univ. of London (Imperial College).

Lehane, B.M., Jardine, R.J., Bond, A.J. & Frank, R. (1993), "Mechanisms of shaft friction in sand from instrumented pile tests", J. Geotech. Engng, ASCE, **119**(1) pp 19-35.

Lehane, B.M. & Jardine, R.J. (1994a), "Displacement pile behaviour in glacial clay", Can. Geotech. J., **31**(1), pp 79-90.

Lehane, B.M. & Jardine, R.J. (1994b), "Displacement-pile behaviour in a soft marine clay", Can. Geotech. J., **31**(2), pp181-191.

Lehane, B.M. & Jardine, R.J. (1994c), "Shaft capacity of driven piles in sand : A new design approach", Proc. Conf. Behaviour of Offshore Structures, pp 23-36.

Lehane, B.M., Jardine, R.J., Bond, A.J. & Chow, F.C. (1994), "The development of shaft resistance on displacement piles in clay", Proc 13th Int. Conf. Soil Mech. & Fndn Engng, New Delhi, 2, pp 473-476.

Design of Axially Loaded Piles – European Practice, De Cock & Legrand (eds)
© 1997 Balkema, Rotterdam, ISBN 90 5410 873 8

Some comparisons of safety for axially loaded piles

R. Frank
CERMES (ENPC-LCPC), Cité Descartes, Champs-sur-Marne, France

ABSTRACT : The various factors of safety for designing axially loaded piles in ultimate limit state situations are reviewed, as given in Eurocode 7-1 (CEN 1994) and in the French code *Fascicule* 62-V (MELT 1993). The overall safeties of both codes are compared, whether the bearing resistance is determined from ground test results (PMT rules) or from load test results. The results are found to be quite consistent. Finally, a proposal is made for replacing cases B and C (for fundamental combinations) of Eurocode 7-1 by a unique case, which is easy to establish for the bearing resistance of axially loaded piles.

1 INTRODUCTION

For designing axially loaded piles with regard to the ultimate bearing resistance, Eurocode 7 - Part 1 (CEN 1994) contains all the factors of safety to be applied, when calculations rules from ground test results are available or when pile load tests are performed. The document describes how to go from the measured value(s) of the ultimate bearing resistance to the characteristic value and, from the characteristic value, to the design value. The factors to apply to the loads in ultimate limit states (ULS) is also known.

In the Codes of practice of most countries the entire route for designing piles with regard to axial loads is also given.

Thus, it is possible, for the axial ultimate bearing resistance problem, to compare the overall safety brought by the Eurocodes with the one of national practices.

In the following, the comparisons are made with the French code of practice *Fascicule* 62-V (MELT 1993). Piles in compression only are examined and no group effect is taken into account, for simplicity reasons.

2 BASIC RELATIONS

The basic relations of Eurocode 7-1 for ultimate bearing resistance are :

$$R_{ck} = R_{cm}/\xi \qquad (1)$$

$$R_c = R_{ck}/\gamma_t$$

$$\text{or } R_c = R_{bk}/\gamma_b + R_{sk}/\gamma_s \qquad (2)$$

where R_{cm} = measured value of ultimate bearing resistance, R_{ck} = characteristic value, and R_c = design value, with R_{bk} = characteristic base resistance and R_{sk}, = characteristic shaft resistance, ξ = factors to apply on measured values, γ_t , or γ_b and γ_s = partial factors of safety on the characteristic pile resistance, or on the base and shaft resistances, respectively.

On the other hand, the applied characteristic loads F are multiplied by factors γ_F in order to establish the compression design load F_c :

$$F_c = \gamma_F F \quad (3)$$

The basic condition to fulfil for all ultimate limit states (ULS) is :

$$F_{c(ULS)} \leq R_c \quad (4)$$

Relations 1 through 4 lead to :

$$F \leq R_{cm}/ \gamma_F \cdot \gamma_t \cdot \xi = R_{cm}/\mathbf{FS} \quad (5)$$

FS = $\gamma_F \cdot \gamma_t \cdot \xi$ can be called the global factor of safety. Note that it is understood in relation (5) that γ_F synthesises loads of different nature, if relevant, and that γ_t synthesises γ_b and γ_s , in case different factors are used on base and shaft.

3 VALUES OF γ FACTORS

3.1 *Eurocode 7-1*

For Eurocode 7-1, in fundamental (persistent and transient) ULS situations, the values of γ_F for cases B and C and γ_t , or γ_b and γ_s , for case C are given in Tables 1 and 2, respectively. For case B, γ_t , or γ_b and γ_s = 1.00. In accidental situations all these factors are equal to 1.00.

3.2 *Fascicule* 62-V

For fundamental ultimate limit states (persistent and transient situations), the partial load factors of *Fascicule* 62-V are very similar to case B of Eurocode 7. In particular : 1.35 on unfavourable permanent loads ; 1.00 on favourable permanent loads and 1.5 on the basic variable action (except for well known service loads, or loads having a specific character). The combination of loads for accidental situations is the same as in Eurocode 7 (for more details, see Frank 1994).

In *Fascicule 62*-V, the same factor is used for the base resistance and for the shaft resistance. This value is :

$\gamma_b = \gamma_s = \gamma_t = 1.4$

Table 1. Eurocode 7-1 : Partial factors γ_F - ULS fundamental situations.

	Actions		
	Permanent		Variable
Case	Unfavou-rable	Favou-rable	Unfavou-rable
B	(1.35)	(1.00)	(1.50)
C	(1.00)	(1.00)	(1.30)

Table 2. Eurocode 7-1 : Partial factors γ_b, γ_s and γ_t - ULS fundamental situations, case C.

Component factors	γ_b	γ_s	γ_t
Driven piles	(1.3)	(1.3)	(1.3)
Bored piles	(1.6)	(1.3)	(1.5)
CFA (Continuous flight auger) pile	(1.45)	(1.3)	(1.4)

Note that for case B, γ_t , γ_b and γ_s = 1.00.

for the ULS fundamental combinations;

$\gamma_b = \gamma_s = \gamma_t = 1.2$

for the ULS accidental combinations.

4 ASSESSMENT OF CHARACTERISTIC VALUES OF BEARING RESISTANCE

Two different cases must be distinguished (see 7.6.3 in Eurocode 7-1) :
i) the design is made on the basis of ground tests results ;
ii) the design is made on the basis of pile load test results.

4.1 *Design from ground test results*

When ground tests results are used, the requirement of Eurocode 7-1 is stated under clause (4)P of § 7.6.3.3 :

"The characteristic values q_{bk} and q_{sik} shall be derived from calculation rules based on established correlations between the results of static load tests and the results of field or laboratory ground tests. These calculation rules shall be devised such that ultimate bearing resistances using the characteristic values q_{bk} and q_{sik} do not exceed the measured ultimate bearing resistances used for establishing the correlation divided by [1.5], on average." (where. q_{bk} = characteristic value of the resistance per unit area of the base and q_{sik} = characteristic value of the resistance per unit area of the shaft in layer i).

In other words ξ must be equal to 1.5 'on average' : $\xi_{ave} = 1.5$

In *Fascicule* 62-V, calculation rules for two field ground tests are given : Ménard pressuremeter tests (PMT) and cone penetration tests (CPT). Details of these rules are described in the national report for France by Bustamante and Frank (1997). Both set of rules come from correlations with the results of numerous static load tests (originally developed by Bustamante and Gianeselli 1981). Note also that static load tests are performed, in France, in a very consistent manner with the procedure of ISSMFE (1985) recommended by Eurocode 7-1.

For comparison with Eurocode 7-1, the key problem for all the calculation rules is to know the value of ξ_{ave}. For the PMT rules contained in *Fascicule* 62-V, a preliminary assessment has been made on a limited number of pile load tests (Renault 1996). This assessment leads to ξ_{ave} = 1.25, without distinguishing between driven or bored piles for the time being (see Figure 1).

4.2 *Design from pile load test results*

The values of ξ proposed by Eurocode 7-1 are given in Table 3. The minimum bearing resistance obtained from the application of both conditions a) and b) is to be used.

In the case of *Fascicule* 62-V :

- if only one single pile load test is performed :

$$\xi = 1.2 \qquad (6)$$

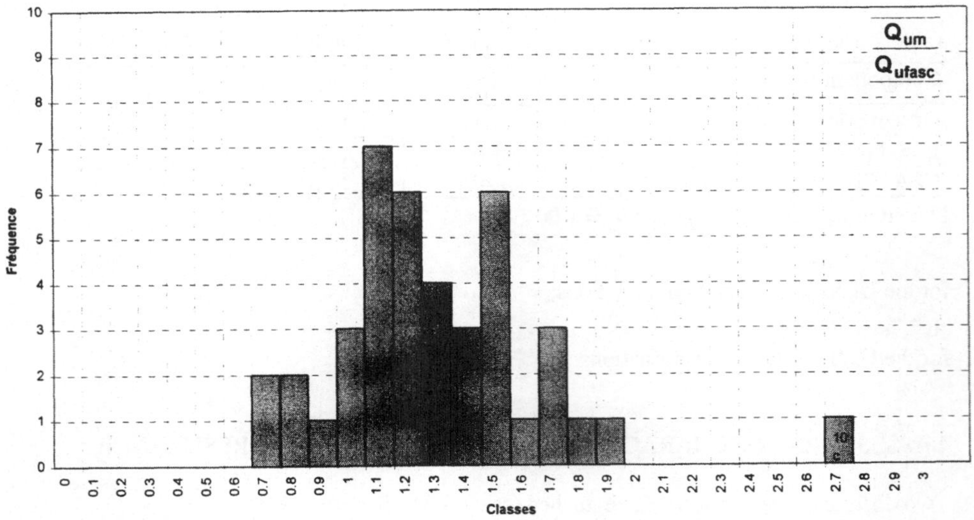

Figure 1. Values of $\xi = R_{cm}/R_{ck}$ for 42 pile load tests in the case of the PMT rules of *Fascicule* 62-V (Renault 1996).

Table 3. Eurocode 7-1 : Factors ξ to derive R_{ck}

Number of load tests	1	2	> 2
a) Factor ξ on mean R_{cm}	(1.5)	(1.35)	(1.3)
b) Factor ξ on lowest R_{cm}	(1.5)	(1.25)	(1.1)

Table 4. *Fascicule* 62-V : Values of factor ξ'.

Number of load tests	2	3	4	5
ξ'	0.55	0.20	0.07	0.00

- if several load tests are performed :

$$R_{ck} = R_{min} (R_{min}/R_{max})^{\xi'} \quad (7)$$

where R_{min} and R_{max} are the minimum and maximum measured values and ξ' is given in Table 4.

5 COMPARISONS OF GLOBAL SAFETY

In the following, the values of FS = $\gamma_F.\gamma_t.\xi$ (Equation 5) is established for Eurocode 7-1 and for *Fascicule* 62-V. For the ULS fundamental combinations all the actions are considered to be unfavourable.

5.1 *Design from ground test results*

In the case of *Fascicule* 62-V, the PMT design rules are used. For Eurocode 7-1, as in

Fascicule 62-V, the same factors γ_t are used for the base and shaft resistances.

i) fundamental combinations :

Eurocode 7-1 :

Case B : 100% of permanent loads :

FS = 1.35 x 1.0 x 1.5 = 2.03 (8)

67% of permanent loads and 33% of variable loads :

FS = 1.4 x 1.0 x 1.5 = 2.10 (9)

50% of permanent loads and 50% of variable loads :

FS = 1.425x 1.0 x 1.5 = 2.14 (10)

Case C :

driven piles, 100% of permanent loads :

FS = 1.0 x 1.3 x 1.5 = 1.95 (11)

67% of permanent loads and 33% of variable loads :

FS = 1.1 x 1.3 x 1.5 = 2.15 (12)

50% of permanent loads and 50% of variable loads :

FS = 1.15 x 1.3 x 1.5 = 2.24 (13)

bored piles, 100% of permanent loads :

FS = 1.0 x 1.5 x 1.5 = 2.25 (14)

67% of permanent loads and 33% of variable loads :

FS = 1.1 x 1.5 x 1.5 = 2.48 (15)

50% of permanent loads and 50% of variable loads :

FS = 1.15 x 1.5 x 1.5 = 2.59 (16)

Fascicule 62-V :

100% of permanent loads :

FS = 1.35 x 1.4 x 1.25 = 2.36 (17)

67% of permanent loads and 33% of variable loads :

FS = 1.4 x 1.4 x 1.25 = 2.45 (18)

50% of permanent loads and 50% of variable loads :

FS = 1.425x 1.4 x 1.25 = 2.49 (19)

ii) accidental combinations (all γ_F = 1.00) :

Eurocode 7-1 :

FS = 1.0 x 1.0 x 1.5 = 1.50 (20)

Fascicule 62-V :

FS = 1.0 x 1.2 x 1.25 = 1.50 (21)

It can be seen that the overall safety of both codes is very similar. On average, *Fascicule* 62-V (PMT rules) is slightly more conservative for driven piles under fundamental combinations (up to + 16 %). These comparisons will have to be confirmed when the assessment of ξ_{ave} will have been completed.

5.2 *Design from pile load test results*

A simple (theoretical) case

A CFA (continuous flight auger) pile is considered and the results of only one pile load test is available, without separate measurement of base and shaft resistances (i.e. for Eurocode 7-1, as in *Fascicule* 62-V, the factor γ_t is used).

i) Fundamental combinations and permanent loads only :

Eurocode 7-1 :

Case B : FS = 1.35 x 1.0 x 1.5 = 2.03 (22)

Case C : FS = 1.0 x 1.4 x 1.5 = 2.10 (23)

Fascicule 62-V :

FS = 1.35 x 1.4 x 1.2 = 2.27 (24)

ii) Accidental combinations :

Eurocode 7-1 :

FS = 1.0 x 1.0 x 1.5 = 1.50 (25)

Fascicule 62-V :

FS = 1.0 x 1.2 x 1.2 = 1.44 (26)

Here also, the overall safety of both codes is very similar, *Fascicule* 62-V being slightly more conservative for fundamental combinations (+ 8%).

Comparison on real cases

The two examples come from a real project for which the application of Eurocodes was requested. The project included several important bridges and viaducts on piles. The design of the pile foundations can be summarised in the following way (Baguelin and Frank 1995).

Bored piles, results of one load test

i) Fundamental combinations and 50% of permanent loads and 50% of variable loads :

Eurocode 7-1 :

Case B : FS = 1.425x 1.0 x 1.5 = 2.14 (27)

Case C : FS = 1.15 x 1.5 x 1.5 = 2.59 (28)

Fascicule 62-V :

FS = 1.425x 1.4 x 1.2 = 2.39 (29)

ii) Accidental combinations :

Eurocode 7-1 :

FS = 1.0 x 1.0 x 1.5 = 1.50 (30)

Fascicule 62-V :

FS = 1.0 x 1.2 x 1.2 = 1.44 (31)

Driven piles, results of 4 pile load tests

i) Fundamental combinations and 67% of permanent loads and 33% of variable loads :

Eurocode 7-1 :

Case B : FS = 1.4 x 1.0 x 1.3 = 1.82 (32)

Case C : FS = 1.1 x 1.3 x 1.3 = 1.86 (33)

Fascicule 62-V :

FS = 1.4 x 1.4 x 1.14 = 2.23 (34)

ii) Accidental combinations ($\gamma_F = 1$) :

Eurocode 7-1 :

FS = 1.0 x 1.0 x 1.3 = 1.30 (35)

Fascicule 62-V :

FS = 1.0 x 1.2 x 1.14 = 1.37 (36)

Again, for the real cases examined here, both codes are in good agreement, *Fascicule* 62-V being more conservative for fundamental combinations in the case of 4 load tests (+ 20%).

6 A UNIQUE CASE REPLACING CASES B AND C

It appears that, in the case of axially loaded piles, cases B and C of the fundamental ULS of

Eurocode 7-1 could be replaced by a unique case U, which would have the following features :

- use for γ_F the values of case B, see Table 1 ;
- use for γ_b and γ_s , or γ_t , values slightly lower than for case C, but respecting for γ_b and γ_t the hierarchy of the piles (see Table 2) :

γ(driven) $\leq$ γ(CFA)$\leq$ γ(bored)

- as a consequence, the value of ξ_{ave} for the design from the ground test results, and the values of ξ in Table 3 for the design from load test results should be lower, perhaps near those of *Fascicule* 62-V.

The following values are proposed for the time being :

$\gamma_{t(driven)} = 1.2$; $\gamma_{t(CFA)} = 1.3$ and $\gamma_{t(bored)} = 1.4$

ξ_{ave}= 1.2 for ground test results , ξ = 1.2 when the results of one pile load test are available and ξ = 1.1 when the results of more than two pile load tests are available.

The application of this unique case U to the above examples is now given.

6.1 *Design from ground test results*

driven piles, 100% of permanent loads :

FS = 1.35 x 1.2 x 1.2 = 1.94 (37)

67% of permanent loads and 33% of variable loads :

FS = 1.4 x 1.2 x 1.2 = 2.02 (38)

50% of permanent loads and 50% of variable loads :

FS = 1.425x 1.2 x 1.2 = 2.05 (39)

bored piles, 100% of permanent loads :

FS = 1.35 x 1.4 x 1.2 = 2.27 (40)

67% of permanent loads and 33% of variable loads :

FS = 1.4 x 1.4 x 1.2 = 2.35 (41)

50% of permanent loads and 50% of variable loads :

FS = 1.425x 1.4 x 1.2 = 2.39 (42)

These results are, in most cases, very near the ones of Equations (8) through (16) and never differ more than 10 %. Incidentally, the PMT rules of *Fascicule* 62-V (Equations 17 to 19) are now more conservative, on average, in all cases of fundamental combinations (+ 22% for driven piles and + 4% for bored piles).

6.2 *Design from ground test results*

Theoretical example : CFA pile, one pile load test

FS = 1.35 x 1.3 x 1.2 = 2.11 (43)

This result is to be compared with the ones of Equations (22) and (23). It is quite satisfactory. *Fascicule* 62-V (Equation 24) is 8% more conservative.

Real case : bored piles, one pile load test, 50% permanent and 50% variable loads :

FS = 1.425 x 1.4 x 1.2 = 2.39 (44)

This result is to be compared with the ones of Equations (27), and (28). It is quite satisfactory. It is exactly the same as for *Fascicule* 62-V (Equation 29).

Real case : driven piles, 4 load tests, 67% permanent and 33% variable loads :

FS = 1.4 x 1.2 X 1.1 = 1.85 (45)

This result is to be compared with the ones of Equations (32) and (33). It is quite satisfactory. *Fascicule* 62-V (Equation 34) remains some 20 % more conservative.

7 CONCLUSIONS

Though the approaches may seem different and there are some differences in the values of partial factors of safety, the overall safeties of Eurocode 7-1 and *Fascicule* 62-V are quite similar, when designing axially loaded piles with regard to the ultimate bearing resistance.

The design rules from field test results (PMT and CPT) given in *Fascicule* 62-V need to be further compared to the results of pile load tests in order to check their compatibility with the requirements of Eurocode 7-1.

For the problem of bearing resistance of axially loaded piles, a unique case for ULS fundamental combinations, replacing cases B and C of Eurocode 7-1, can easily be found.

REFERENCES

Baguelin F. & R. Frank 1995. *Private communication.*

Bustamante M. & R. Frank 1997. Design of axially loaded piles in France : National Report. *Proc. Int. Seminar Design of axially loaded piles, European practice, Brussels, 17-18 April 1997.*

Bustamante, M. & L. Gianeselli 1981. Prévision de la capacité portante des pieux isolés sous charge verticale. Règles pressiométriques et pénétrométriques. *Bull. Liaison Labo. P. et Ch.* 113 : 83-108 (in French).

CEN (Comité Européen de Normalisation) 1994. *ENV 1997-1 Geotechnical design - General rules. 30 October 1994, CEN/TC 250/SC7.* CEN : Brussels.

Frank, R. 1994. The new Eurocode and the new French code for the design of deep foundations. Keynote lecture, *Proc. FHWA Int. Conf. Design and Construction of Deep Foundations, Orlando, Florida, 6-8 December 1994, Vol. 1* : 279-304. FHWA : Washington, D.C.

ISSMFE Subcommittee on Field and Laboratory Testing 1985. Axial pile loading test - Part 1 : Static loading. *Geotechnical Testing J.* 8(2) : 79-90.

MELT 1993. *Règles techniques de conception et de calcul des fondations des ouvrages de génie civil, CCTG, Fascicule N°62 - Titre V* (in French). Ministère de l'Equipement du Logement et des Transports : Paris.

Renault J. 1996. *Private communication.*

Design of Axially Loaded Piles – European Practice, De Cock & Legrand (eds)
© 1997 Balkema, Rotterdam, ISBN 90 5410 873 8

Pile dynamic testing, driving formulae, monitoring and quality control: Background for discussion

A. Holeyman
U.C.L., Louvain-la-Neuve, Belgium

ABSTRACT : Background information is provided for discussion on pile dynamic testing, driving formulae, monitoring and quality control as those practices relate to the design of piles. A general framework to classify pile dynamic testing is presented. Topics for discussion are suggested.

1. PILE DYNAMIC TESTING

Dynamic testing of piles has become part of many modern civil engineering projects as a result of the increased availability and performance of testing equipment and interpretation procedures (Holeyman, 1992). Two major classes of methods can be distinguished on the basis of energy level and intent, as presented in Figure 1 : high-strain dynamic testing methods primarily intended to provide bearing capacity information and low-strain dynamic testing methods primarily intended to provide information on pile integrity. Amongst the dynamic testing methods, the impact method is the most closely related to pile design considerations. The most common high-strain dynamic testing involves dropping a mass on the head of a pile and is addressed by ASTM Standard D-4945-89 (1989).

The load-bearing behaviour may be summarised by an allowable load or described by a complete load-settlement curve generally derived from distributed shaft and toe resistance terms. The interpretation following that procedure is based on the wave-equation theory and both force and velocity signals measured at the pile head during impact. The dynamic nature of the load and the time consumption for the shock wave to travel down and back up allows one to relate time to position down the pile and resolve soil reactions distributed along the shaft and at the toe.

Resolution of the shaft resistance terms versus depth (depth resolution) is afforded by the sharp increase of the force at the wave front and by the short length or duration of the original wave form. The sharpness of the wave relative to the pile characteristics can be used as a criterion to separate different types of "dynamic" pile tests. Table 1 provides a summary of key attributes of several known pile test types. Of particular significance to this discussion is the relative wave length Λ, which represents the length of the force pulse in terms of the double length (2L) of the pile. It can be noted from Table 1 that integrity testing is typically characterised by a relative wave length of 0.1, which provides for the sharpest depth resolution available. The dynamic bearing capacity test is typically characterised by a relative wave length of 1, which still allows for depth resolution while providing high-strain testing.

Longer-duration impacts, such as generated by the Dynatest (Gonin et al., 1984) or the Statnamic Test (Bermingham and Janes, 1989), are characterised by a relative wave length Λ of 10 or higher and, therefore, do not allow for depth resolution. It is suggested that, although those tests resort to inertial actions on masses to generate their extended force pulse, they be referred to as "kinetic tests" mainly because the inertial forces within the pile are small compared to the current force being applied and because the interpretation of these tests does not significantly benefit from the use of the wave equation framework.

PILE DYNAMIC TESTING METHODS

HIGH-STRAIN PRIMARILY FOR BEARING CAPACITY	LOW-STRAIN PRIMARILY FOR INTEGRITY
IMPACT	**IMPACT**
Using - Pile driving equipment - Pile testing equipment	*Using* - Hand held hammer - Explosion - Piezocrystals
Measuring - Strain and velocity	*Measuring* - Velocity and optionally force
1997 State-of-the-practice	*1997 State-of-the-practice*
VIBRATION	**VIBRATION**
Using - Pile vibratory driving equipment	*Using* - Pile testing vibrator
Measuring - Strain and velocity	*Measuring* - Force and velocity
Not presently developed	*Used 1965 - 1985*
PROLONGED IMPACT	**SONIC LOGGING**
Using - Soft springs with heavy hammer - Explosives slowly burning in engineered chamber	- Crosshole - Single hole - Parallel seismic using piezocrystals/hand held hammer and measuring pressure at depth
Measuring - Head displacement and force	
1997 State-of-the-art	*Requires non-standard pile set-up*

Fig. 1 - Pile dynamic testing methods

Figure 2 provides a representation of the pile tests available in terms of relative wave length Λ and of acceleration. Figure 2 also presents typical times expressed in terms of relative wave lengths required to reach 90% consolidation around a pile in sand, silt and clay. This diagram allows, in the writer's opinion, the separation between dynamic, kinetic, and static testing. Compared to static tests, one is faced with the difficulty in kinetic tests of sorting out the velocity dependency on the soil resistance, and in dynamic tests of resolving dynamic effects with, however, the advantage of depth resolution.

Table 1. Typical Key Attributes of Different Types of Pile Tests

	Integrity Testing	High-Strain Dynamic Testing	Kinetic Testing	Static Testing
Mass of Hammer	0.5 - 5 kg	2,000 - 10,000 kg	2,000 - 5,000 kg	N/A
Pile Peak Strain	2 - 10 μstr	500 - 1,000 μstr	1,000 μstr	1,000 μstr
Pile Peak Velocity	10 - 40 mm/s	2,000 - 4,000 mm/s	500 mm/s	10^{-3} mm/s
Peak Force	2 - 20 kN	2,000 - 10,000 kN	2,000 - 10,000 kN	2,000 - 10,000 kN
Force Duration	0.5 - 2 ms	5 - 20 ms	50 - 200 ms	10^{7} ms
Pile Acceleration	50 g	500 g	0.5 - 1 g	10^{-14} g
Pile Displacement	0.01 mm	10 - 30 mm	50 mm	> 20 mm
Relative Wave Length Λ	0.1	1.0	10	10^{8}

Primary difficulties and limitations associated with high-strain testing are the conversion of the dynamically mobilised resistance measured during the test into static resistance and the limited transient displacement enforced by the impact. Conversion of dynamic resistance into static resistance is rendered difficult in part because of the following effects:

- Inertial and radiation-damping effects, which are frequency-dependent,
- Differences in the deformation pattern along the shaft and at the base between dynamic and static loading,
- Effect of pore-pressure generation and dissipation, and
- Dependence of the soil's modulus and shear strength on velocity.

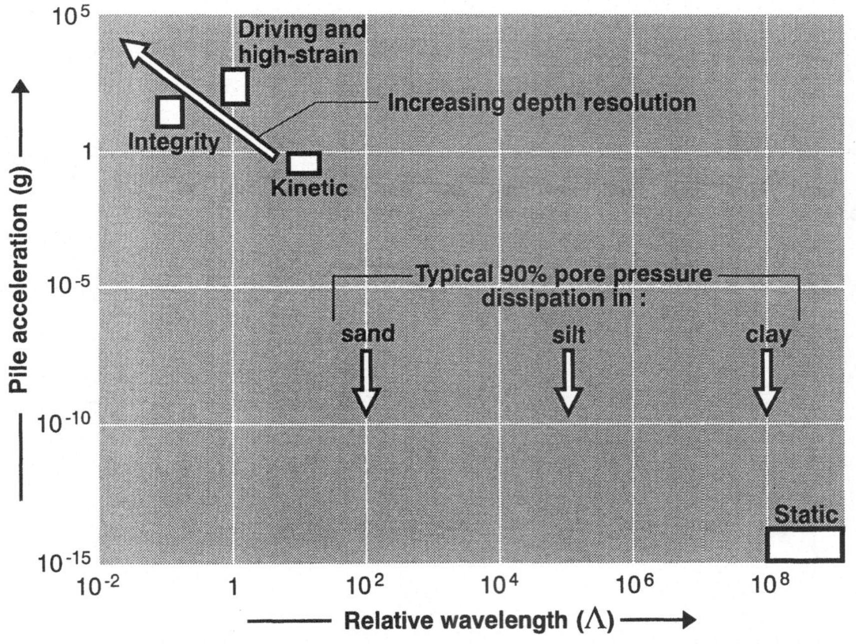

Fig. 2 - Sharpness and duration of force pulse for different pile tests

For driven piles monitored during driving, one must also contend with the effects of cyclic pore pressure generation and soil set-up (or relaxation). Also, and less often mentioned, reliability problems of measurements, especially of the force for cast-in-place piles, and velocity and displacement in general must be contented with. Finally, the development, commercial success, and persistence of early simplistic models, which still represent the bulk of the practice, have deterred most end users from addressing the complexity of the phenomena at hand.

High-strain vibration, although easily implementable in practice, has not seen many applications. Vibrators are regularly used to install sheet piles; however, in that case, axial capacity is not usually a primary concern. Also, vibrations imply cyclic loading, which generates an additional difficulty in the interpretation because of pore-pressure generation and fatigue effects (Holeyman and Legrand, 1996).

2. DRIVING FORMULAE

Pile driving formulae can be viewed as a particular case of using installation monitoring to derive information on the performance of piles. Generally based on energy considerations, those formulae strive to provide a direct relationship between the set and the total downward ultimate bearing capacity (some say "driving resistance" or resistance during driving). Driving formulae usually take into account the rated or observed energy of the hammer, some loss of energy within the driving system, and an assumed or measured rebound of the pile.

Most driving formulae are a particular case of the following general expression :

$$\eta_i \, \eta_c \, Mgh = \frac{1}{2} Q_D \, s_{el} + Q_D \cdot s$$

in which
- η_i is the efficiency of impact
- η_c is the efficiency of drop
- M is the mass of the hammer
- g is gravity (9.81 m s^{-2})
- h is the stroke of the hammer
- Q_D is the pile driving resistance
- s_{el} is the transient displacement or "elastic" rebound.
- s is the set

Driving formulae still appearing in special provisions of some specifications include those of Eytelwein (or Dutch formula), Janbu (or Danish formula), Hiley and Delmag (after Crandall). Their parameters are summarised in the following table :

	η_i	η_c	s_{el}
Dutch	$\frac{1}{1+\mu}$	1	0
Danish	1	0.7 to 1	$\left(\frac{2\eta_c \, Mgh \, L}{A_p \, E}\right)^{1/2}$
Hiley	$\frac{1+e_r^2 \, \mu}{1+\mu}$	0.75 to 1	$C_1 + \frac{Q_D \, L}{A_p \, E} + C_3$
Delmag	$\frac{1}{1+\mu}$	1	0.6 10^{-3} L

with L = length of pile [m]
A_p = section of pile [m^2]
E = modulus of deformation of pile [Mpa]
M_p = mass of pile [Mkg]
μ = M_p/M
e_r = coefficient of restitution [-]
C_1 and C_3 = Hiley constants obtained from tables ($0 < C_1 + C_3 < 12\ 10^{-3}$ [m])

The large factors of safety typically used (4 to 12) to adopt the allowable value of the pile resistance form the interpreted driving ultimate resistance underline the low reliability of that approach as a design tool.

3. INSTALLATION MONITORING

Besides driven piles discussed above, it appears that little information is available to relate installation parameters to the bearing capacity of a pile. The writer anticipates that efforts pursued in that direction could provide, if not the bearing capacity, at least some confirmation of the installation dependent portion of the bearing capacity. More specifically, methods should be developed and promoted to reward and recognise increases in bearing capacity that depend on monitored processes and workmanship (e.g. lower soil relaxation around excavated piles).

Installation monitoring, as perceived by the writer, appears to be considered as a component of the quality control of the installation, more than a design related tool.

4. QUALITY CONTROL

The quality and resulting performance of piles can be controlled from "cradle to grave". Components of a quality control program relate to materials, installation and fabrication process, and the finished product. In present piling practice, materials are systematically controlled, installation and fabrication often controlled, and the finished product less often controlled. As shown in Fig. 1, low-strain dynamic testing can provide means to verify the integrity of the finished product.

The most common low-strain dynamic testing involves hitting the pile head using a hand-held hammer and monitoring the pile head to obtain its transient velocity, and optionally the impact force. This test is well documented, but is not, to the writer's knowledge, the object of a national standard. The primary objective of the low-strain dynamic test is to assess the integrity of the pile as a structural member. Anomalies that impair the integrity of a pile and that are expected to be identified by integrity tests include the presence of material of poorer quality than expected (locally and overall) and variations in the cross section of the shaft (e.g., crack, necking, and bulb). Additionally, some idea of the pile and soil behaviour at low-strain may be inferred. Because the primary information offered by the test is the manner in which waves travel and are reflected within the pile material, pile material strain during those integrity tests has a typical maximum of only 2 to 10 μstr.

Primary difficulties associated with low-strain integrity testing are:

- Test repeatability (improved to some degree by signal averaging),
- Elimination of spurious vibrations (in hammer and Rayleigh wave effects),
- Discrimination between soil resistance and shaft impedance effects,
- Difficulty in identifying gradual changes in shaft section,
- Masking of potential necking below bulb,
- Historical distrust of engineering community towards results stemmed from early days, and
- Lack of one simple, quantitative, and rational interpretation method.

Other low-strain methods are used to investigate the integrity of piles, although not exclusively relying on the transmission of longitudinal waves. These are the Parallel Seismic Testing,

Crosshole Sonic Logging, and Single hole Sonic Logging (Stain, 1982). These three methods require the provision of casings outside or within the pile shaft.

Parallel Seismic Testing is typically used when the pile head is not accessible. A bore-hole is drilled immediately adjacent and parallel to the pile, and a slotted tube is installed. The boring is usually drilled to within 1 m of the shaft and at least 3 to 5 m deeper than the presumed pile depth. The cased hole is filled with water, and a hydrophone is lowered down the hole to monitor, at regular depth intervals (typically 0.5 m), the water pressure wave resulting from the impacts imparted on a structural element directly connected to the pile head. Wave arrival time delays are plotted versus depth in order to identify the deep foundation bottom.

Crosshole and single-hole sonic logging are typically used to evaluate the concrete condition of drilled shafts and slurry walls. Casing within the pile generally consists of water-filled tubes attached to the reinforcement cage before the casting of concrete. Ultrasonic pulses are generated by a piezoelectric motion generator (source), and the resulting water pressure waves are recorded by a hydrophone (receiver). Pulses have a typical duration of 50 microseconds (μs) and result in a concrete strain on the order of 0.1 μstr. Crosshole logging is performed by simultaneously lowering source and receiver into separate tubes; single hole logging is performed by lowering a source/receiver assembly, separated by a fixed depth interval, into a single hole. Wave arrival time delays and amplitudes are interpreted with a view to identifying zones with poor quality concrete, voids, intrusions, and breaks.

Difficulties and present limitations associated with seismic and sonic logging are:

- Planning requirement and interference with construction process,
- Control of casing positions,
- Quality of mechanical contact between tube and concrete,
- Defect must fully separate receiver from source (i.e., defect boundary must ideally intercept casing to be detected), and
- Qualitative more than quantitative interpretation.

5. SUGGESTIONS FOR DISCUSSION

As a preliminary approach to be adjusted based on a more complete review of questions raised at the Seminar, the following topics are suggested for discussion :

5.1 Pile dynamic testing

- Dependence of mobilised load on energy level
- Need for correlation with static loading tests
- Use for non driven piles
- European standard.

5.2 Pile Driving Formulae

- Standard for set measurement upon retap
- Soil set up factors
- Use of complete driving history versus last blows to assess bearing capacity
- Automatic set recording systems

5.3 Installation monitoring

- Availability of monitoring systems for various pile types (e.g. screwed, continuous flight auger (CFA), grouting, ...)
- Correlation between installation parameters and bearing capacity
- Case histories documenting different performance of a given pile type resulting from different installation records

5.4 Quality control

- Limitations of integrity tests
- Characterisation of anomalies in term of nature, volume, shape
- Addressing identified anomalies (repair, discount of capacity, etc...)
- European standards for low-strain dynamic tests

6. REFERENCES

ASTM D4945-89. 1984. Standard Test Method for High-Strain Dynamic Testing of Piles, Annual Book of American Society for Testing and Materials, Volume 4.08 (7p.)

Bermingham, P., and Janes, M. 1989. An Innovative Approach to Load Testing of High Capacity Piles, Proc. of the International Conf. on Piling and Deep Foundations, London, May, pp. 409-413

Gonin, H., Coelus, G. & Leonard, M. 1984. Theory and Performance of a New Dynamic Method of Pile Testing, Proc. 2nd Int. Conf. Application of Stress-Wave Theory to Piles, Stockholm, pp. 403-410

Holeyman, A. 1992. Technology of Pile Dynamic Testing, Keynote lecture, Proc. of the 4th International Conference on the Application of Stress wave theory to piles, The Hague, September 21-24, pp. 195-215

Holeyman, A. and Legrand, C., 1996, A Method to Predict the Drivability of Vibratory Driven Piles, Proceedings of the 5th International Conference on the Application of Stress-Wave Theory to Piles, Orlando, September, pp. 1101-1112.

Stain, R.T. 1982. Integrity Testing, Civil Engineering (London), April, pp. 53-59 & May, pp. 71-73.

National reports on design practice for axially loaded piles

Design of Axially Loaded Piles – European Practice, De Cock & Legrand (eds)
© 1997 Balkema, Rotterdam, ISBN 90 5410 873 8

Design of axially loaded piles – Belgian practice

A. Holeyman *(U.C.L., Louvain-la-Neuve)*, C. Bauduin *(Besix)*, M. Bottiau *(Socofonda)*,
P. Debacker *(Federal Public Buildings Agency)*, F. De Cock *(Franki Foundations)*,
E. Dupont *(Fundex)*, J. L. Hilde *(Technical Control Office Seco)*,
C. Legrand & N. Huybrechts *(CSTC/WTCB, Belgian Building Research Institute)*,
P. Mengé *(Ministry of the Flemish Community, Ghent University)*, J. P. Miller *(Tractebel)*,
G. Simon *(Ministry of Equipment and Transport of the Walloon Region)*

PREAMBLE : This report has been put together by combining individual contributions provided by the authors listed above, reviewing assembled drafts in plenary meetings, and allowing the General Reporter to edit the report for general consistency. The sole intent of the report is to provide our foreign colleagues with a pragmatic representation of the current piling design practice in Belgium. It is therefore explicitly stated that this report should not be construed as superseding work products presently elaborated by several professional bodies such as Committees, Ministries, and Institutes. As no final decisions have been made regarding the Belgian National Application Document (NAD) for EC7 so far, references to the NAD contained in this report have to be considered as tentative.

1 REGIONAL GEOLOGY

The Belgian territory is rather flat with a continuous transition from a plain at the North Sea and the Dutch border to the highlands of the Ardennes, the highest point being situated at Botrange (694 m above sea level). The geology of the Tertiary and Quaternary formations in Belgium is characterised by a Southeast Northwest oriented epirogenetic axis [34], which follows the valleys of the rivers Haine, Sambre, Meuse and Vesdre (Figure 1), and which divides Belgium into approximately two equal parts.

In the North, the stratigraphy has been governed by fluctuations in the coastal line. Consequently the bedrock is covered by alternating Tertiary clay, sand and (occasionally) gravel sediments, with thickness up to hundreds of meters. The Quaternary Pleistocene formations have been heavily influenced by the glacial periods, giving rise to the formation of marine, coastal, river, lake or wind deposits of sand, clay, peat and silt (loess). Holocene erosion and river sedimentation, as well as human activities, have further influenced the actual subsurface. In the South of the epirogenetic axis, the bedrock is often found at rather shallow depths, overlain by colluvium layers consisting of weathered rock and river sediments.

As a result of the geological history, one can find in the North a wide variety in stratigraphy, with complicated and heterogeneous soil layer patterns. It is not therefore surprising that the North of Belgium (like the Netherlands) has to face serious foundation problems, requiring particular foundations such as piling or ground improvement. In accordance with those geological conditions, depths for deep foundations generally range between 10 and 25 meters, and more typically between 13 and 18 meters.

2 PRACTICE FOR SOIL INVESTIGATION

Table 1 provides a summary of the Belgian practice for soil investigation, based on differences identified from the standpoint of usage, major applications, pile design objectives, and results used for pile design in common practice.

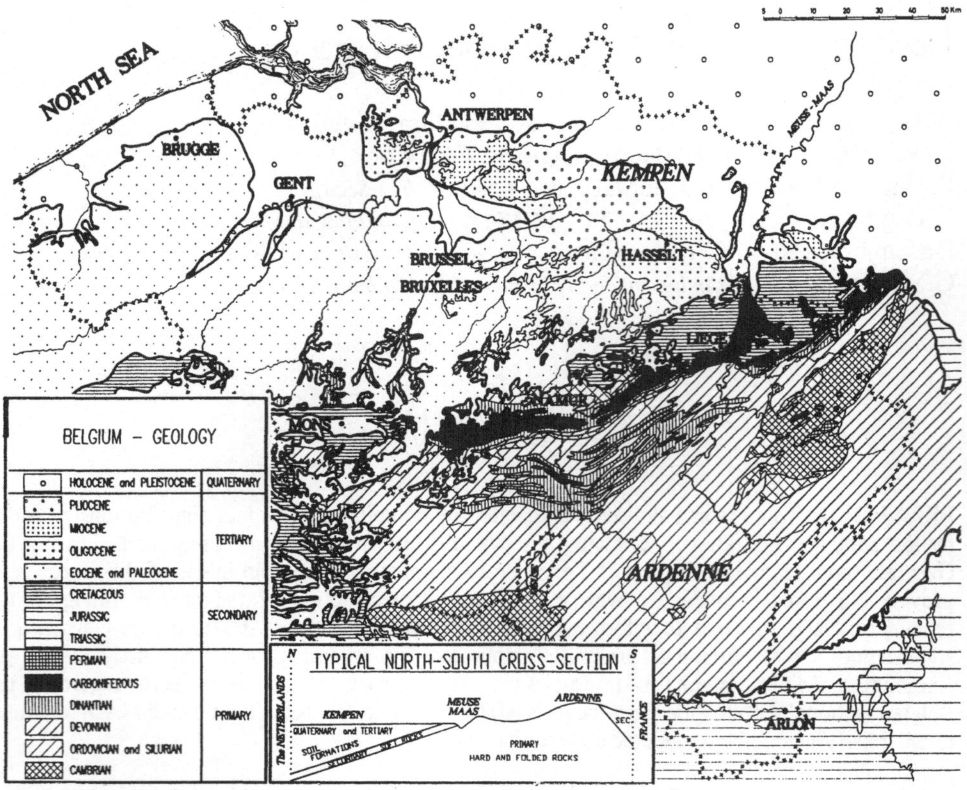

Figure 1. - Geological Map of Belgium (After de Béthune [21b])

The scope of the soil investigation performed for a particular project at a given site depends on a large number of project and site dependent factors. The project dependent factors such as the magnitude and distribution of loads, settlement sensitivity, and Owner requirements are mainly discussed in Section 5.1 relating to the general design philosophy supporting a piling project. This section 2 attempts to present the Belgian general practice for soil investigation as far as it relates to site dependent factors.

As a result of Belgium's high population density and intensive development, an extensive soil data base has been generated over the years. That data base consists not only in informal libraries of soil investigation files in private firms but also in formally accessible libraries or publications. Two main official sources of information are noteworthy: the library of the Belgian Geological Survey which contains logs of borings (copies of borelogs provided by boring companies) and the published Geotechnical Maps. The geotechnical maps currently cover parts of the most developed areas of the country (Cities of Antwerp, Brussels, Charleroi, Ghent, Mons, and Liège). Geological maps covering Belgian territory can also be consulted.

Based on this wealth of information and experience allowing for correlations, most soil investigations consist exclusively in CPT tests where feasible, i.e. where the CPT tests can be performed to a depth allowing the piling project to be designed. This is particularly the case for the very heterogeneous quaternary soil layers encountered in large areas in Belgium, where soil profiling is essential. For the reasons explained in Section 1, sites located in the northern part of Belgium, where piles are required, generally fulfil those conditions.

The use of the CPT was developed in Belgium by De Beer and Verdeyen from 1945 on. De Beer did pioneering work in developing his own interpretation and calculation methods based on the simple cone with closing nut (M4). In addition to interpretation methods providing a friction angle and the compression index, a particular method was published to derive the axial bearing capacity of a pile, based on the results of a CPT test (see Section 5.4.). The history, equipment and use of the CPT in Belgium are discussed in detail in Nuyens e.a. [31].

Table 1. Belgian Practice for Soil Investigation for Axially loaded Pile Design

Type	Sounding/ Testing	Usage	Major applications	Pile design objectives	Results used for pile design (see note below)
IN SITU (Belgian Practice First Choice)	CPT (Cone Penetration Test using 10 cm^2 cones M4, M1, M2 and E1)	Extensive	- Northern part of Belgium (M and E) - Southern part of Belgium (M exclusively)	Soil profiling, identification of potential end bearing layer, quantification of base and shaft resistance	q_c and f_s vs. depth q_c and Q_{st} vs. depth
	PMT (Pressure meter Test Menard type)	Rare to occasional, but growing	- Southern part of Belgium - Settlement sensitive structures	Quantification of base and shaft resistance, settlement	p_l, p_f, E_m
	Destructive Boring with parameters logging	Occasional	- Combined with CPT - Southern part of Belgium - Combined with non-destructive borings for calibration	Soil profiling and classification (to complete that inferred from CPT, where applicable), control of continuity and quality potential end bearing layer	Penetration and rotation velocity, crowding force, torque, vibration, pressure of flushing fluid
	Non-destructive Boring	Occasional	- Combined with CPT - Southern part of Belgium	Same as above + Quantification of base and shaft resistance	Results of LABORATORY Testing (see below) performed on collected samples
	DMT (Dilatometer Test)	Experimental	Research	Quantification of shaft resistance	c_u , K_D
	Vane	Rare	Clayey sites, friction piles	Quantification of base and shaft resistance	c_u
	Geophysical	Occasional, growing	South of Belgium (seismic and resistivity surveys)	Identification of end bearing layer	V_p, V_s
LABORATORY (in combination with IN SITU testing and when IN SITU testing is not feasible)	Identification	Occasional	General	General	Grain size, Atterberg limits, Organic content, CaCO3 content, w, γ_d, γ, γ_s
	Oedometer	Occasional	Compressible layer below pile toe	Pile group settlement analysis	C_c
	Unconfined compression	Rare	Rock	Quantification of shaft and base resistance	q_u
	Triaxial Compression	Exceptional	Soil and rock	Base and shaft resistance	c' , φ'
	Vane	Exceptional	Clayey layers	Base and shaft resistance	c_u

Note : ISSMFE Symbols used for soil properties, specialised literature symbols used for test results.

Investigation methods other than CPT testing are performed to complement CPT tests where warranted or where CPT testing is deemed unfeasible. The major components of those alternate piling investigations will still include in situ testing, leaving a minor role to laboratory testing. Information complementary to the normal CPT practice is warranted when investigation is performed far away from prior developments or when settlements must be specifically evaluated.

Table 2. Piling technologies used in Belgium

GROUP I	PILES EXECUTED WITH HIGH SOIL DISPLACEMENT	
IMPACT DRIVEN PILES		
	Type A:	Prefabricated (concrete and timber) piles
	Type B:	Steel tube piles, close ended or open ended with plugging
	Type C:	Cast-in-situ piles (with or without enlarged base)
SCREWED PILES		
	Type A:	Cast-in-situ piles with screw shaped shaft
	Type B:	Cast-in-situ piles with smooth shaft
	Type C:	Prefabricated (concrete or steel tube) piles
JACKED PILES		
	Type A:	Prefabricated concrete elements
	Type B:	Steel elements
GROUP II	PILES EXECUTED WITH LIMITED SOIL DISPLACEMENT OR LOW SOIL RELAXATION	
IMPACT DRIVEN OR VIBRATED PILES		
	Type A:	Prefabricated concrete piles with enlarged base
	Type B:	Steel beams with or without partial plugging
	Type C:	Steel open ended tubes
	Type D:	Steel beams with enlarged base
	Type E:	Steel piles with grouting (low or high pressure)
DRILLED PILES		
	Type A:	Cast-in-situ piles with special provisions to limit soil relaxation
GROUP III	PILES EXECUTED BY EXTENSIVE EXCAVATION OF THE SOIL	
DRILLED PILES		
	Type A:	With prefabricated concrete shaft
	Type B:	With steel tube shaft
	Type C:	Cast-in-situ piles (a) executed with temporary casing, (b) executed under thixotropic liquid, (c) continuous flight auger (CFA)
	Type D:	With post grouting (low or high pressure)
GROUP IV	MICROPILES	

More sophisticated soil mechanical parameters may be needed when pile design is enhanced by means of finite element programs. As these calculation methods are getting more popular in geotechnical practice, in situ and laboratory tests to define specific parameters such as the shear modulus may also become a larger part of the investigation practice.

3 PILING TECHNOLOGY

Table 2 provides a classification of piling technologies available in Belgium, according to installation process and its potential resulting influence on design.

The evolution of the piling techniques used in Belgium has been originally mainly influenced by the historical development of the Franki-type rammed driven pile with dry concrete. The original system has evolved with the years while its Belgian and foreign competition has developed alternative systems of impact driven piles with a shaft concreted with plastic concrete and with or without enlarged base. These systems are still widely used in Belgium, amongst driven piles.

Cast-in-situ piles are the predominant type. Precast piles are used where the soil geotechnical conditions are homogeneous enough and usually for limited bearing capacities or special applications. In recent years, major concerns have arisen around the problems of noise and vibrations, and vibration-free systems were extensively developed. One of the particularities of Belgium is the coexistence of different types of soil-displacement screwed piles which are well suited to our soil conditions.

Driven piles (Group I in Table 2, and Group II to some extent) are thus preferred in many cases, specially in weak subsurface conditions where soil failure governs the design. When a hard layer is encountered (intermediate or bearing layer), piles with partial or extensive soil excavation (Groups II and III) are generally preferred, specially when pile embedment into the hard layer is required. The classification provided in Table 2 also lists other types of piles which can be used more marginally or for special purposes.

4. NATIONAL RELEVANT DOCUMENTS

To this day, there is no single Belgian standard (norm or code) available to officially regulate the national piling practice. In the absence of truly relevant national documents, several owners and engineers have developed their own specifications or recommendations.

For the public construction market, the calculation and the choice of the pile system is always governed by the specifications of the different Administrations. This means that the relevant documents for the public market depend on the owner. These particular piling specifications from each Administration change from time to time, taking into account new technologies or/and relevant calculation criteria. For each tender on the public market one can ask for the general provisions of the piling specifications of the relevant Administration (e.g. Public Works, Federal Public Buildings Agency, National Railways, etc.) while special provisions for the project are normally part of the tender documents.

In the private construction market, every private consultant, small or large, uses his own piling specifications. Beside the lack of standards it is perhaps interesting to mention the so-called "Type Specifications 104", dated 1973, which are often used as a basis by administrations and consultants to draft their own specifications. On the other hand, the Federal Public Buildings Agency is drafting its own technical specifications "STS 21" on pile foundations [55], as an accompaniment to Type Specifications 104.

Belgium is also working on its National Application Document (NAD) for pile foundations, in order to implement the Eurocode 7 in Belgium. This document is expected to be available in the short-term future.

5 NATIONAL DESIGN METHODS

5.1 *General Philosophy*

5.1.1 Background

The general philosophy supporting the design of piles within a construction project stems from historical and structural factors influencing the organisation of the Belgian profession in general and of the project in particular. Those factors provide an imperfect framework that yet facilitates the achievement of the goal of pile design: "identify, within the physical (mainly geotechnical) and human environment of the site, the most adequate foundation system taking into account the loads and the deformability of the structure".

As a result of the paucity of national standards, the Belgian approach to pile design and construction is characterised by a truly integrated process combining the responsible contributions of key members of the piling project team: the Owner, the Civil Engineer, the General Contractor, the Piling Subcontractor, and, when adopted, the Technical Controller. It is common practice that the Engineer recommends the pile type(s) and specifies the so-called "nominal" load(s) Q_{SP} (more strictly called "specified" load) based on actions, pile layout, and identification of the bearing layer. It is also common practice that the Piling Contractor assumes the responsibility for the performance of the piles and for their design as far as toe level and soil bearing capacity are concerned. His assessment of the allowable value of the pile resistance (R_{ca} or R_{ta}) is however reviewed by the Engineer and the Technical Controller. The advantage of that division of responsibility is that the project benefits from the piling specialist's (subcontractor) know-how.

Because Belgian piling specialists remain competitively involved and rewarded in the design process, they have developed an engineering and problem solving expertise that is respected. Conflicts between "smart engineer" and "dumb contractor" are also thereby reduced. In the case a Technical Controller is assigned to the project, the project is covered by an umbrella insurance, which then further enhances the co-operative atmosphere of the design team, and permits more creative divisions of responsibility.

It should also be noted that civil Engineering has been and is being taught and practised in Belgium as an integrated field of engineering revolving around the construction process. There is no professional segmentation of the civil engineering practice, such as sanctioned abroad for example by titles distinguishing "Structural Engineers" from "Geotechnical Engineers". The structure of the civil engineering profession in Belgium also tends to facilitate communication between engineers employed by key members of the design team.

5.1.2 Basis of design .

In the vast majority of cases, piling solutions are designed on the basis of a geotechnical investigation, performed as discussed in Section 2, with the objectives to identify possible bearing levels and provide quantitative data to determine pile ultimate capacities. Methods to derive pile capacity from the geotechnical investigation data are often imposed in the special provisions of the project specifications. Although the methods specified are generally uniform, thereby confirming the existence of a body of generally accepted design principles (GADP), some differences can be noted between sets of specifications regarding the numerical value of coefficients.

The scope of the investigation programme depends on the size of the project and requirements from the Owner. However one has to acknowledge a strong tendency from the owners to limit the cost of the geotechnical investigation, fuelled by the lack of incentive attached to the present GADP.

Pile design methods generally accepted in Belgium are characterised by the semi-empirical, yet direct transformation of soil bearing parameters measured using in situ testing or sounding, as discussed in more detail in Section 5.4. Mostly used in Belgium are design methods based on the CPT test: the unit base resistance is obtained from a scaling procedure of the cone resistance diagram [15, 38, 53] while the pile shaft friction is obtained from the CPT total friction, local friction, and/or cone resistance diagrams. The De Beer method has been validated and further calibrated thanks to an extensive experimental basis spanning 30 years of full-scale co-operative research, as described in Section 8.

Because the basic CPT-based design method provides the ultimate base and shaft resistance of jacked or driven compression displacement piles, installation coefficients have to be introduced to account for the installation effects of each pile type with reference to the displacement pile types which were used to validate the basic design method. Tension piles are also designed based on CPT tests, but with a higher degree of conservatism, reflecting the limited tension pile test data base . When PMT is used in Belgium, the French design approach [52, 54] is followed.

As discussed in more detail in Section 5.3, static load testing of piles is rarely used as a design tool and generally confined to the control of designed piles. Safety and serviceability issues are discussed in Sections 5.7 and 5.8, respectively.

5.2 *Definitions and symbols*

s_b, s_h	Pile settlement; at the toe and at the head, respectively
$Q^m_{s_b}$, $Q^{s_h}_m$	Load corresponding to a pile settlement s_b (subsript) at the toe and s_h (superscript) at the head, respectively; m indicating the value measured from a static load test.
Q_{SP} or Q_n	Specified (or "nominal" per Belgian usage) pile "load". Used as reference in the piling contract and for control testing, the "nominal" load is specified generally by the engineer based on an envelope analysis of the reactions needed from isolated piles and piles in groups, when considering "normal" actions and the identification of an economical reaction module. "Normal" actions include all loads except for accidental and exceptional loads; exceptional wind but no snow is however considered if less favourable; negative skin friction if applicable should be accounted for. In all rigor, the "nominal" load Q_n is a reference loading level on the pile (i.e. a reference action) while the "specified" load Q_{SP} can be viewed as a reference value to establish the required performance of the pile (i.e. a reference value of the pile resistance, which could be labelled R_{SP}).
R^m_{ca}, R^c_{ca}	Allowable pile resistance in compression obtained from a pile load test and from a calculation, respectively. A satisfactory load test implies $R^m_{ca} \geq Q_{SP}$; a satisfactory design for an individual pile, as generally guaranteed by the Piling Contractor, implies $R^c_{ca} \geq Q_n$.
R_{bu}, R_{ba}	pile *base* resistance; ultimate and allowable value, respectively
R_{su}, R_{sa}	idem for pile *shaft* resistance
R_{cu}, R_{ca}	idem for total pile resistance in *compression*
R_{tu}, R_{ta}	idem for total pile resistance in *tension*
q_b, q_s	unit values for pile base resistance and shaft resistance, respectively
$q^{(m)}_{bu}$, $q^{(m)}_{su}$	ultimate unit pile base resistance and pile shaft resistance, respectively, directly derived from the q_c values in the natural ground conditions
q_c, f_s, Q_{st}	CPT values : cone resistance, local unit side friction, and total side friction, respectively
d	diameter of the CPT sounding tube and/or cone
α_b, ε_b	empirical factors for pile base resistance
α_s, β_s, ε_s, ξ_f, η_p	empirical factors for pile shaft resistance
A_b, X_b, D_b	respectively the cross section, perimeter, and diameter of the pile base
A_s, X_s, D_s	respectively the cross section, perimeter, and diameter of the pile shaft

5.3 *Static load tests*

A clear distinction between design load tests and control load tests is made in Belgium. Because of the proven reliability of in-situ tests based design methods and of the high cost and time consumption of static load tests, loading test piles with a view to design production piles is used only in extreme projects where the large quantity of piles is able to leverage out the benefit of an improved design. The Belgian practice of pile load testing is therefore primarily motivated by the need to control the conformity and intrinsic quality of the piles and should in principle, be discussed in Section 7.2. However because control static load tests also provide assurance of the design and may occasionally warrant its fine-tuning, procedures and interpretation of results are discussed in this section.

The three main documents available in Belgium covering the execution of a static pile load test place the emphasis on the Control type tests. These documents are the Recommendations of the former National Committee on Pile Foundations [49], Draft provisions of the Federal Public Buildings Agency (STS 21) [55], and Specifications of the Flemish Community [50]. A waiting period of 1 to 12 weeks must be respected between the installation of the pile and the execution of a static load test, depending on soil type and pile material. The loading procedure belongs to the maintained load type (ML) and uses the specified load as a reference. The maximum load of 1.5 or 2.0 Q_{SP} is achieved after 6 to 10 load increments, followed by complete unloading achieved after 3 or 4 steps. The loads are maintained for at least half an hour and as long as the pile head settles more than 0.05 mm per half-hour.

The load that is allowed on a pile depends on several criteria relating not only to the amplitude of settlement at given load levels but also to the shape of its load-settlement curve. Alternatively, the load test data can be interpreted to evidence a creep load [54]. The Belgian settlement criteria to assess the reference value of the pile resistance R^m_{ref} on the basis of one measured load-settlement curve are as follows:

$$R^m_{ref} = \min\left(Q^m_{0.010D_b}, \frac{Q^m_{0.017D_b}}{1.35}, \frac{Q^m_{0.025D_b}}{1.70}\right), \quad \text{according to [49],}$$

$$= \min\left(Q_m^{0.0075D_b}, \frac{Q_m^{0.015D_b}}{1.5}\right), \quad \text{according to [55], and}$$

$$= \min\left(Q^m_{3mm}, \frac{Q^m_{6mm}}{1.5}\right), \quad \text{according to [50]}$$

Settlement criteria specified for the pile toe are converted to pile head settlement criteria, allowing for the shortening of the pile shaft (either determined experimentally or calculated from the elastic compression assuming a given transfer of the load to the pile toe). In addition, a pile is not acceptable according to [49] and [55] if its load-settlement curve does not fulfil the following shape criterion: the measured settlement at any load step may not exceed by more than 3 mm the value obtained for the same load step by linear interpolation between any couple of other measured points.

The ultimate load R^m_{cu} is usually assessed based on a 10% settlement criterion for the pile toe. In case that condition cannot not be experimentally evidenced, it is inferred from an extrapolation of the load settlement curve using procedures suggested by Van der Veen [34b] or Chin. The value of R^m_{ca} is deduced either from R^m_{ref} or from R^m_{cu} after taking into account effects not included in the testing situation such as downdrag. In case the testing condition is deemed representative of the service conditions, one usually adopts $R^m_{ca} = R^m_{ref}$.

5.4 Design by calculation on basis of soil ground test results

<u>5.4.1 CPT-related direct method for the ultimate resistance design of compression piles</u>

At least 90 % of all pile design in Belgium is based on semi-empirical formulae, directly assessing both base and shaft resistance of compression piles from CPT data in the natural ground conditions (i.e. before pile installation). The formulae include pile and soil depending empirical factors, which are calibrated based on various research programs (see Section 8). The method itself is conceptually the same for most common pile types (displacement as well as bored piles) and for both non-cohesive and cohesive soils.

It should be noted that in most research work undertaken to date, the simple mechanical cone (M4) or the electrical cone (E1) have been used for the soil investigation, and thus also for calibrating the calculation of the ultimate pile resistance with the measured pile resistance. All empirical factors given hereunder are in principle applicable to M4 or E1 tests only. In practice however, many engineers are unaware of this limitation with the result that in many cases, correction factors for other CPT methods are not considered in the design.

In order to improve that situation, the Belgian NAD is anticipated to establish the electrical cone E1 as the reference cone and to require that data from other cones be transformed to E1-values. Studies on the influence of the CPT method on test results suggest that conversion factors should depend not only on cone type and penetration mode, but also on soil type. $q_{c,M}$ values have been noted to be approximately 35 % higher than $q_{c,E}$ values measured in OC Clay, but have been observed to be approximately 10 % lower than $q_{c,E}$ values measured in sand. Conversion factors for intermediate soils are expected to belong to the range covered by those two soil types. Total side friction (Q_{st}) does not appear to significantly depend on the CPT method.

Basic formulae

The pile ultimate base resistance R_{bu} is deduced from the CPT data by :

$$R_{bu} = \beta \times q_{bu} \times A_b = \beta \times \alpha_b \times \varepsilon_b \times q_{bu}^{(m)} \times A_b \qquad (1)$$

with :

β = a shape factor introduced for non circular nor square-shaped bases; (e.g. for barrettes); $\beta = \frac{1+0.3B/L}{1.3}$ with B = width and L = length of rectangular base

α_b = an empirical factor taking into account the method of installation of the pile and soil type;

ε_b = a parameter referring to the scale dependant soil shear strength characteristics (e.g. in case of fissured clay)

$q_{bu}^{(m)}$ = ultimate unit pile base resistance derived from the CPT results in the natural ground conditions

A_b = the nominal pile base cross-sectional area.

Estimation of the ultimate shaft friction R_{su} is based on one of the following CPT values : the total side friction Q_{st} (easiest and most common method); the cone resistance q_c; and/or the local unit side friction f_s.

The total pile shaft resistance R_{su} can be directly evaluated by proportioning the pile shaft resistance to the CPT total side friction increment ΔQ_{st} in the relevant shaft bearing layer(s) :

$$R_{su} = \frac{X_s}{\pi d} \times \xi_f \times \Delta Q_{st} \quad \text{or} \quad = \frac{X_s}{\pi d} \times \Sigma\, \xi_{fi} \times \Delta Q_{sti} \qquad (2)$$

with :

ξ_f = an overall empirical factor ($=\alpha_s \,.\, \beta_s \,.\, \varepsilon_s$) introducing the effects of pile installation method (α_s), of the nature of the shaft's material and roughness (β_s) and soil structure scale effects (ε_s);

X_s resp. πd = the perimeter of the pile shaft and of the sounding rod, respectively.

The pile shaft friction can also be evaluated from a semi-empirical correlation between the ultimate unit shaft friction q_{su} and the cone resistance values q_c:

$$q_{su} = \eta_p \times q_c \quad \text{or further detailed as :} \quad q_{su} = \xi_f \times \eta_p^* \times q_c \qquad (3a)\&(3b)$$

and thus :

$$R_{su} = X_s \times \Sigma H_i \times \eta_{pi} \times q_{ci} = X_s \times \Sigma H_i \times \xi_{fi} \times \eta_{pi}^* \times q_{ci} \qquad (4)$$

wherein η_p = an overall empirical factor depending on both soil *and* pile type. For clarity, the correlation $\eta_p = q_{su}/q_c$ can be split into (1) a pure soil parameter η_p^* equal to the ratio of q_c and the average unit side friction $q_{su}^{(m)}$, and (2) a pile/soil dependant empirical factor ξf as already defined in equation (2). Values for $q_{su}^{(m)}$ and η_p^* have been suggested for Belgian practice in [48], and are summarised in Table 3 for ease of reference.

A third method relates the pile shaft friction to the directly measured local unit side friction f_s by

$$q_{su} = \alpha_{fs} \times f_s \qquad (5)$$

Table 3. $q_{su}^{(m)}$ and η_p^* values (in Eq. 3b and 4) commonly used in Belgian Practice, after [48]

CLAY	q_c [Mpa]	0.075	0.2	0.5	1.0	1.5	2.0	2.5	3.0	≥ 3.0
	$q_{su}^{(m)}$ [kPa]	5	10	18	31	44	58	70	82	q_c[kPa] /36.6
SAND	q_c	≤ 10 MPa	10 < q_c < 20 MPa							> 20 MPa
	$q_{su}^{(m)}$	q_c / 150	Linear interpolation between q_c / 200 and q_c / 150							q_c / 200

Again, one can expect that a_{fs} depends on pile and soil type, and should hence be defined by calibration with static load tests. This method is not widely used for direct pile design (except in the Begemann concept for cyclic loaded piles; See section in §5.4.6.2), because little calibration data or available in Belgium and because experience has revealed the high sensibility of the f_s values from cone type and cone wear.

Calculation of ultimate unit base resistance $q_{bu}^{(m)}$

One fundamental aspect of the Belgian pile design is the introduction of the so-called "scale effect" for the pile base resistance. The scale effect aims to take into account that the base resistance of a pile is defined by the failure pattern, which extends over a certain height below and above the pile toe, this height being related to the pile base diameter. The approach aims at transforming the CPT diagram (generally obtained with a 3.6 cm diameter cone) into the CPT diagram that would be obtained with a sounding rod having a diameter equal to that of the pile base.

While in foreign countries this scale effect is calculated by rather simple mathematical approaches (smoothing and averaging the q_c-values over a certain range such as in France and The Netherlands, a more analytical method has been developed in Belgium in the 70's by De Beer [15] and then been widely introduced in the Belgian design practice. The method and later modifications have also been reported in ECSMFE and ICSMFE [48] proceedings by De Beer and Van Impe among others. It has been observed that the method aims to predict the limit load ($Q_{0.025Db}$) near the upper and lower boundaries of the bearing layer but provides the conventional rupture load ($Q_{0.10Db}$) at large depths in that layer.

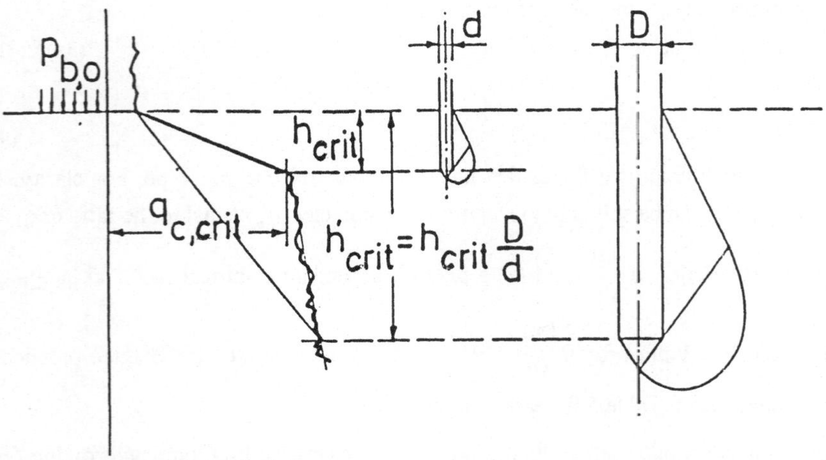

Figure 2. Scale effect principle

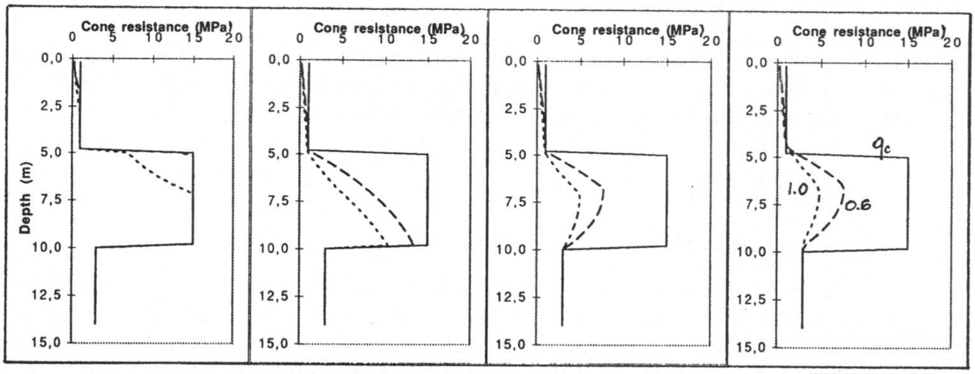

Figure 3. Step by step illustration of De Beer procedure : (a) homogeneous values, (b) downward values, (c) upward values and (d) blended values; for 0.6 and 1.0 m diameter base, respectively.

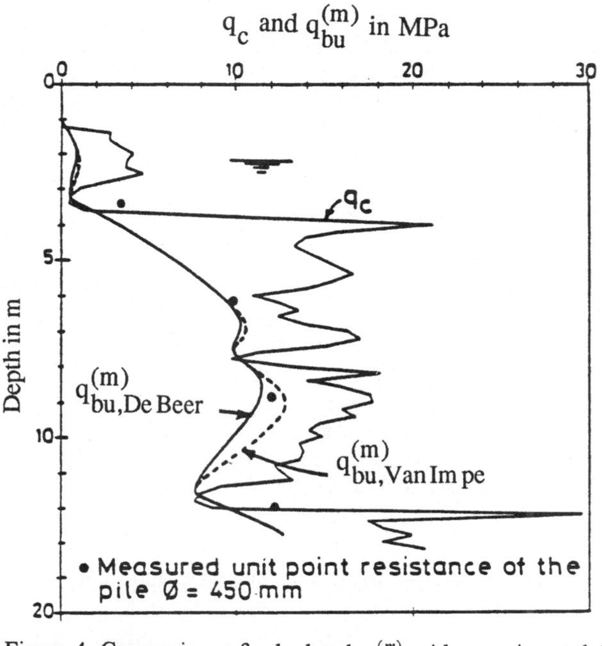

Figure 4. Comparison of calculated $q_{bu}^{(m)}$ with experimental data

The De Beer method is based on a thorough application of the principles of the scale effect, when transitioning from a soft to a hard soil layer as shown in Figure 2.

This application of the scale effect is done in 4 steps, designated by the terms (a) homogeneous values, (b) descending or downward values, (c) upward values and (d) mixed or blended values. These final mixed values $q_{bu}^{(m)}$ are the basis values for the further base resistance calculation of the pile. To demonstrate the procedure, step by step results of a De Beer calculation are given in figure 3 for a simplified soil profile. A practical calculation example is provided in figure 4, showing $q_{bu}^{(m)}$ for a given displacement pile, as calculated according to the basic De Beer method as well as the slightly modified method as proposed by Van Impe and as measured at several depths. The example design provided in section 6 provides an other illustration of the De Beer method [15]

Installation factors in design formulae

Installation empirical factors for a given pile in a given soil type should be deduced from calibration with static pile load tests. It is most often assumed in Belgian specifications, that for traditional piles of the displacement type, all empirical factors = 1.0, so that :

$$R_{cu} = R_{bu} + R_{su} = q_{bu}^{(m)} \times A_b + \Delta Q_{st} \times \frac{X_s}{\pi d} \qquad (6)$$

More refined factors are however discussed hereafter.

The ε_b factor in equation (1) has been introduced to take into account the scale effect of the size of the failure mechanism of a pile base relative to the failure mechanism of the CPT cone, as recognised in the stiff fissured OC Boom clay [16]. Although size dependent shear resistance might also exist in other soil types, that phenomenon has been explicitly introduced only in stiff OC clay, where, one applies in Belgium :

$$0.476 \le \varepsilon_b \approx 1 - 0.01\,(D_b / d - 1) \qquad (7)$$

Table 4. Commonly used α_b factors (in equation (1)) for various pile types

Pile type	α_b factor for	
	sand	stiff OC clay
Group I - High soil displacement		
Impact driven piles	0.8-1.15 $^{(1)}$	0.8-1.0 $^{(1)}$
Screwed piles	0.8 -1.0 $^{(2)}$	1.0
Jacked piles (smooth)	1.0	1.0
Group II - Low soil displacement or low relaxation		
Impact driven	see [20,21&46]	see [20,21 & 46]
Drilled with special provisions	0.6-0.8	0.8
Group III - Soil excavation		
Cast-in-situ bored piles (large diameter and CFA)	0.33-0.67	0.8

$^{(1)}$: highest value for expanded base with semi-dry concrete only; for cast-in-situ with plastic concrete, function of the diameter of the bottom plate relative to diameter of driving tube;

$^{(2)}$: depending on the allowance or not for vertical soil displacement near the pile base

Table 5. Commonly used ξ_f factors (in Eq. (2) and (4)) for various pile types and shaft material

Pile type	ξ_f factor for	
	sand	stiff OC clay
Group I - High soil displacement		
Shaft in compacted semi-dry concrete	1.6	1.15
Shaft in plastic concrete or prefabricated concrete	0.8-1.0 $^{(1)}$	0.65-1.0
Screwed piles - plastic concrete	0.8-1.25 $^{(2)}$	0.8-1.25 $^{(2)}$
Steel shaft	0.6	0.45-0.65
Group II - Low soil displacement or low relaxation		
Impact driven - steel shaft	see [20,21 & 46]	see [20,21 & 46]
Drilled with special provisions - wet concrete	0.6-0.8	0.65-0.85
Group III - Soil excavation		
Cast-in-situ bored piles (large diameter and CFA)	0.4 - 0.6	0.5

$^{(1)}$: for cast-in-situ with plastic concrete, function of diameter of the bottom plate relative to diameter of driving tube;

$^{(2)}$: highest values for screw shaped shaft.

The α_b, ξ_f and η_p factors in equations (1), (2), and (4) have been deduced from static pile load tests in various research projects (see Section 8). Tables 4 and 5 are giving an overview of values commonly used in the Belgian design practice for sand and stiff OC Clay, Intermediate values between those listed are adopted for intermediate soil types. For certain pile types, installation factors that have not yet been calibrated, based on an objectively conducted full scale load test program, require the input of some judgement and are therefore the subject of some debate. For the η_p factors, Van Impe [42] has summarised the values resulting from Belgian research work. In some cases, the factors prescribed in e.g. the Dutch code (NEN 6743) or the French regulations [52, 54] are applied.

Extended research work, on the other hand, has been performed on the bearing resistance of steel H-beam piles, with or without base or shaft enlargements [21,21a & 46]. That research indicates that basic formulae similar to formula (1) for the base resistance and to formula (2) or (4) for the shaft resistance can be used. However, the design is based on the most conservative of two possible rupture models : (1) the H-beam penetrates into the soil like a knife, without any plug formation; and (2) a partial plug (in granular soils) or a full plug (in cohesive soils) is formed in the space between the flanges of the H-section.

5.4.2 Other methods for ultimate resistance design of piles on basis of in situ ground tests

Beside the widely used direct method on basis of CPT results, as detailed above, other methods are occasionally used in Belgium. CPT-based design is sometimes performed according to codes or recommendations from neighbouring countries, (e.g. Dutch codes NEN6740 and NEN6743 and French DTU 13.2 or Fascicule 62). PMT-based design most likely follows the French methods (DTU 13.2 or Fascicule 62) as well. When dynamic penetration tests (DPT) or standard penetration tests (SPT) have been used, the geotechnical engineer generally transforms the test data into more familiar CPT values to apply the methods detailed in Section 5.4.1, but may occasionally refer to Bustamante (1993).

Micropiles are most commonly designed according to the LCPC method, published by Bustamante e.a. [3]. As the design charts are elaborated in terms of PMT values, CPT q_c data are converted into limit pressure p_l values. The conversion is often based on the comparative research work by Van Wambeke [44], and may be simplified into a 3-6-9 rule :

$q_c / p_\ell \approx 3.0$ for clays, ≈ 6.0 for silts, and ≈ 9.0 for sands.

Although currently being used in Belgium for e.g. tension piles, very little experimental data is available for *post-grouted* piles. Design most likely is performed using the French recommendations [52,54].

5.4.3 Design methods based on laboratory tests

Calculations of pile bearing resistance on basis of shear parameters (ϕ, c or c_u) using static formulae are usually not used. Their exceptional use is limited to the calculation or verification of the shaft resistance and in that particular case, for instance :

- to define a lower limit of the shaft resistance for bored piles in granular soils;
- to calculate the shaft friction in cases where the in situ performed soil tests are insufficient, or doubtful, or where they might not be representative of the soil conditions (e.g. in case of deep excavations after soil testing);
- to calculate the shaft friction of tension piles (see below).

5.4.4 Design methods for tension piles on basis of ground test results

For straight-sided tension piles, the uplift capacity (or tensile resistance) R_{tu} is mostly calculated

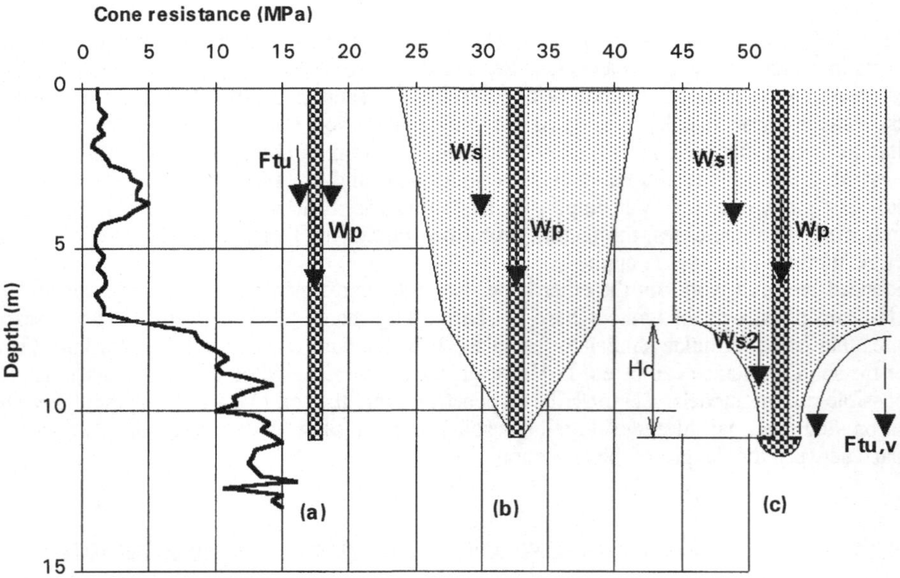

Figure 5. Pile uplift failure mechanisms

by assuming a slip failure along the shaft-soil interface (fig. 5a). The shear resistance along this interface is generally calculated by direct methods on basis of the CPT data, applying the same formulae and factors as those detailed above for the friction on compression piles. An occasional alternative consists in applying a capped Mohr-Coulomb failure criterion assuming drained and/or undrained soil conditions. In case of small D_s/L_p ratios, the uplift of the block of ground co-operating with the pile, has also to be checked (see figure 5b).

For piles which are strongly embedded at their lower end, the governing failure mechanism will be rather similar to that of a plate anchor. In this case, an internal slip line will occur, at least over the depth of embedment into the bearing layer. Conditions for such a slip line are generally fulfilled for piles with cast in situ expanded base (e.g. Franki type) and for piles installed with lateral displacement of the soil (e.g. helicoidal screw piles), both conditions being combined with a rough shaft-soil interface.

For granular soils, the tension resistance of such a piles is often calculated on basis of the analytical methods, elaborated by Lousberg e.a. [29]. The assumed slip line is given in figure 5c. The height of the trumpet shaped slip line along which the shear resistance is taken into account is limited to the minimum 4 D_b and the embedment height H_c into the dense layer.

5.4.5 Particular load cases

Downdrag

The downdrag F_n generally is calculated following the method of Zeevaert [47], further detailed by De Beer [11]. For a given critical height h_c over which the downdrag is estimated to occur, the downdrag caused by a surcharge p_o on a pile with perimeter X_s and an area of influence A amounts to : $F_n = F_{n,o} + F_{n,\gamma} = A_o p_o \left(1 - e^{-m_o h_c}\right) + A_\gamma \gamma_k h_c \left(1 - \frac{1 - e^{-m_\gamma h_c}}{m_\gamma h_c}\right)$

with : $m_O = \frac{K_O \tan\phi X_s}{A_O}$ and $m_\gamma = \frac{K_O \tan\phi X_s}{A_\gamma}$

The influence areas A_O and A_γ are calculated following the hypothesis of influence cones that extend over a top area of $A_O = \pi h_c^2 / 4$ and $A_\gamma = \pi h_c^2 / 16$ respectively.

Cyclic loading

Particularly for piles supporting pylons for high voltage lines, specifications often require a verification of the side friction on the piles under cyclic loading by the method suggested by Begemann (1969). The side friction resistance is then calculated on basis of the local friction as directly measured in the CPT, following formula (5) : $q_{su.} = \alpha_{fs} \times f_s$. On the other hand, it is proposed by Begemann [2] for alternating loading (compression/tension), to reduce the friction over the middle half of the embedded length of the shaft by a factor of 3.0.

5.5 *Driving Formulae*

In Belgium, piles are not designed based on driving formulae. However the final blow counts can be used to check and fine-tune the penetration required by design as follows.

During driving the first pile at the very location or at least in the close vicinity of a CPT test, the set is measured at the proposed level. This set is then imposed within a narrow margin (typically 20 %) when driving the neighbouring piles. Driving is thus to be continued until each pile is placed in the same layer as the test pile and at such a depth that the set criterion is fulfilled. For calculating the set (or penetration per blow), the mean value over the last 10 or 25 cm or the mean value of 5 consecutive observations of 10 blows is taken.

For Franki-type piles, the set recorded at base level governs the volume of dry concreted required to form the expanded base.

5.6 *Wave equation analysis*

The deduction of the static bearing capacity from dynamic measurements is still considered by several Belgian engineers as controversial because of two considerations: (1) the dynamic loading behaviour is not necessarily representative of the static one, and (2) the displacement induced during driving or dynamic testing is much smaller than what is recognised to yield significant data about the ultimate bearing capacity of a pile. Development of pile dynamic testing in Belgium is however further discussed in Sections 7 and 8.

5.7 *Factors of Safety*

5.7.1 General Concept

The factor of safety that establishes the ratio between the calculated ultimate bearing capacity and the allowable load is meant to cover a sufficient margin of safety with regards to failure but also encompasses uncertainties attached to e.g. subsurface conditions, calculation methods, quality of the piles, and actual working load. In the Belgian practice, factors of safety are applied within a deterministic framework, i.e. they are globally applied to components of the bearing capacity.

When bearing capacity is evaluated from CPT tests, Belgian engineers generally use a factor of safety of 2 on the end bearing term and a factor of safety of 3 on the shaft resistance. In the case of a PMT based evaluation, French prescriptions are followed. There exists a current trend to assign identical factors of safety to all pile types after allowing installation coefficients to account for differences in the ultimate bearing capacity between different pile types. The working load of piles is generally assessed based on the most conservative CPT test within a given zone of the site, thereby neglecting the favourable effects due to load redistribution amongst piles and discounting the design advantage offered by a potentially more extensive geotechnical investigation.

5.7.2 Factors of safety with respect to ultimate bearing resistance

In the deterministic method, global factors of safety are used :

$R^c_{ca} = R_{bu} / S_b + R_{su} / S_s$ with : $S_b = 2.0$ and $S_S = 3.0$

One verifies that : R^c_{ca} (for all CPT's individually) $\geq Q_n$

In the case a downdrag $F_n \leq R_{bu}$ is expected, one generally uses :

$$R^c_{ca} = \frac{R_{bu} - F_n}{S_b} + \frac{R^+_{su}}{S_s}$$ with

R^+_{su} : pile shaft resistance accrued below the neutral point.

$R^c_{ta} = W'_p / S_w + F_{su} / S_s$ with : $S_W = 1.0$ to 1.3

and $S_S = 3.0$ to 5.0, depending on project type

One verifies that : R^c_{ta} (for all CPT's individually) $\geq Q_n\uparrow$

Alternatively, De Beer e.a. [19] suggested for displacement piles, using the results of several CPT's within a design zone:

$$R_{ca,1} = \frac{1}{\gamma_t}\left[\frac{R_{bu,\max}}{S'_b} + \frac{R_{su,\max}}{S'_s}\right]$$ with : $\gamma_t = 1.4$, $S'_b = 1.5$ and $S'_s = 1.3$ and $R^c_{ca} = R_{ca,1} \leq R_{ca,2}$

$$R_{ca,2} = \frac{1}{\gamma_t}\left[R_{bu,\min} + R_{su,\min}\right]$$

That approach has been extended by Van Impe (1986) who suggested $\gamma_t = 1.7$ for bored piles and a statistical determination of S'_b and S'_s .

Partial factoring is still to be decided in the Belgian NAD.

5.7.3 Structural Safety

Belgian norms available for the design of structural elements made of reinforced concrete are not applied to design pile shafts. It is the practice to conduct that calculation on the basis of an equivalent composite cross section equal to the concrete section plus 14 times the steel cross sectional area and characterised by an allowable compressive stress. That allowable compressive stress varies little between current specifications (typically 5 to 7 MPa), but is always lower than that allowed to design reinforced concrete columns, on the grounds that the quality of concrete can not be enforced and exposed as easily.

5.8 Serviceability

In regular present Belgian practice, serviceability conditions are not usually explicitly analysed. Experience has indeed shown that for piles designed under usual conditions and for regular structures, the factors of safety indicated in Section 5.7 are conservative enough to satisfy the service states. A settlement analysis is explicitly performed however for pile groups located above potentially compressible layers, for settlement sensitive structures, or for marginal and challenging subsurface conditions. That analysis is conducted first for a single pile using mobilisation curves for the shaft and the end bearing reactions derived from pile load tests performed under similar conditions, and later refined using the results of control load tests performed on the site. Alternatively, mobilisation curves can be derived from the results of CPT tests, as suggested by Verbrugge [56]. Settlement of the pile group is then evaluated using stress distribution theories (e.g. Buisman) and a linear or logarithmic stress-strain relationship for the soil.

6 PARTICULAR EXAMPLE

The following example illustrates the widely used procedure described in Section 5.4.1.
Pile : Cast-in-situ pile (group I type C), wet concrete shaft, D = 520 mm
Pile base depth : 10.5 m
Soil : Loam (0 m - 8 m) and sand (8 m - 12 m)
CPT : M4

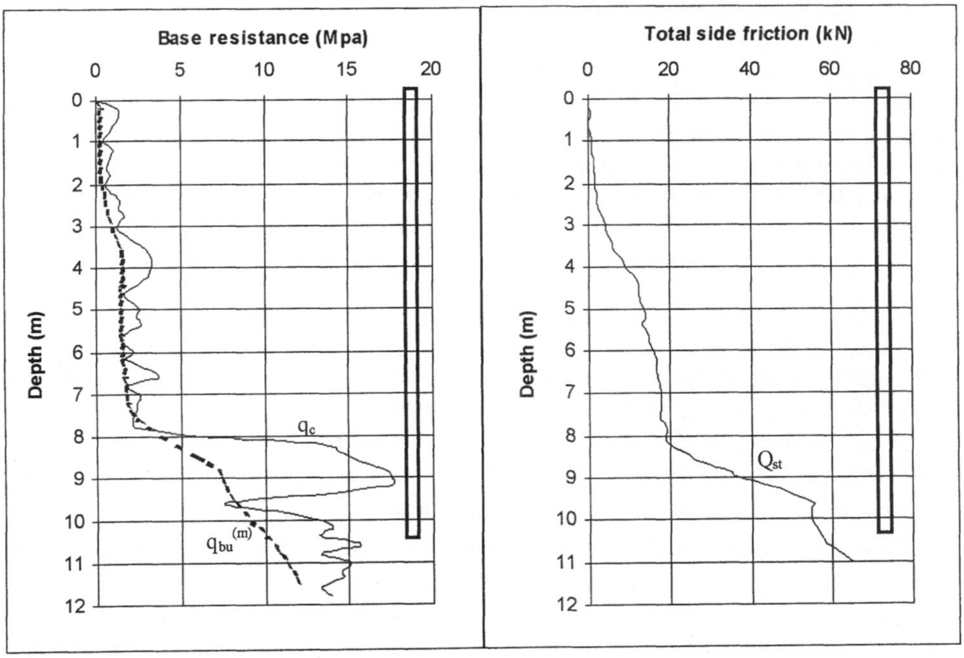

Pile base resistance	Pile shaft resistance
$q_{bu}^{(m)}$ = 10.3 MPa (De Beer method)	ΔQ_{st} = 55 kN
$\alpha_b = 1.0$ $\varepsilon_b = 1.0$	$\xi_f = 1.0$
$R_{bu} = 1.0 \times 1.0 \times \frac{\pi \times 0.52^2}{4} \times 10300 = 2184$ kN	$R_{su} = 1.0 \times 55 \times \frac{0.520}{0.036} = 801$ kN
$S_b = 2$	$S_s = 3$
$R_{ca} = R_{bu}/S_b + R_{su}/S_s = 1.09 + 0.27 = 1.36$ MN	

7. QUALITY CONTROL AND MONITORING

7.1 Monitoring of pile installation

In addition to controlling the quality of the used materials such as concrete and reinforcement, one can also monitor the different execution parameters during the installation of the pile. To ensure the reliability of the monitoring, some basic data is always recorded, such as date and time of installation, co-ordinates of the pile location, etc. The following paragraphs address specific aspects of the monitoring of impact driven piles on one hand and drilled and screwed piles on the other hand.

The blow count which is normally recorded against the penetration of the pile by the piling foreman can also be registered by a monitoring device. The monitoring of the end of driving is discussed in section 5.5. Monitoring devices (such as Pile Dynamic Analyser (PDA) or blow count recorders versus depth) are not currently specified, although some piling contractors are equipped with those systems.

Drilled and screwed piles are monitored with regards to the drilling or screwing process as well as the concreting process. Depending on the pile system and monitoring system, several of the following parameters are generally recorded : speed of penetration, speed of rotation, depth, rotational torque (usually inferred from the hydraulic oil pressure of the drill table). The concreting which is most often performed using a pump must be controlled by a monitoring device measuring the volume of the used concrete, the pressure applied to the concrete and the pull-out speed. For some pile types used in Belgium, a computer-based monitoring system has already been implemented. When demonstrated to be reliable, that type of monitoring is asked more and more often by the quality control department of owners and consultants.

Soil relaxation resulting from the installation process can be evaluated on the basis of soundings performed alongside the pile.

7.2 Visual inspection, Static Loading Testing, and Core Sampling

Although visual inspection only gives limited information on the top surface and the small portion of the shaft which may be exposed, it is always performed.

In spite of its rare use, the static loading test is still in the Belgium the least disputed method to test the integrity and to verify the bearing capacity, as discussed in more detail in section 5.3. In the case of a control test, a pile is deemed satisfactory when $R^m_{ref} \geq Q_{SP}$. In the event that condition is not fulfilled, further analysis by the engineer and negotiations between the contracting parties ensue, based on the value of R^m_{ca} or R^m_{cu} evidenced form the static load test.

Especially for bored piles, vertical core sampling is sometimes carried out. The sampling provides a continuous control of the quality of the concrete in the pile shaft. Continuing the sampling through and beyond the toe of the pile allows one to examine the tightness of the contact between the base of the pile and the bearing soil layer.

7.3 Non destructive tests

Information on pile integrity is obtained using sonic or gamma-gamma logging, the echo method, and the mechanical admittance method.

Sonic and gamma-gamma logging is usually performed on bored piles using access tubes mounted on the reinforced cage to evaluate the quality of concrete between emitter and receiver.

Depending on the extent and success of the testing program [23], the evaluations expected from the sonic echo and the mechanical impedance methods may include the length of the pile, its cross-section, the extent to which these dimensions vary, the density of the concrete, the

Table 6. Belgian Pile Testing Experience - Part 1: Programs

Test Sites				Summarized soil configuration				Program information		
Abrev.	Name	Reference	Period of tests	Depth (m)	Nature of upper stratum	Depth (m)	Nature of lower stratum(s)	Num -ber	and Type of piles tested	Objectives - Results
ZEL	Zelzate	[12]	1968	0 - 21	loose loamy sand	21 - 26	Boom-clay	1	Tension bored	Skin friction resistance under tension
ANT I	Antwerp	[12, 48]	1968	0 - 11	loose sand	> 11	Boom-clay	5	Tension driven cast-in-situ	Skin friction resistance and bulb effect under tension
ANT II	Antwerp (North)	[48]	?	0 - 8	soft clay and peat, and loose sand	> 8	dense sand	1	Tension driven cast-in-situ	Skin friction resistance and bulb effect under tension
OST	Ostend	[48]	?	0 - 13	soft clay and peat, locally sandy	> 13	dense sand	1	Tension driven cast-in-situ	Skin friction resistance and bulb effect under tension
ZWI	Zwijnaarde (Ghent)	[14, 45, 48]	1969/70	0 - 13	loose sand to sandy silt	> 13	Medium dense sand	4	Driven cast-in-situ	Scale effect for base resistance (medium dense sand)
KO I	Kontich I	[8, 16, 48]	1975/76	0 - 3	sandy loam (silt)	> 3	tertiary Boom-clay	12	Driven cast-in-situ, and prefabricated	Scale effect for base resistance (fissured clay), shaft friction
KA I	Kallo I	[17, 18, 48]	1977/78	0 - 8	soft clay and peat	> 8	Dense sand	7	Driven cast-in-situ	Scale effect for base resistance (dense sand)
KA II	Kallo II	[20, 21, 48]	1977/81	0 - 8	soft clay and peat	> 8	Dense sand	12	Driven H steel pile	Plugging, influence of pile length, enlargement in sand, prediction by stress wave analysis.
KO II	Kontich II	[20, 21, 48]	1977/81	0 - 3	sandy loam (silt)	> 3	tertiary Boom-clay	12	Driven H steel pile	Plugging, influence of pile length, enlargement in clay, prediction by stress wave analysis.
KA III	Kallo III	[38, 48]	1982	0 - 6	soft clay and peat	> 6	Dense sand	4	Bored and driven steel tube	Comparison bored and driven piles in sand (all same geometry)
GRB	Groot-Bijgaarden	[48]	1983/85	0 - 8	sandy loam and clay	> 8	Fine sand/ dense sand	6	Driven cast-in-situ	Influence of base plate enlargement on base resistance
ZWE	Zwevegem (Kortrijk)	[29]	1984	0 - 15	tertiary Ieper-clay	> 15	compact sand	2	Screwed (auger) cast-in-situ	Design of cast-in-situ displacement screwed piles in clay
GH I	Ghent I	[40]	1985	0 - 14	silt & silty clay	> 14	Dense clayey sand	2	Screwed (auger) cast-in-situ	Design of cast-in-situ displacement screwed piles in sand
KA IV	Kallo IV	[21a]	1986	0 - 5	soft clay	>5	Dense sand	3	Driven H steel pile	Influence of geometry and position of lagging, prediction by stress wave analysis.
GH II	Ghent II	[23, 26, 30, 36, 37, 38, 40, 41]	1987	0-10	silt & silty clay	> 10	Dense clayey sand	12	Screwed (auger) cast-in-situ, CFA & driven precast concrete	Comparison different screwed and CFA piles in sand. Prediction by dynamic loading tests.
GEE	Geel	[32]	1988	0 - 8	loose sand, silty sand	> 8	sand	3	CFA	Design of CFA piles- Use of DMT
NSC	North Sea Coast	[24]	1990	0 - 10	soft clay	10 - 16	clayey sand	2	Screwed (auger) cast-in-situ	Design of cast-in-situ displacement screwed piles in sand. Prediction by dynamic loading test
KOE	Koekelare	[4, 22]	1992	0 - 5	silty/clayey sand	> 5	tertiary Ypresian clay	10	Screwed (auger) cast-in-situ and screwed steel tube	Design displacement screwed piles in clay. Prediction by dynamic loading test. Use of DMT
LIM	Limelette	[28]	1995-96	0 - 9	loam	> 9	dense sand	4	Driven (cast-in-situ, steel tube, precast concrete)	Comparison of driving techniques in sand; induced vibrations, prediction by dynamic loading test.
VIL	Vilvoorde	[7]	1995	0-7.5	sandy silt	> 7.5	sand	3	Screwed cast-in-situ Omega and driven precast concrete	Design of Omega-piles, comparison with precast

Table 6 (continued). Belgian Piles Testing Experience - Part 2: Piles data (1)									
Test site	Test pile	Description	C, T (1)	Ø Shaft (m)	Ø base (m)	Base depth	Q_{max} (MN)	S_{max} (mm)	Q_{dyn} (MN)
ZEL	1	Bored with bucket with casing & under bentonite	T	0.80	0.80	26.4	2.90	40.66	nr
ANT I	B	Franki (normal expanded base, rammed shaft)	T	0.45	> 0.45	8.3	0.96	3.98	nr
ANT I	C	Franki (normal expanded base, rammed shaft)	T	0.36	> 0.36	8.3	0.90	6.24	nr
ANT I	D	Franki (normal expanded base, rammed shaft)	T	0.45	0.45	8.3	0.90	3.94	nr
ANT I	E	Franki (normal expanded base, rammed shaft)	T	0.36	0.45	8.3	0.90	6.28	nr
ANT I	G	Steel tube pile	T	0.32	0.32	8.3	0.90	17.63	nr
ANT II	1	Franki (normal expanded base, rammed shaft)	T	0.52	0.72	10.1	2.48	57.00	nr
OST	1	Franki (normal expanded base, rammed shaft)	T	0.52	0.72	10.1	3.04	35.00	nr
GRB	1	Vibrex pile + very enlarged steel plate	C	0.52	0.72	20.0	2.44	11.49	nr
GRB	2	Vibrex pile + normal steel plate	C	0.52	0.52	10.0	2.44	64.95	nr
GRB	3	Vibrex pile + normal enlarged steel plate	C	0.52	0.62	10.0	2.20	66.09	nr
GRB	4	Super-Vibrex pile + steel plate + cast in-situ enlarged base	C	0.52	0.52	10.0	2.44	30.14	nr
GRB	5	Vibrex pile + very enlarged steel plate	C	0.52	0.72	10.0	2.20	72.47	nr
GRB	6	Closed-end steel tube pile	C	0.51	0.52	10.0	1.46	55.15	nr
ZWE	1	Atlas screw (auger) cast-in-situ pile	C	0.36/0.46	0.46	12.7	1.58	12.00	nr
ZWE	2	Atlas screw (auger) cast-in-situ pile	C	0.36/0.46	0.46	12.7	1.76	36.00	nr
GH I	1	Atlas screw (auger) cast-in-situ pile	C	0.36/0.46	0.46	13.5	2.76	75.00	nr
GH I	2	Atlas screw (auger) cast-in-situ pile	C	0.36/0.46	0.46	13.5	3.00	75.00	nr
KA IV	1	H Steel without lagging	C	0.36x0.41	0.36x0.41	14	3.50	112.00	2.78
KA IV	2	H Steel with lagging near the toe	C	0.36x0.41	1.08x0.41	14	4.35	110.00	3.51
KA IV	3	H Steel with lagging and plate near the toe	C	0.36x0.41	1.08x0.41	14	5.50	167.00	4.97
GH II	3	Fundex screw (auger) cast-in-situ pile	C	0.38	0.45	13.0	2.04	14.00	nr
GH II	5	Continuous flight auger cast-in-situ pile	C	0.45	0.45	14.5	1.72	15.00	nr
GH II	6	Atlas screw (auger) cast-in-situ pile	C	0.36/0.46	0.46	12.5	2.01	13.00	nr
GH II	8	Fundex screw (auger) cast-in-situ pile	C	0.38	0.45	13.0	1.86	14.00	nr
GH II	9	Continuous flight auger cast-in-situ pile	C	0.45	0.45	14.5	1.62	15.00	nr
GH II	10	Atlas screw (auger) cast-in-situ pile	C	0.36/0.46	0.46	12.5	2.02	13.00	nr
GH II	11	Driven precast concrete pile	C	0.32x0.32	0.32x0.32	13.3	2.80	16.00	nr
GH II	12	Fundex screw (auger) cast-in-situ pile	C	0.38	0.45	13.0	nr	nr	1.81..2.60
GH II	13	Continuous flight auger cast-in-situ pile	C	0.45	0.45	14.5	nr	nr	0.87..1.66
GH II	14	Atlas screw (auger) cast-in-situ pile	C	0.36/0.46	0.46	13.0	nr	nr	1.24..2.60
GH II	15	Driven precast concrete pile	C	0.32x0.32	0.32x0.32	13.4	nr	nr	1.64..2.80
NSC	1	Atlas screw (auger) cast-in-situ pile	C	0.36/0.46	0.46	12.0	0.68	3.30	1.25
GEE	33	Continuous flight auger cast-in-situ pile	C	0.35	0.35	10.5	1.78	2.67	nr
GEE	55	Continuous flight auger cast-in-situ pile	C	0.35	0.35	10.5	1.55	1.97	nr
GEE	99	Continuous flight auger cast-in-situ pile	C	0.35	0.35	10.5	1.39	6.27	nr
KOE	1	Steel tube screw (auger) pile	C	0.35	0.65	13.0	nr	nr	nr
KOE	2	Atlas screw (auger) cast-in-situ pile	C	0.36/0.50	0.50	13.0	nr	nr	nr
KOE	3	Atlas screw (auger) cast-in-situ pile	C	0.51/0.65	0.65	13.0	nr	nr	nr
KOE	8	Steel tube screw (auger) pile	C	0.35	0.65	13.0	1.30	80.03	nr
KOE	9	Atlas screw (auger) cast-in-situ pile	C	0.36/0.50	0.50	13.0	1.95	60.04	nr
KOE	10	Atlas screw (auger) cast-in-situ pile	C	0.51/0.65	0.65	13.0	2.50	64.32	nr
KOE	15	Atlas screw (auger) cast-in-situ pile	C	0.36/0.50	0.50	13.0	2.05	26.04	nr
KOE	16	Atlas screw (auger) cast-in-situ pile	C	0.51/0.65	0.65	13.0	2.35	29.63	nr
KOE	21	Atlas screw (auger) cast-in-situ pile	T	0.36/0.50	0.50	13.0	nr	nr	?
KOE	23	Atlas screw (auger) cast-in-situ pile	T	0.51/0.65	0.65	13.0	nr	nr	?
LIM	5	Driven steel tube	C	0.30	0.30	9.5	1.22	40.64	nr
LIM	8	Driven precast concrete pile	C	0.29x0.29	0.29x0.29	9.5	1.74	34.42	nr
LIM	12	Driven cast-in-situ with steel plate	C	0.40	0.40	9.5	2.82	52.39	nr
LIM	16	Driven cast-in-situ with enlarged base (Franki)	C	0.40	0.52	9.5	3.07	174.83	nr
VIL	3	Screwed cast-in-situ Omega	C	0.41	0.41	14.0	2.08	18.66	nr
VIL	4	Driven precast concrete pile	C	0.35	0.35	7.2	1.1	76.6	nr
VIL	5	Screwed cast-in-situ Omega	C	0.41	0.41	14	1.9	55.93	nr

(1) C = compression, T = tension

Table 6 (continued). Belgian Piles Testing Experience - Part 2: Piles data (2)									
Test site	**Test pile**	**Description**	**C, T (1)**	**Ø Shaft (m)**	**Ø base (m)**	**Base depth (m)**	**Q_{max} (MN)**	**S_{max} (mm)**	**Q_{dyn} (MN)**
ZWI	3	Franki (overexpanded base, shaft friction eliminated)	C	nr	1.43	12.3	3.60	13.40	nr
ZWI	4	Franki (overexpanded base, rammed shaft)	C	0.52	1.47	12.3	3.60	9.10	nr
ZWI	5	Franki (overexpanded base, no shaft friction)	C	nr	1.58	7.3	2.38	51.00	nr
ZWI	6	Franki (overexpanded base, rammed shaft)	C	0.52	1.33	7.3	3.58	46.00	nr
KO I	1	Franki (overexpanded base, shaft friction eliminated)	C	nr	1.40	11.4	2.16	133.99	nr
KO I	2	Franki (overexpanded base, vibrated shaft)	C	0.41	1.45	11.4	3.12	211.85	nr
KO I	3	Franki (overexpanded base, rammed shaft)	C	0.41	1.45	11.4	3.44	160.06	nr
KO I	4	Franki (overexpanded base, vibrated shaft)	C	0.41	1.45	11.4	3.28	173.40	nr
KO I	5	Franki (overexpanded base, shaft friction eliminated)	C	nr	1.40	11.4	2.32	94.85	nr
KO I	6	Franki (normal expanded base, shaft friction eliminated)	C	nr	0.62	10.6	0.62	105.64	nr
KO I	7	Franki (normal expanded base, vibrated shaft)	C	0.41	0.64	10.6	1.52	40.60	nr
KO I	8	Franki (normal expanded base, rammed shaft)	C	0.41	0.64	10.6	2.16	192.37	nr
KO I	9	Vibrated shaft, steel plate	C	0.41	0.41	10.0	0.88	184.37	nr
KO I	10	Franki (no expanded base, rammed shaft)	C	0.41	0.41	10.0	1.20	198.69	nr
KO I	11	Driven prefabricated concrete pile	C	0.41	0.41	10.0	1.04	173.13	nr
KO I	12	Driven steel tube pile (closed end)	C	0.41	0.41	10.0	0.72	189.40	nr
KA I	1	Franki (normal expanded base, vibrated shaft)	C	0.52	0.90	9.7	6.18	96.39	nr
KA I	2	Franki (normal expanded base, shaft friction eliminated)	C	nr	0.54	9.7	2.87	103.33	nr
KA I	3	Franki (normal expanded base, vibrated shaft)	C	0.41	0.62	9.8	3.35	110.61	nr
KA I	4	Franki (normal expanded base, shaft friction eliminated)	C	nr	0.82	9.8	5.21	84.30	nr
KA I	5	Driven steel tube pile (closed end)	C	0.41	0.41	9.3	1.80	68.02	nr
KA I	6	Driven steel tube pile (closed end), enlargement 1.6m above basis, 1.6m length	C	0.41	0.41	11.4	4.74	81.35	nr
KA I	7	Vibrated shaft, enlarged steel plate at the basis	C	0.41	0.54	9.4	2.87	68.07	nr
KA III	A	Driven steel tube pile (closed end)	C	0.60	0.60	11.0	5.45	263.00	nr
KA III	B	Driven steel tube pile (closed end)	C	0.60	0.60	11.0	6.00	300.00	nr
KA III	C	Bored with bucket under bentonite	C	0.60	0.60	11.0	5.10	320.00	nr
KA III	D	Bored with bucket under bentonite	C	0.60	0.60	11.0	4.30	290.00	nr
KO II	1	H steel pile without lagging	C	0.39x 0.38	0.39x 0.38	18.0	2.28	48.00	2.11
KO II	2	H steel pile with local constant enlargement near the toe	C	0.39x 0.38	0.81x0.38	14.5	1.52	97.00	1.85
KO II	3	H steel pile with local variable enlargement near the toe	C	0.39x 0.38	0.81x0.38	15.5	nr	nr	1.62
KO II	4	H steel pile with local variable enlargement near the toe	C	0.39x 0.38	0.81x0.38	19.0	nr	nr	1.17
KO II	5	H steel pile with enlargement (plate) near the toe	C	0.39x 0.38	0.55x0.55	19.3	nr	nr	1.51
KO II	6	H steel pile with local variable enlargement near the toe	C	0.39x 0.38	0.80x0.80	14.5	nr	nr	1.50
KO II	7	H steel pile with local variable enlargement near the toe	C	0.39x 0.38	0.39x 0.38	19.0	nr	nr	0.50
KO II	8	H steel pile without lagging	C	0.39x 0.38	0.39x 0.38	50.0	6.51	119.00	6.85
KO II	9	H steel pile with steel plate at the bottom	C	0.39x 0.38	0.39x 0.38	18.5	nr	nr	0.40
KO II	10	H steel pile with steel plate near the bottom	C	0.39x 0.38	0.39x 0.38	18.5	nr	nr	0.31
KO II	11	H steel pile with local constant enlargement near the toe	C	0.39x 0.38	1.13x0.39	18.0	2.75	94.00	3.86
KO II	12	H steel pile with local constant enlargement near the toe	C	0.39x 0.38	1.13x0.39	18.5	nr	nr	3.20
KA II	1	H steel pile without lagging	C	0.39x 0.38	0.39x 0.38	18.0	3.10	13.50	3.09
KA II	2	H steel pile with local constant enlargement near the toe	C	0.39x 0.38	0.39x 0.38	15.0	6.50	41.00	6.54
KA II	3	H steel pile with local variable enlargement near the toe	C	0.39x 0.38	0.81x0.38	14.5	5.30	57.00	4.56
KA II	4	H steel pile with local variable enlargement near the toe	C	0.39x 0.38	0.81x0.38	18.0	nr	nr	3.49
KA II	5	H steel pile with enlargement (plate) near the toe	C	0.39x 0.38	0.81x0.38	18.5	nr	nr	3.20
KA II	6	H steel pile with local variable enlargement near the toe	C	0.39x 0.38	0.55x0.55	14.2	7.40	85.00	5.54
KA II	7	H steel pile with local variable enlargement near the toe	C	0.39x 0.38	0.80x0.80	18.7	nr	nr	4.53
KA II	8	H steel pile with local variable enlargement near the toe	C	0.39x 0.38	0.39x 0.38	18.0	nr	nr	4.53
KA II	9	H steel pile with steel plate at the bottom	C	0.39x 0.38	0.39x 0.38	19.0	nr	nr	-
KA II	10	H steel pile with steel plate near the bottom	C	0.39x 0.38	0.39x 0.38	19.0	nr	nr	4.06
KA II	11	H steel pile with local constant enlargement near the toe	C	0.39x 0.38	0.39x 0.38	19.0	nr	nr	5.15
KA II	12	H steel pile with local constant enlargement near the toe	C	0.39x 0.38	1.13x0.39	18.5	nr	nr	4.58

(1) C = compression, T = tension

propagation velocity of stress waves in the pile and the soil, and the pile toe condition in the bearing layer.

Belgian experience of the sonic echo method has evidenced however several limitations in the case of cast-in-situ concrete piles (driven, screwed, vibrated, injected or bored) which often have a very irregular lateral surface. A limitation has been found when one encounters several discontinuities in a particular pile : the number of echoes which may be partially superimposed is thereby increased and makes the interpretation of the graphs more difficult. Another limitation has been identified when heavy damping of the signal due to the corrugated texture of the shaft prohibits in some cases the interpretation of the test. It has also been observed that the wave speed travelling in screwed piles shafts is lower than the concrete bar wave speed.

The mechanical admittance method is used when quantification of the pile cross-sectional area and of the pile-soil interaction parameters is needed, in addition to information regarding the integrity of the pile.

7.4 Dynamic load tests

Dynamic load tests with measurement of the strain and velocity of the pile head are increasingly used to evaluate the behaviour of piles [25b, 26b]. Few dynamic tests have been performed on non-displacement (bored) piles however, and some judge that there is not enough experimental data to confirm the feasibility of the method in such cases.

In spite of these arguments, deductions are made using available methods based on the wave equation, including the "Case" and "Capwap" type approaches. Studies of the "Case" method in Belgium [25] tend to show that the result depends strongly on the shape of the impacting force diagram (role of helmet) and on the level of energy.

For the "Capwap-type" procedure, Belgian experience has found a reasonable degree of reliability for the prediction of the ultimate skin friction and of the loading curve at the base, up to the mobilised load [25]. The ultimate failure load, if required, is then a matter of extrapolation as in the case of a loading test not carried out to failure. Tests conducted in Belgium indicate that the maximum transient displacement at the base can exceed that corresponding to the limit load.

8 PARTICULAR NATIONAL EXPERIENCES

De Beer's method [13] finds its roots in theoretical and laboratory experimental research work [7,8] dealing with the interpretation of the CPT test. The method was further enhanced using experimental full scale research conducted on displacement piles at different sites in Belgium (Zwijnaarde, Kontich, Kallo I and III). Table 6 provides information on those test programs as well as on other research programs undertaken for other pile types. Three test sites were used to study the behaviour of H steel piles : two in Kallo II and IV (sand) and another in Kontich II (clay).

Other comparative tests were undertaken in Groot-Bijgaarden with Vibrex-type piles having different lengths and base types. More recently, a test site in Limelette with different driven piles has been undertaken in order to compare driving techniques, performance and induced vibration.

Tensile piles have not been frequently tested (test sites of Ostend, Antwerp I and II, Zelzate focused on bored piles and cast-in-situ driven piles with an enlarged base). The question of the prediction of their ultimate tensile load remains open in many cases.

For more than fifteen years, the market of the continuous flight auger cast-in-situ pile and the screwed (auger) cast-in-situ pile has been rapidly increasing in Belgium, and significant research work has been undertaken on such piles in different soil conditions (test sites of Zwevegem, Ghent I and II, North Sea Coast, Geel, Koekelare and Vilvoorde).

During the last decade, research work has been undertaken in order to improve prediction of pile behaviour by means of stress wave analysis (Kontich II and Kallo II), dynamic loading test (Ghent II, North Sea Coast, Koekelare) and the use of the dilatometer test, DMT (Geel and Koekelare).

In total, more than 100 piles have been scientifically tested during the last 30 years in Belgium in order to develop and improve design methods for axially loaded piles (compression or tension). All this research work has been accomplished thanks to the financial aid of the National Institute for Scientific Research in Industry and Agriculture, the services for funding industrial research in the three Belgian Regions, the Ministry of Public Works, the Belgian Geotechnical Institute, the Civil Engineering Department of the Université Catholique de Louvain, the Laboratory of Soils Mechanics of the University of Ghent, the Belgian Building Research Institute, and Belgian piling contractors.

9 REFERENCES

[1] Aboutaha, M., De Roeck, G. & Van Impe, W. 1993. Bored versus displacement piles in sand-experimental study. *Proceedings of the 2nd International Seminar on Deep Foundations on Bored and Auger Piles :* p 157-162, Ghent.

[2] Begemann, H.K.S. 1969. The Dutch static penetration test with the adhesion jacket cone. Chapter III Tension Piles. *LGM Mededelingen,* Delft.

[3] Bustamante, M. & Doix, B. 1985. Une méthode pour le calcul des tirants et des micropieux injectés. *Bulletin de liaison des Laboratoires des Ponts et Chaussées*, n° 140, p. 75-92, Paris.

[4] Bustamante, M. & Gianeselli, L. 1993. Design of auger piles from in situ tests. *Proceedings of the 2nd International Seminar on Deep Foundations on Bored and Auger Piles (BAP II).* p 21-34, Ghent.

[5] Bottiau, M.& Massarch K.R. 1991. Quality aspects of reinforced augercast piles. *Proceedings of the 4th International Conference on Piling and Deep Foundations (DFI).* p 41-49, Stresa.

[6] Bottiau, M. 1993. Influence of the concreting on the bearing capacity. *Proceedings of the 2nd International Seminar on Deep Foundations on Bored and Auger Piles (BAP II).* p 179-184, Ghent.

[7] Bottiau, M. 1995. Comparative assessment of the bearing capacity out of CPT results. *International Symposium on Cone Penetration Testing (CPT95)*, vol.2, p 399-406, Linköping.

[8] Carpentier, R. 1976. Studie van het grensdraagvermogen van palen geheid in stijve en gescheurde kleilagen. Bepaling van het schaaleffekt. *Thesis presented at the State University of Ghent to obtain the degree of doctor in applied sciences*, Ghent.

[9] De Beer, E. 1963. The scale effect in the transposition of the results of deep-sounding tests on the ultimate bearing capacity of piles and caisson foundations. *Géotechnique*, vol. 13, No 1, p39-75, London.

[10] De Beer, E. 1965. De invloed van de dwarsafmetingen van een paal op de puntweerstand. *De Ingenieur*, p13-21, No 3, Brussels.

[11] De Beer, E. 1968. Quelques problèmes concernant les fondations sur pieux dans des zones portuaires. *Proceedings of the 5th International Harbour Congres,* Antwerp.

[12] De Beer & E., Wallays, M. 1969. Mesures expérimentales de la résistance à l'extraction de pieux. *Proceedings of the VIIth ISSMFE Conference, Speciality Session No 8 on negative skin friction and settlements on piled foundations*, Mexico.

[13] De Beer, E. 1970. Experimental determination of the shape factors and the bearing capacity factors of sand. *Géotechnique*, vol. 20, No 4, p387-411, London.

[14] De Beer & E., Wallays, M. 1972. Pieux Franki à bases surélargies. *La Technique des Travaux.* Liège.

[15] De Beer, E. 1971-1972. Méthodes de déduction de la capacité portante d'un pieu à partir des résultats des essais de pénétration. *Annales des Travaux Publics de Belgique*, No 4 (p 191-268), 5 (p321-353) & 6 (p 351-405), Brussels.

[16] De Beer, E., Lousberg, E., Wallays, M., Carpentier, R., De Jaeger, J. & Paquay, J. 1977. Bearing capacity of displacement piles in stiff fissured clays. *I.R.S.I.A.- I.W.O.N.L., Comptes Rendus de Recherches - Verslagen over Navorsingen*, No 39, Brussels.

[17] De Beer, E., Lousberg, E., De Jonghe, A., Carpentier, R. & Wallays, M. 1979a. Analysis of the results of loading tests performed on displacement piles of different types and sizes penetrating at relatively small depth into a very dense sand layer. *Proceedings of the Conference on "Recent Developments in the Design and Construction of Piles"*, ICE, London.

[18] De Beer, E., Lousberg, E., De Jonghe, A., Carpentier & R., Wallays, M. 1979b. Prediction of the bearing capacity of displacement piles penetrating into a very dense sand layer from the results of CPT tests. *Proceedings of the 7th ESSMFE Conference*, vol. 3, Brighton.

[19] De Beer, E., Lousberg, E., De Jonghe, A., Carpentier & Wallays, M. 1981.Partial safety factors in pile bearing capacity. *Proceedings of the 10th ISSMFE Conference*, vol. 1, p 105-110, Stockholm.

[20] De Beer, E., De Jonghe, A., Carpentier, R., Hever, M. & Scholtes, P. 1981. H steel piles in dense sand. *Proceedings of the 10th ISSMFE Conference*, vol. 2, p. 693-698, Stockholm.

[21] De Beer, E., Scholtes, P. & Carpentier, R. 1982. Draagvermogen van staal liggerpalen. *Tijdschrift van Openbare werken van Belgïe*, No 3, 4 &5, Brussels.

[21a] De Beer, E., Lousberg, E., and Weber, L., 1989, Amélioration de la capacité portante de pieux métalliques HP, *Annales des Travaux Publics de Belgique*, No.6, pp. 455-524, Brussels.

[21b] de Bethune, P. 1961. Atlas de Belgique, Planche 8 - Géologie, Bruxelles; Académie Royale de Belgique.

[22] De Cock, F., Van Impe, W. & Peiffer, H. 1993. Atlas screw piles and tube screw piles in stiff tertiary clays - Assessment of pile performance and pile capacity on basis of instrumented loading tests. *Proceedings of the 2nd International Seminar on Deep Foundations on Bored and Auger Piles (BAP II).* p 359-368, Ghent.

[23] De Jaeger, J., Van den Broeck, M., Holeyman, A. & Legrand, C. 1988. Integrity tests of various types of piles. *Proceedings of the 1^{rst} International Seminar on Deep Foundations on Bored and Auger Piles (BAP I)*, p 577-586, Ghent.

[24] Gilles, P., Tancré, E., Lousberg, E. & De Cock, F. 1991. Determination of the bearing cpacity of screwed-in piles by dynamic piles testing. *Proceedings of the 4th International Conference on Piling and Deep Foundations (DFI).* p 571-578, Stresa.

[25] Holeyman, A. 1984. Contribution à l'étude du comportement dynamique non-linéaire des pieux lors de leur battage. *Doctoral thesis presented at the Université Libre de Bruxelles to obtain the degree of doctor in applied sciences*, ICC, Brussels.

[25b] Holeyman, A., 1987. Théorie des essais dynamiques, Proceedings of the "Journée d'études du Groupement Belge de l'ISSMFE, Brussels, pp. II-50 to II-98.

[26] Holeyman, A., Legrand, C., Lousberg, E. & D'Haenens, A. 1988. Comparative dynamic pile testing in Belgium. *3rd International conference on "Application of Stress-wave Theory to Piles.* P 542-554, Ottawa, Canada.

[26b] Holeyman, A., 1992. Pile Dynamic Testing - State of the Art, 4th International Conference on "Application of stress-wave Theory to Piles, The Hague, Balkema.

[27] Legrand, C. 1987. Essais dynamiques. Intégrité et capacité portante. Données de base du programme d'essais. *Proceedings of the "Journées d'Etudes du Groupement Belge de l'ISSMFE"*, Brussels.

[28] Legrand, C & Huybrechts N. 1997. Recherche en Géotechnique. Etude comparative de différents essais de réception de pieux de fondation. *BBRI report*, Brussels.

[29] Lousberg, E., Jacquemart, J. & Zech, Y. 1974. Ancrages verticaux dans un sol pulvérulent. Etude théorique et essais en laboratoire. *Annales des Travaux Publics de Belgique*, n° 2, Brussels.

[30] Lousberg, E. & D'Haenens, A. 1987. Essais de mise en dynamiques - Dynamische belastingsproeven - Dynamic loading tests. *Proceedings of the "Journées d'Etudes du Groupement Belge de l'ISSMFE"*, Brussels.

[31] Nuyens, J., De Cock, F., Legrand, C., Maertens, J., Menge, P., Van Alboom, G., Van Den Broeck, M. & Welter, P. 1995. National Report 10 - CPT in Belgium in 1995. *International Symposium on Cone Penetration Testing (CPT95)*, vol.1, p. 17-28 , Linköping.

[32] Peiffer, H., Van Impe, W., Cortvrindt & Van den Broeck M. 1991 Analysis of CFA-pile behaviour with DMT-resuslts at Geel test site. *Proceedings of the 4nd International Conference on Piling and Deep Foundations (DFI)*. p 571-578, Stresa.

[33] Peiffer, H., Van Impe, W., Cortvrindt, G. & Bottiau, M. 1993. Evaluation of the influence of pile execution parameters on the soil condition around the pile. *Proceedings of the 2nd International Seminar on Deep Foundations on Bored and Auger Piles (BAP II)*. p 217-222, Ghent.

[34] Silence, P. 1992. Keynote Paper: Luxembourg, Belgium and Holland. *Proceedings of the Conference "Piling - European practice and worldwide trends"*, p.1-5, ICE, London.

[34b] Van der Veen, 1953., The Bearing capacity of a pile, Proc. 4th ICSMFE, Vol. 2, p. 76-84.

[35] Van Impe, W. 1985. The bearing capacity of screwed piles in cohesive layers. *Proceedings of the 11th ISSMFE Conference*, vol. 3, p 1493-1497, San Francisco.

[36] Van Impe, W. 1986. Evaluation of deformation and bearing capacity parameters of foundations from CPT-results. *Proceedings of the 4th International Geotechnical Seminar. Field Instrumentation and In-situ Measurements*, p 51-70, Singapore.

[37] Van Impe, W. & Van den Broeck, M. 1987. Discussion on some loading test results on cast in situ auger piles. *Proceedings of the 8th Asian Regional Conference on Soil Mechanics and Foundation Engineering*, Kyoto, Japan.

[38] Van Impe, W., De Beer, E. & Lousberg, E. 1988. Prediction of the single bearing capacity in granular soils out of CPT-results. *International Symposium on Penetration Testing (ISOPT I), Speciality session*, p 1-34, Orlando.

[39] Van Impe, W., Van den Broeck, M. & Thooft, K. 1988. End and shaft bearing capacity of piles evaluated separately out of static pile loading tests results. *Proceedings of the 1rst International Seminar on Deep Foundations on Bored and Auger Piles (BAP I)*, p 489-498, Ghent.

[40] Van Impe, W. 1988. Considerations on auger pile design. *Proceedings of the 1^{rst} International Seminar on Deep Foundations on Bored and Auger Piles (BAP I)*, p 193-218, Ghent.

[41] Van Impe, W. 1989 Auger piles, planning and use. *Conference of the Association of Finnish Civil Engineers*, Helsinki.

[42] Van Impe, W. 1991. Deformation of deep foundations. General report. P*roceedings of the 13th ISSMFE Conference*, Florence.

[43] Van Impe, W. 1996. Experiences with a new type of displacement screw-pile. *Proceedings of the 6th International Conference on Piling and Deep Foundations (DFI)*, Bombay.

[44] Van Wambeke, A. & D'Hemricourt, J. 1982. Correlation between the results of static or dynamic probings and pressuremeter tests. *Proceedings of the Second European Symposium on Penetration Testing*, p. 941-944, Amsterdam.

[45] Wallays, M. 1972. Pieux Franki à bases surélargies. *Annales de l'Institut Technique du Bâtiment et de Travaux Publics*, No 300, p129-163, Paris.

[46] Weber, L. 1985. Tests on steel H bearing piles. *Belgian Geotechnical Volume published for the 1985 Golden Jubilee of the International Society for Soil Mechanics and Foundation Engineering* (ISSMFE), p. 19-25, Brussels.

[46b]Verbrugge, J.-C., 1981, *Evaluation du tassement des pieux à partir de l'essai de pénétration statique,* Revue Française de Géotechnique, No 15, pp. 75-82.

[47] Zeevaert, L. 1969. Reduction of point capacity of piles because of negative friction. *Proceedings of the first Panamerican Conference on Soil Mechanics and Foundation Engineering*, Vol. III Mexico.

[48] Belgian Geotechnical Volume published for the *1985 Golden Jubilee of the International Society for Soil Mechanics and Foundation Engineering* (ISSMFE), by Belcotec and Belgian Member Society of ISSMFE. Brussels, 1985.

[49] Commission National Fondations sur Pieux, recommandations provisoires des groupes de travail "Reconnaissance de sol et contrôles de reception.

[50] Ministerie van de Vlaamse Gemeenschap, Dept. LIN, AOSO. 1996. Bestekteksten voor kunstwerken, bijgevoegd artikel nr. 201 : Funderingen met betonnen palen.

[51] ISSMFE Subcommittee on Field and Laboratory Testing. 1985. "Axial Pile Loading Test - Part 1 : Static Loading," *Geotechnical Testing Journal, GTJODJ*, Vol. 8, No. 2, pp. 79-90.

[52] Ministère de l'équipment, du logement et des transports, Fascicule n° 62 - Titre V, Règles techniques de conception et de calcul des fondations des ouvrages de génie civil, Paris.

[53] Nederlands Normalisatie Instituut, NEN 6743. 1993. Geotechniek - Berekeningsmethode voor funderingen op palen - Drukpalen.

[54] Normalisation française, DTU 13.2. 1992. Fondations profondes pour le bâtiment.

[55] Regie der gebouwen, "Typebestek 104 index 21 (STS 21). 1990. Diepe funderingen deel A : Funderingspalen vervaardigd door grondverdringing, experimentele uitgave.

Design of Axially Loaded Piles – European Practice, De Cock & Legrand (eds)
© 1997 Balkema, Rotterdam, ISBN 90 5410 873 8

Design of axially loaded piles – Czech practice

J. Feda
Institute of Theoretical and Applied Mechanics, Academy of Sciences, Prague

P. Juranka & V. Pachta
SG-Geotechnika, a.s., Prague

J. Krásný
Faculty of Science, Charles University, Prague

J. Masopust
Zalkládáni Group, a.s., Prague

ABSTRACT : On the background of regional geology, piling technology and design methods in the Czech Republic are dealt with. Soil investigation methods, quality control and monitoring are mentioned too.

1 REGIONAL GEOLOGY OF THE CZECH REPUBLIC

The diversity and complexity of the geology of the Czech Republic is due to the fact that it is composed of two different basic geological units. The greater part of the Czech Republic (Bohemia and Northern and Western Moravia) belongs to the Bohemian Massif while the Easternmost and Southern Moravia are formed by the Carpathians (Fig. 1). These two basic geological units differ considerably in the ages of their rocks and decisive tectogenetic processes and in their palaeogeographic development, structural features and morphology. On the ground surface, the boundary between these two main geological units runs in a SW-NE direction along the line connecting the towns of Znojmo, Přerov and Ostrava. At depth, however, the structures of the Bohemian Massif extend farther to the SE, thus underlying the Carpathians.

The Bohemian Massif is an old geologically heterogeneous block extending from the Czech territory into the neighbouring countries of Austria, Germany and Poland (Fig. 1). Having been affected by older orogeneses such as the Precambrian and Caledonian, the tectonic development of the Bohemian Massif culminated in the formation of the Variscan mountain chain during the late Palaeozoic. Thick continental Permocarboniferous sequences consisting mostly of arkoses, mudstones and shales and containing economically important coal-seams were deposited as filling in intermontane depressions in the newly formed Variscan orogen. Consolidated by the Variscan folding, the Bohemian Massif turned into cratogen. Later it was covered only by the epicontinental Cretaceous sea and by lakes in the Cretaceous and the Tertiary periods. After the Permocarboniferous period the Bohemian Massif was peneplanized until the beginning of the Neogene. During the Neogene and Quaternary, the general uplift of the whole Bohemian Massif and the intensive tectonic activity, causing horsts and grabens, together with the volcanic activity led to the geomorphological rejuvenation of the Bohemian Massif. The depressions were filled in by Neogene sandy-clayey and organogenic (coal seams and diatomaceous clays) deposits up to some hundred metres thick. During the Pleistocene, sandy and gravely deposits of fluvial and glaciofluvial origin were deposited over extensive areas.

The Carpathians belong to the Alpine-Carpathian mountain chain. They had a geological history that was comparable with that of the Bohemian Massif only until the Palaeozoic era. After the beginning of the Mesozoic, when the extensive Alpine-Carpathian geosyncline arose in the southern and eastern vicinity of the Bohemian Massif, their different geological development started. The thick sedimentary complexes of the Carpathians were folded and thrusted from the central part northwards and north-westwards over the more rigid units in their foreland (including Bohemian Massif) mainly during the late Cretaceous, the Oligocene and the Neogene. Within the Czech territory only the outer parts of the West Carpathians occur: the Neogene Carpathian Foredeep, the Flysch belt and the important intermontane Vienna basin.

Due to the aforementioned complex geological history, foundation soils in the Czech Republic vary from residual soils (weathered crystalline and sedimentary rocks - Fig. 1, No. 3), often restricting the use of driven piles, to large areas formed by Neogene stiff clayey sediments (either of marine or lacustrine origin, with abundant fissuration - No. 7). Sandy material is often encountered locally (Cretaceous sediments of Northern Bohemia - No. 6: sandstone's underlain by clays and claystone's or Flysch sandstone's), as are limestone's (No. 4). About 10 % of the country is covered by wind-blown sediments (loess, loessial loam, silts). In many areas, the ground water is aggressive. Alluvial soils represent geomaterials of high compressibility. Glacial sediments are rare (the Quaternary glaciation stopped at the boundary of the mountains of Northern Bohemia and Moravia).

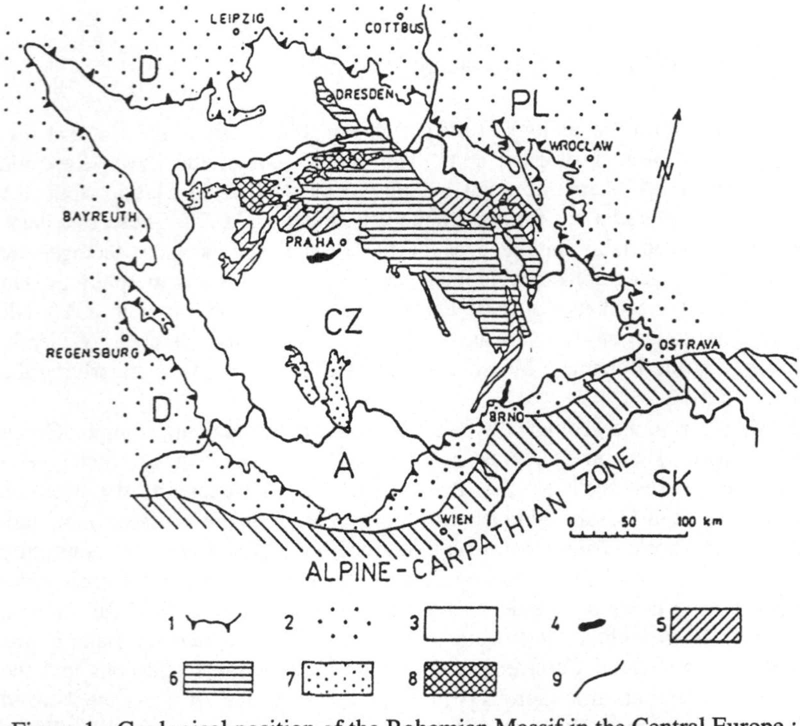

Figure 1 - Geological position of the Bohemian Massif in the Central Europe :
1 - surface boundaries of the Bohemian Massif, 2 - Post-Variscan deposits outside the superficial boundaries of the Bohemian Massif, 3-4 - basement of the Bohemian Massif intensively folded by Variscan tectogenesis, 3 - crystalline, Precambrian and Palaezoicmostly noncarbonate ("hard") rocks, 4 - Silurian and/or Devonian karstified limestones, 5-8 - Post-Variscan cover of the Bohemian Massif, 5 - Permocarboniferous basins, 6 - Bohemian Cretaceous Basin, 7 -Tertiary basins (mostly Neogene, in Southern Bohemia also Cretaceous deposits), 8 - Tertiary volcanic rocks, 9 - frontiers between states.

2 PILING CONDITIONS AND THE COMMON PRACTICE OF SOIL INVESTIGATION

The engineering geological features of the Czech Republic are very varied being complex and regionally differentiated. Within the cratogen block of the Bohemian Massif, Precambric and older Palaeozoic well-diagenetically solidified rocks prevail (metamorphic, folded sediments).

The platform cover of the cratogen block consists of sedimentary filling of the Central Bohemian, Western Bohemian and Eastern Bohemian Permocarbon basins or marine Upper Cretaceous sediments in Central and Eastern Bohemia. They mostly form fairly resistant rocks In contrast, the young Tertiary deposits of the Krušné Hory graben or Budějovice basin, where pellitic sediments prevail, fluctuate between weak rocks and soils and often require deep foundations for more sensitive and demanding structures.

The Eastern part of the Czech Republic is less favourable for foundations because of the softer, diagenetically less solidified rocks of the Foredeep of the Alpine orogen and the flysch rocks of the Carpathian arch on the border with Slovakia.

Quaternary sediments by their nature and mainly by their relatively small thickness do not require the application of the deep foundations. Exceptions are some local accumulations of eolian sediments or thicker alluvial deposits along the banks of the larger rivers.

On the other hand, exogenic factors have exerted an unfavourable effect on the shallow foundation soil zone, often marked already on the Palaeozoic peneplane. Dominating is the effect of the periglacial climate of the Pleistocene. This is manifested by the mechanical or combined alteration of the shallow zone which varies in intensity with depth, eventually combining with the cryoturbation phenomena.

Deep foundations, especially with piles, will be sometimes chosen even if they are not essential for the current subsoil conditions. This is the situation where subsoil properties in the shallow zone cannot be reliably identified or are excessively variable, the control of the ground water inflow into the foundation pit would be too difficult or expensive or where the water table lowering could endanger the stability of structures or the capacity of water resources.

As a consequence of the urbanisation and industrialisation of the Czech Republic, more and more frequently one has to use sites disturbed by the human activity. These include various waste deposits, landfills, mining tips etc. whose heterogeneity and hardly definable properties (mostly unfavourable) call for the use of deep foundations.

Piles are current in the Czech Republic. Owing to the distinct interruption in the exploration for ore deposits at the turn of the sixties and seventies releasing high capacity of the drilling rigs for use in construction large diameter bored piles came into use, either as point bearing or floating piles for foundation medium depths (10 -20 m).

Micro piles, irrespective of the foundation soil, are used to remediate the subsoil of statically endangered bearing elements of old structures or for the underpinning of neighbouring structures endangered by excavations for foundations.

Engineering geological subsoil investigation is governed by the standard ČSN 730090 "Geological investigation for building purposes". Details are given in "Rules for the execution of the engineering geological and hydrogeological investigation" of the former Czech Geological Institute which has been revised several times. It is accepted as a standard.

Subsoil investigations follow the principle of progressing in stages: for deep foundations, starting from the initial incomplete information, the exploration is gradually focused on the collection and improvement of the information about the properties of the deeper layers. As the information and the plans for the project develop, the foundation concept emerges.

Knowledge of the geological conditions in the Czech Republic is rather extensive, especially in the urban and industrial agglomerations and in the mining areas. For over forty years the central archive (Geofond) has been gathering the results of the majority of site investigations, mostly in the form of maps and soundings.

Archive information and the knowledge of the terrain often suffice for the design of common shallow foundations. For deep foundations one has, as a rule, to perform complex identification and verification investigations requiring both field and at desk work.

The underlying geological structure is verified almost exclusively by core boring. To identify the important irregular geotechnical boundaries, geophysical methods are used (geoelectrical, seismic and georadar methods).

The state of granular soils is identified by means of standard penetration procedures. In weak decomposed rocks, pressuremeters or loading tests in the boreholes are used.

In the course of sounding, soil and rock samples are taken for laboratory testing. The set of physical and mechanical properties is recorded for evaluation of the pile-soil interaction.

Some important aspects of the soil investigation can be found in the standard ČSN 731002 "Pile foundations".

The data concerning the ground water level, its fluctuation and aggressivity, are evidently a part of the investigation. In some cases, one has to determine, using hydrodynamical testing, the effect of the depression cone due to pumping on the neighbouring structures.

3 PILING TECHNOLOGY

In the Czech Republic, piled foundations represent the basic and most widespread deep foundation method for industrial, residential and engineering structures. Among the wide range of pile systems and methods the most frequently used are bored piles. This results from their relative universality as far as the foundation soil is concerned which is suitable for the variable foundation conditions of the Czech Republic (see above).

The major advance of piling technology started in the sixties due to the intensive building of large numbers of precast high-rise apartment blocks. Until the eighties, the volume of pile foundations represented more than 20 % of the total volume of foundations and more than 45 % for the apartment blocks alone. Gradually piles were used for bridges and industrial structures. In the fifties driven reinforced piles prevailed. Since the end of sixties bored piles started to be used almost exclusively (which was affected by the importing of special pile systems like Callweld, Terradril, Benoto and Poclain). Early in the seventies Franki and vibropressure VUIS piles were introduced. Both systems are to some extent still used. Presently about 70 % of the total pile production in the Czech Republic are bored piles (CFA piles included) with diameters d = 500 to 2200 mm, Franki piles form about 20 % and VUIS piles 10 %. Driven piles and other technologies are not in use.

Circular bored piles dominate the Czech market. The rotary method is normally used with steel casing and, for pile diameter d > 1500 mm, with clayey suspension. Typical bored piles are telescopic, cased through the upper (granular) soils and encased in the deeper weak rocks (typical conditions). They are usually 8-15 m long, occasionally longer (the longest recent piles were L = 27 m). Raked piles are not used - from the statical point of view raking is considered to be useless. Boring is effected by special foreign boring machines like Soil-Mec, Callweld, Wirth, Delmag, Casagrande and Czech machines G-1300 and G-1500. Typically drilling buckets and augers are used, to a lesser degree also a grab and drop chisel for the Benoto system. Piles with enlarged shaft or underreamed base are not used. Concreting is effected mostly by tremie pipes, less by concrete pumps. Accounting for the frequent considerable aggressivity of ground water secondary protection of pile shafts is used usually by including a PVC membrane of 1.1 mm thickness. Recently CFA piles were performed with d = 400 and 600 mm. Piles are usually reinforced by steel reinforcement cages reaching down to the pile toe, however, piles of plain concrete provided with connecting steel bars in the head only are also accepted.

For large bridges even elements of trench walls (barrette piles) are used carried out under bentonite slurry protection. Their thickness ranges from 0.6 to (exceptionally) 0.8 m.

Franki piles have been in use since the early seventies (when the first KPF machines came from Poland). They are carried out with d = 420 and 520 mm and L < 14 m, vertically or raked to 8:1. Later on further mobile machines were produced in the Czech Republic. Franki piles represent about 20 % of all piles. They are mostly applied for foundations on clayey landfills and thick layers of clayey sands. They are not permitted in urban areas owing to the considerable vibrations.

The Research Institute of Engineering Structures (VUIS) in Bratislava developed in the mid sixties vibropressure VUIS piles, with d = 380 and 420 mm. A special casing penetrates (via induced vibrations) the soil followed by a simultaneous concreting and extraction of the casing. Modifications VUIS-B, VUIS-P and VUIS-Y possess a conical casing enlargement of the pile head. The pile is "bored" by open-end casing vibrated down into the soil and extracted full of penetrated soil. Thus the borehole is carried out either to the full length or to the depth were its wall is stable. If unstable, the casing is provided with a sacrificial shoe and vibrated to the required depth. Then the reinforcing cage is inserted and a pressure vessel full of concrete is fastened to the casing head. Under the effect of pressurised air introduced into the vessel and simultaneously extracting the casing, concreting is effected. In suitable conditions, VUIS piles are produced in lengths of 8-12 m and, even nowadays, represent about 10 % of the total pile production.

4 NATIONAL RELEVANT DOCUMENTS

The basic document for the design, execution, control and testing of piles and pile foundations of plain, reinforced and prestressed concrete, steel and wood is the Czech Standard ČSN 731002 "Pile foundations" valid since 1989. It is a revision of a standard with the same name valid since 1967. It is, however, not applicable to Franki and VUIS piles, to underground (slurry-trench, diaphragm) walls and to micropiles. With regard to the actual situation in the Czech Republic, it is, consequently, the standard for bored piles. The Standard contains just the basic requirements for the design, execution, control and testing of piles. It is linked with ČSN 731001 "Foundation soil of shallow foundations" which is the basic geotechnical standard of the Czech Republic containing soil and rock classifications for geotechnical purposes. It is also linked with ČSN 731010 "Terminology and symbols for foundations of structures", ČSN 730031 "Structures and foundations. Basic requirements for the design" and ČSN 732400 "Execution and control of concrete structures". ČSN 731002 is accompanied by a separate and extensive commentary (Pochman et al., 1989) explaining individual paragraphs and recommending some calculation procedures (some of them are exploited in the parts 5.5 and 5.6 of this report). ČSN 731002 is not obligatory and other procedures, at least of the same level as that in the Standard, may be used.

For Franki and VUIS piles no standard exists. The production companies, however, provide technological rules and catalogues. Presently a translation of the European standard EN 1536/1996 "Execution of special geotechnical work: Bored piles" is under preparation. The Czech Republic intends to accept this standard in the course of 1997 when it will become a member of CEN.

Design, calculation and production of piles and experiences thus gained are dealt with in many papers published at many conferences and symposia on soil mechanics and foundation engineering organised in the Czech Republic in the course of the past 20 - 30 years. The most important are the Brno annual conferences with the international participation organised since 1970.

Piles are the subject of some publications the most important being that of Bažant (1979), Feda (1977) and Masopust (1994) who takes into account the experiences gained through the design and execution of piles in the Czech Republic during the past 25 years.

5. DETAILED DESCRIPTION OF THE NATIONAL DESIGN METHODS

5.1 *General philosophy*

The fundamental type of pile, as already mentioned, is the bored pile which is the most universal and therefore the most suitable for the foundation conditions of the Czech Republic. Research into the theory and design of bored piles was retarded and the axial bearing capacity was calculated using the classical static formulae (e.g. according to ČSN 731002 valid since 1967 by the Caquot-Kerisel method etc.). Numerous loading tests (formerly more frequent than now) did not agree satisfactorily with the calculations. The investigation of the real behaviour of piles using the static loading tests (especially instrumented) showed the substantial effect of the pile technology. Thus the basic principles of the execution of those piles were established and the share of the natural and technological effects was set up. These experiences were incorporated into the producers' technologies and in the former ČSN 731004 (now invalid), one of the first standards in Europe dealing with the design and construction of piles and attempting to reflect the mechanism of the real pile behaviour.

The bearing capacity of a pile depends on the :

- Natural effects, i.e. the kind and quality of the foundation soil.
- Technological effects, i.e. the effects depending on the complete process of the pile installation.
- The effect of the mutual interaction of piles in a group.

The design load of a pile foundation may be arrived at

- either by the first group of limit states, i.e. by the ultimate limit state
- or by the second group of limit states, i.e. by the serviceability limit state.

Ultimate limit state calculations are founded on the Rankine theory of limit equilibrium of granular materials. The ultimate load of a pile is calculated like the bearing capacity of a shallow foundation with the depth effect. The skin friction follows the Coulomb law with the normal stress resulting from a modified coefficient of earth pressure. Šimek and Sedlecký introduced into the design a coefficient of technological effects γ_r The ultimate limit state calculation does not enable the load-settlement curve to be constructed and thus cannot model the pile-structure interaction. It is rarely used.

The limit state of serviceability is characterised by the construction of a load-settlement curve for a bored pile. Either the elasticity theory (Poulos 1972) or quasi-elastic constitutive relations are used (Desai 1974; Ellison et al. 1971). The methods are modified by means of back analyses of loading tests and records of the behaviour of structures (Bažant and Masopust 1978, 1981). Recently, the construction of the load-settlement curve of a bored pile has been based on a non-linear pile settlement theory (Masopust & Muehl 1990).

5.2 *Definitions*

According to ČSN 731002 a pile is a foundation element designed according to the theories of the bearing capacity of piles.

A differentiation is made between small - (d = 0.2 - 0.6 m) and large - (d = 0.61 - 2.5 m) diameter piles. In the first case, min L/d = 5, in the latter case 3, min L = 2 m. Pile inclination is not mentioned.

Piles are classified according to :

1. Execution method:
   - precast (driven, vibrated, jacked, screwed, etc.) - displacement piles;
   - bored (borehole produced by boring, predriving, using a grab, by vibropressure etc.) nondisplacement piles.

2. The mutual relation (single and group piles).
3. The diameter (large - and small - diameter).
4. The inclination (vertical and raked).
5. The load transmission (compression, tension, bending, torsion).
6. The material (wooden, steel, concrete, reinforced concrete).
7. The shape of the shaft and toe (cylindrical, conical, underreamed).

The minimum axial distance should amount to 2.5 d (small diameter piles) or 1.5 d (large diameter piles), but $\geq d + 0.5$ m; for Franki piles 3.5 d.

The axial bearing capacity usually depends on the geotechnical conditions and allowable settlement, with respect to the dimensions of the shaft and toe, depth of embedment into the bearing soil layer, loading mode, pile material and technology.

The vertical design bearing capacity of a pile U_{vd} according to ČSN 731002 should be determined by :

- a static loading test (experimental bearing capacity) U_{ve};
- using the tabulated values for the vertical bearing capacity $U_{v,tab}$
- applying a static calculation based on the strength and deformation parameters of soils and rocks;
- calculating the failure of the pile shaft.

The vertical component of the extreme design load V_d (due to the structure) at the pile head cannot surpass the vertical pile design bearing capacity U_{vd}, i.e.

$$U_{vd} \geq V_d \qquad (1)$$

5.3 *Design on the basis of static load tests*

Static load tests are:

- preliminary (design) tests,
- verification tests,
- control tests.

Preliminary tests - usually a part of the engineering geological soil investigation of the site - are carried out especially in the case of :

- abnormally loaded piles;
- piles of unusual technology;
- complicated structures in exceptionally complex geotechnical conditions (3rd category);
- situations where it is legitimate to assume that the test results will radically decrease the expense of the foundation work.

Verification loading tests are performed usually before or at the beginning of piling if the number of piles > 1500. If their number < 1500, they are carried out (following agreement between the designer and the client) if the geotechnical conditions encountered differ from those assumed.

Control loading tests are ordered, as a rule, by the client (investor) at the end of piling or if the quality of the piles is in doubt. At least 3 piles selected randomly should be tested. Because of the working function of the tested piles, the designer must be involved in their selection.

Static loading tests result in identification of the vertical experimental bearing capacity U_{ve} which depends on the allowable deformation of the structure to be piled. From the pile settlement curve one may deduce (Fig. 2) :

- the ultimate bearing capacity U_y defined by an asymptote of the pile-settlement curve (fig. 2a);
- the bearing capacity U_{pr} at the proportionality limit corresponding to the larger curvature change of the load-settlement diagram (fig. 2b);
- the bearing capacity at the deformation limit U_{def} corresponding to the stabilised pile head settlement equal to 0.1 d (fig. 2c);

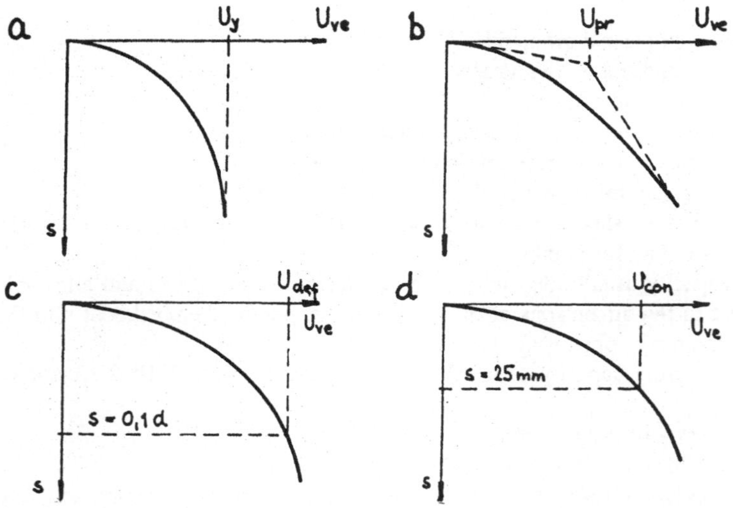

Figure 2 - The determination of U_{ve} from the load-settlement curve : a - ultimate bearing capacity U_y; b - bearing capacity at the proportionality limit U_{pr}; c - bearing capacity at the deformation limit U_{def}; d - conventional bearing capacity U_{con}.

$U_{vd} \geq U_{ve}/\gamma_{re}$ (2)

where γ_{re} (the coefficient of working conditions) is equal to :

1.4 - 1.5 for U_y and U_{pr};
1.25 -1.3 for Ud_{ef};
1 - 1.1 for U_{con}.

- the conventional bearing capacity U_{con} corresponding to the stabilised pile head settlement of 25 mm or to that admitted by the designer.

The vertical design bearing capacity is obtained from the experimental capacity, U_{ve}, using the relation

5.4 *Tabulated values of the vertical bearing capacity* $U_{v,tab}$.

Tabulated values of the vertical design bearing capacity $U_{v,tab}$ may be used for simple structures and for the first geotechnical category (first category: simple structures + simple foundation conditions; second category: either simple structures in complex foundation conditions or complex structures in simple foundation conditions; third category: complex structures in complex foundation conditions - see ČSN 731001), especially for bored piles. Tables 2-6 are based on a set of 226 static loading tests of bored piles with d = 0.6 to 1.5 m (Masopust 1978). For foundation soils classified after ČSN 731001 (R1- sound crystalline rocks; R2- sound sedimentary rocks; R3 -sound claystones, moderately weathered R1 and slightly weathered R2 rocks; R4 - sound siltstones and claystones but slightly cemented, some shales, strongly weathered R1, R2 and moderately weathered R3 rocks; R5 - very little cemented sandstone's, claystones, completely weathered R1, R2, strongly weathered R3 and moderately weathered R4 rocks; R6- completely weathered R1 to R5 rocks; ID - relative density; IC - consistency index), Tab. 1 shows the frequency of the analysed 226 loading tests. As a basis for comparison, the load at the total pile settlement of s = 10 mm was accepted. Some values of the load at s = 25 mm were also recorded.

The extreme design load V_d must fulfil the condition

$V_d <= 0,8\ U_{v,tab}$ (3)

Table 1 - Number of loading tests in various types of soils and rocks

Group	Class according to ČSN 73 1001	Number of tests for s = 10 mm	Number of tests for s = 25 mm
Rocks and weak rocks	R1	-	-
	R2	17*	-
	R3	33*	-
	R4	43	9
	R5	22	14
Gravels and sands	$I_D = 0.5$	9	9
	$I_D = 0.7$	15	9
	$I_D = 1$	39	9
Fine soils	$I_C = 0.5$	5	5
	$I_C \geq 1$	43	18
* Settlement s < 10 mm			

Table 2 - Vertical tabulated bearing capacity $U_{v,tab}$ of piles bored in rocks R1 to R3

Embedment length l_f in m in the rock (class R1-R3)	Bearing capacity of piles $U_{v,tab}$ in kN in rocks class R1 to R3 and for pile diameter d in m			
	0.3	0.5	1.0	1.5
0.0 - 0.5	200	600	2300	6000
1.5	300	720	2500	6000

Table 3 - Vertical tabulated bearing capacity $U_{v,tab}$ of piles bored in rocks R4 to R6

Embedment length l_f in m in the rock R4 -R6	Bearing capacity of piles $U_{v,tab}$ in kN in rocks class R4 to R6 and for pile diameter d in m			
	0.3	0.5	1.0	1.5
0.0 - 0.5	100	300	1 000	2 000
1.5	150	400	1 250	2 200
3.0	200	500	1 500	2 600

Table 4 - Vertical tabulated bearing capacity Uv,tab of piles bored in gravels G1 to G4

Embedment length l_k in m in sands G1 to G4	Bearing capacity of piles $U_{v,tab}$ in kN in sands G1 to G4 and for pile diameter d in m											
	0.30			0.50			1.00			1.50		
	for relative density I_D											
	0.33	0.66	1.0	0.33	0.66	1.0	0.33	0.66	1.0	0.33	0.66	1.0
1.0-1.5	35	70	200	100	200	600	400	800	2300	820	1600	5000
3.0	80	160	380	160	330	870	520	1050	2800	1000	2000	5600
5.0	110	220	500	220	420	1060	630	1300	3200	1100	2300	6300
10.0	180	370	800	320	650	1500	840	1700	4000	1450	3050	8000

Note : G1: well graded, G2 : poorly graded, G3 - G4 : with various fines contents.

Table 5 - Vertical tabulated bearing capacity $U_{v,tab}$ of piles bored in sands S1 to S5

Embedment length l_k in m in sands S1 to S4	Bearing capacity of piles $U_{v,tab}$ in kN in sands S1 to S5 and for pile diameter d in m											
	0.30			0.50			1.00			1.50		
	for relative density I_D											
	0.33	0.66	1.0	0.33	0.66	1.0	0.33	0.66	1.0	0.33	0.66	1.0
1.0-1.5	20	50	175	60	120	500	240	480	1900	520	900	4200
3.0	35	110	275	85	235	680	300	700	2300	580	1300	4800
5.0	50	160	370	100	320	820	340	870	2500	650	1600	5300
10.0	70	280	570	140	520	1100	400	1200	3000	750	2200	6500

Note: S1: well graded, S2 : poorly graded, S3 - S5 : with various fines contents.

Table 6 - Vertical tabulated bearing capacity Uv,tab of piles bored in fine soils F1 to F6, G5

Embedment length l_k in m in sands S1 to S4	Bearing capacity of piles $U_{v,tab}$ in kN in sands F1 to F6, G5 and for pile diameter d in m											
	0.30			0.50			1.00			1.50		
	for relative density I_D											
	0.5	1.0	1.5	0.5	1.0	1.5	0.5	1.0	1.5	0.5	1.0	1.5
1.0-1.5	25	60	120	60	150	300	230	630	1000	500	1250	2000
3.0	60	130	240	130	260	520	370	860	1500	700	1600	2700
5.0	90	180	340	170	350	700	460	1050	1850	820	1800	3400
10.0	160	320	580	260	550	1100	700	1430	2600	1200	2400	4200

Note: F1 to F6 : loam to low plasticity clay, G5 : clayey gravel.

The value of $U_{v,tab}$ depends, first of all, on the embedment depth in the homogeneous bearing layer - upper non-bearing layers are not considered. For layered bearing soil, the embedment depth equals its total thickness and $U_{v,tab}$ corresponds to the layer with the least bearing capacity. The tabulated bearing capacities are valid if :

- the total pile length ≥ 3 d;
- there is no compressible soil layer beneath the pile toe (to the depth of 3d or 2.5 m);
- the boring is concreted within 36 hours after being finished and cleaned up immediately before concreting, especially if water-filled;
- when a clayey suspension is used, it is concreted at least within 8 hours and close before concreting the filter cake of the shaft and toe is cleaned up;
- neither the casing nor a secondary protection are left in the borehole;
- the pile is not underreamed;
- the pile may be considered to be a single one, i.e. the pile spacing ≥ 1.5 d.

Linear interpolation (as opposed to extrapolation) is allowed between the pile diameters and depths of embedment into the bearing layer.

5.5 *Calculation of the bearing capacity of bored piles for the ultimate limit state*

Design values of the angle of internal friction ϕ_d, cohesion C_{efd} or C_{ud} and unit weight γ_d are used. Design values are equal to the characteristic (standard) values (based on laboratory tests for category 3 or tabulated in ČSN 731001 for category 2), divided by the partial safety factor γ_m. For piles and

- for ϕ, $\gamma_{m\phi} = 1.4$;
- for c, $\gamma_{mc} = 2$;
- for the unit weight and hydrostatic pressure $\gamma_m = 1$.

The basic equation for the design vertical capacity of a pile is

$$U_{vd} = U_{bd} + U_{fd} \geq V_d \tag{4}$$

where U_{bd} is the design bearing capacity of the pile toe and U_{fd} the design bearing capacity of the pile shaft. The vertical design bearing capacity of the pile toe is equal to

$$U_{bd} = A_s\, R_d\, K_1 \tag{5}$$

where A_s is the toe area, R_d is the design bearing capacity of the pile toe given by (with the parameters as defined in figure 3)

$$R_d = 1.2\, c_{d2}\, N_{C2} + (1 + \sin\phi_{d2})\, \frac{\gamma_o d + \gamma_1 h_1 + \gamma_2 h_2}{d + h_1 + h_2}\, (L + d)\, N_{d2} + 0.7\, \gamma_2\, N_{b2}\, b/2 \tag{6}$$

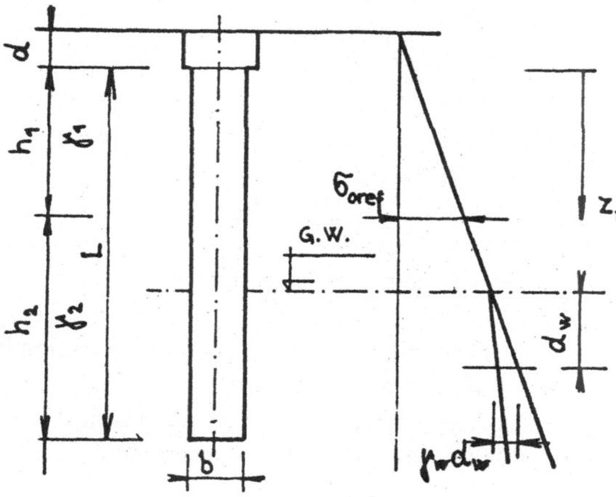

Figure 3 - Design scheme for the vertical design bearing capacity of a bored pile for the ultimate limit state

The coefficients of the bearing capacity are (ČSN 731001) :

$N_c = 2 + \pi$ for $\phi = 0$

$N_c = (N_d - 1)\cot\phi_d$ for $\phi \geq 0$

$N_d = \exp(\pi \tan\phi_d)\tan^2(45 + \phi_d/2)$

$N_b = 1{,}5\,(N_d - 1)\tan\phi_d$

and the coefficient K, indicates the increase of the bearing capacity due to the depth :

$L \leq 2$	$K_1 = 1$
$2 < L \leq 4$	$K_1 = 1{,}05$
$4 < L \leq 6$	$K_1 = 1{,}1$
$L < 6$	$K_1 = 1{,}15$

The design bearing capacity of the pile shaft

$$U_{fd} = \pi \Sigma b_i h_i f_{si}; \qquad (7)$$

where the skin friction

$$f_{si} = \sigma_{xi}\tan(\phi_d/\gamma_{ri}) + C_{efd}/\gamma_{r2} \qquad (8)$$

and the (horizontal) contact stress in the i-layer

$$\sigma_{xi} = K_2\,\sigma_{0ri} \qquad (9)$$

where σ_{0ri} is the original (vertical) geostatic effective stress at the depth $(z_i + d)$, K_2 the coefficient of the lateral earth pressure: for $z \leq 10$ m $K_2 = 1$, for $z > 10$ m $K_2 = 1{,}2$.

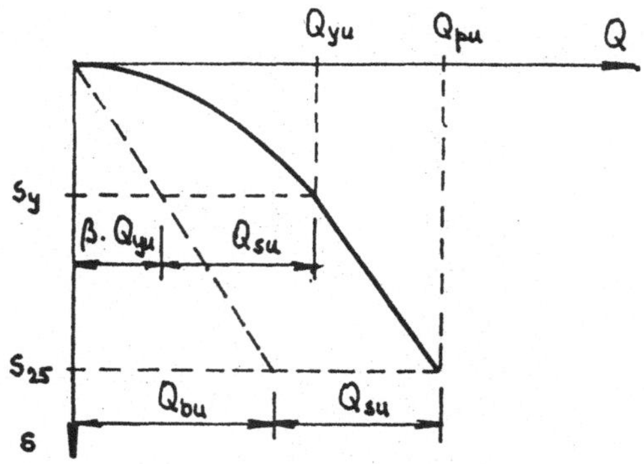

Figure 4 - Limit load-settlement curve of a bored pile

Coefficients of the working conditions of the subsoil γ_{r1} and γ_{r2} are :

for $z \leq 1$ $\gamma_{r2} = 1,3$

$1 < z \leq 2$ $\gamma_{r2} = 1,2$

$2 < z \leq 3$ $\gamma_{r2} = 1,1$

$z < 3$ $\gamma_{r2} = 1,0$

Coefficients γ_{r1} express the technological effects and are equal to

$\gamma_{r1} =$ 1.0 for uncased piles

$\gamma_{r1} =$ 1.1 for uncased piles in cohesive soils and weak rocks

$\gamma_{r1} =$ 1.2 for uncased water-filled borings after pumping, temporary cased borings for the prepact concreting

$\gamma_{r1} =$ 1.25 for borings cased by a clayey suspension, piles with the secondary protection by PVC, PE membranes of the thickness < 0.25 mm

$\gamma_{r1} =$ 1.6 for piles cased by a clayey suspension with secondary protection by the membrane PVC, PE piles of diameter d > 2 m cased by a clayey suspension.

5.6 *Calculation of the bearing capacity of piles for the serviceability limit state*

This procedure requires the shape of the limit load-settlement curve to be found as indicated in Fig. 5.
Piles penetrate, as a rule, a layered foundation soil according to Fig. 5.
The ultimate bearing capacity of the pile shaft Q_{su} may be calculated by means of regression curves deduced from the statistical analysis of the field loading tests of bored piles mentioned in part 5.4 (Masopust 1981). The skin friction is given by the relation

$q_{si} = a - [\, b / (D_i/d_i)]$ (kPa) (10)

where a and b are the regression coefficients in tab. 7. D_i is the depth from the surface to the middle of the respective layer. The total ultimate skin friction equals

$Q_{su} = 0,7\, \pi\, \Sigma\, d_i\, h_i\, q_i$ (11)

The value of the toe pressure at the deformation corresponding to the mobilisation of the skin friction equals

$q_0 = e - [f / (L/d_0)]$ (kPa) (12)

where e and f are the regression coefficients in tab. 7 (d_0 - the toe diameter - fig. 5).
The mean skin friction along the pile shaft is

$q_s = (\Sigma d_i\, h_i\, q_{si}) / (\Sigma d_i\, h_i)$ (13)

The proportion of the total load carried by the pile toe (the toe share as shown in fig. 4)

$\beta = q_0 / (q_0 + 4\, L/d_0\, q_{si})$ (14)

The (full) skin friction is mobilised at the load (fig. 4)

$Q_{yu} = Q_{su} / (1 - \beta)$ (15)

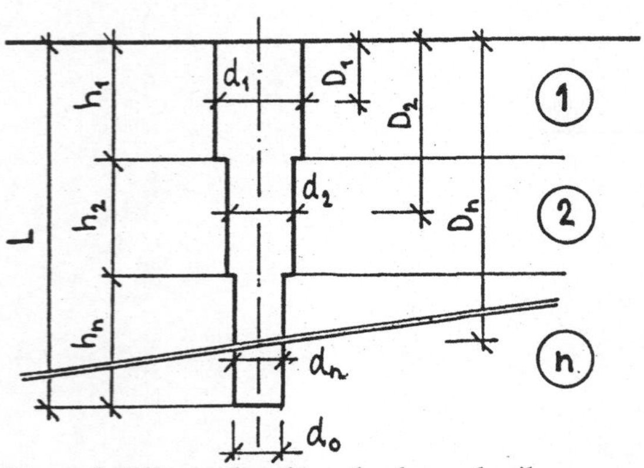

Figure 5 - Pile passing through a layered soil

corresponding to a pile head settlement of (fig. 4)

$$s_y = I\,[Q_{yu} / (d\,Es)] \qquad (16)$$

where I is the influence factor of settlement, E_s the mean value of the secant deformation modulus. The influence factor

$$I = I_l\,R_k\,R_h \qquad (17)$$

Table 7 - Regression coefficients for various soils and rocks

Rock (soil)	a	b	e	f
R3	246.02	225.95	2841.31	1298.96
R4	169.98	139.45	1616.22	1155.34
R5	131.92	94.96	957.61	703.89
$I_D = 0.5$	62.46	16.06	268.11	174.89
$I_D = 0.7$	91.22	48.44	490.34	445.42
$I_D = 1$	154.03	115.88	1596.70	1399.00
$I_C = 0.5$	46.39	20.81	197.74	150.22
$I_C \geq 1$. R6	97.31	108.59	987.60	1084.26

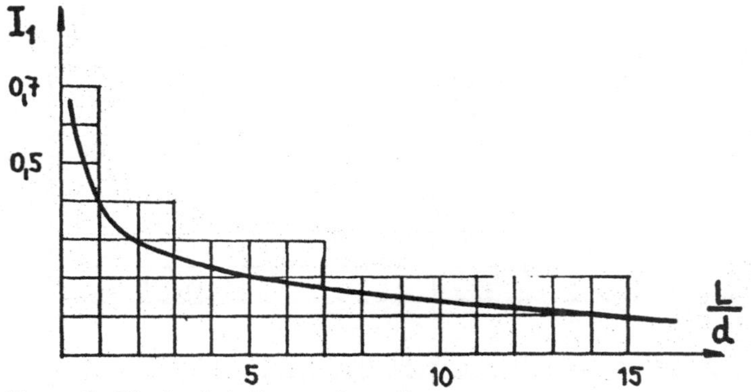

Figure 6 - The basic influence factor I_l

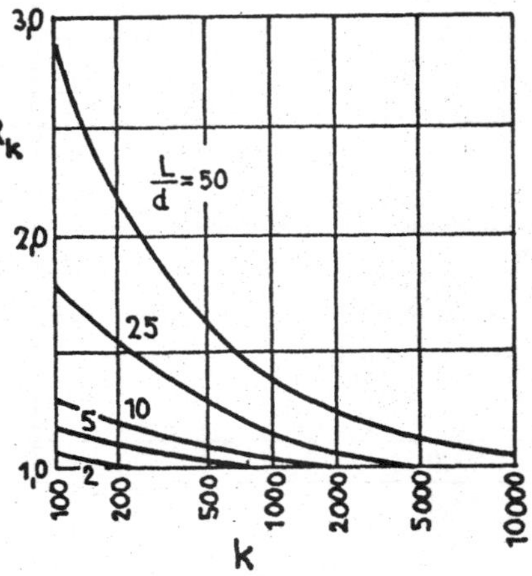

Figure 7 - Coefficient R_k

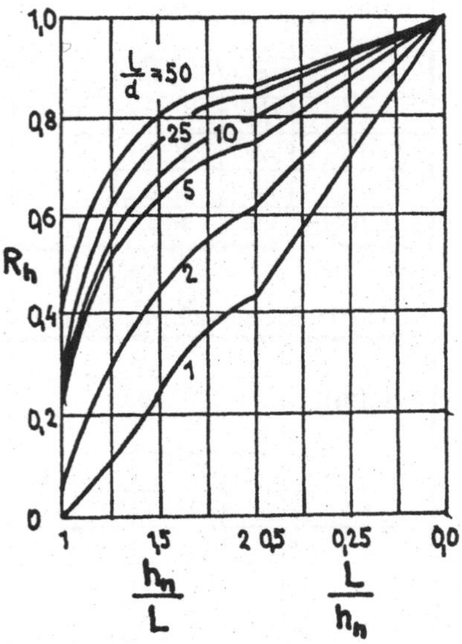

Figure 8 - Coefficient R_h

where I_1 is the basic influence factor according to fig. 6, R_k the coefficient of the pile rigidity according to fig. 7 ($k = E_b/E_s$; E_b deformation modulus of concrete), R_h the coefficient expressing the effect of the depth of the incompressible layer beneath the pile toe in the depth $(h_n - L)$ according to figures. 5 and 8.

Values of the secant moduli E_{si} for different types of soils and pile geometry are recommended in Tables 8, 9 and 10 (Masopust 1980). The mean E_s-value

$$E_s = (\Sigma E_{si}\, h_i) / \Sigma\, h_i \tag{18}$$

The first segment of the load-settlement curve (fig. 4) is a parabola of the second order

$$S = S_y\, (Q/Q_{yu})^2 \tag{19}$$

for the load interval $Q \in (0, Q_{yu})$. The second segment is linear with the coordinates of the ultimate point (S_{25}, Q_{pu}) where

$$Q_{pu} = Q_{su} + Q_{bu} \tag{20}$$

and the pile base (toe) load is

$$Q_{bu} = \beta\, Q_{yu}\, S_{25}/S_y \tag{21}$$

6 AN EXAMPLE DEMONSTRATING THE CALCULATION METHODS

The above calculation methods will be demonstrated by an example of the foundation column

Table 8 - Secant deformation moduli E_s (MPa) for piles in rocks and weak rocks

h (m)	d (m)								
	0.6			1.0			1.5		
	R3	R4	R5	R3	R4	R5	R3	R4	R5
1.5	50.3	28.2	20.0	72.3	35.0	24.7	85.5	33.5	22.3
3.0	64.5	43.1	30.8	105.5	57.3	41.0	138.3	58.8	41.2
5.0	-	58.2	41.3	-	75.3	54.8	-	87.9	63.7
10.0	-	87.5	61.6	-	114.5	83.2	-	133.0	97.0

Table 9 - Secant deformation moduli E_s (MPa) for piles in cohesionless soils

h (m)	d (m)								
	0.6			1.0			1.5		
	I_D								
	0.5	0.7	1	0.5	0.7	1	0.5	0.7	1
1.5	11.0	13.7	28.3	12.8	15.8	30.6	13.0	15.3	29.0
3.0	15.5	20.2	44.5	18.4	25.0	47.8	19.4	24.5	52.5
5.0	18.8	26.6	56.1	22.8	32.5	69.1	24.5	36.0	78.2
10.0	23.8	36.6	72.1	29.8	47.8	93.4	32.6	54.0	107.3

Table 10 - Secant deformation moduli E_s (MPa) for piles in cohesive soils.

h (m)	d (m)					
	0.6		1.0		1.5	
	I_C					
	0.5	≥1	0.5	≥1	0.5	≥1
1.5	6.9	13.2	7.9	13.4	8.6	12.3
3.0	10.0	22.0	12.5	23.9	13.7	23.0
5.0	12.5	31.2	15.9	33.4	18.4	36.7
10.0	15.5	44.3	21.3	51.3	24.6	57.4

of a monolithic concrete skeleton loaded by a design load V = 1.8 MN (extreme design load V_d = 2.25 MN).

Geological log :

0.0 - 0.8: fill - non-loadbearing soil **1** (γ_1 = 18 kNm^{-3})

0.8 - 1.5 : clayey loam soft - non-loadbearing soil **2** (γ_2 = 19.5 kNm^{-3})

1.5 - 5.3 : sand and clayey gravel, water saturated, dense, soil **3** (I_D= 0.7; ϕ_{ef3} = 30°, C_{ef3} =5 kPa, γ_3 = 20 kNm^{-3}, E_{def} = 22 Ma)

5.3 - 6.7 : firm marl, soil **4** (I_C = 1.0; ϕ_{ef4} = 24°, C_{ef4} = 10 kPa, γ_4 = 21 kNm^{-3}, E_{def} = 16 MPa)

6.7 - 12.0 : marlstone weathered to slightly weathered, soil **5** (R5, ϕ_{ef5} = 30°, C_{ef5}, = 20 kPa, γ_5 = 21.5 kNm^{-3}, E_{def} = 40 MPa)

Ground water level in the depth of 2.20 m.

Rotary bored piles are proposed with steel casing (d = 1220 mm) down to a depth of 5.5 m and uncased (d = 1070 mm) further down to the final depth of 8.5 m. The casing is recovered after concreting. The dimensions and stresses assumed in the pile design are shown in figure 9.

6.1 *Ultimate limit state design (part 5.5)*

Toe bearing capacity : ϕ_{d5} = 30/1.4 = 21.42°, C_{d5} = 20/2 = 10 kPa → N_d = 7.37; N_b = 3.75; N_C = 16.24

R_d = 1.0 x 10 x 16.24 + (1 +sin 21.42).(18 x 0.8+ 19,5 x 0.7 + 20 x 3.8 + 21 x 1.4 + 21.5 x 1.8) x 7.37 + 0.7 x 21.5 x 1.07/2 x 3.75 = 194.88 + 1732.09 + 30.19 = 1957.16 kPa

L > 6 m → K_1 1.15; $A_s = \pi.1.07^2/4 = 0.898$ m^2

U_{bd} = 1.15 x 0.898 x 1957.16 = 2 021.16 kN

Skin friction in the layers 3, 4, 5 :

layer **3** : φ_{d3} = 30/1.4 = 21.42°, c_{d3}= 5/2 = 2.5 kPa, σ_{or3} = 20.3 x 4 - 10 x 1.2 = 56.8 kPa,

Z_3 = 3.4 m < 10 m→ K_2 = 1.0, σ_{x3} = 56.8 kPa, γ_{r2} = 1.0

f_{g3} = 56.8.tg (21.42/1.5) + 2.5/1.0 = 16.96 kPa

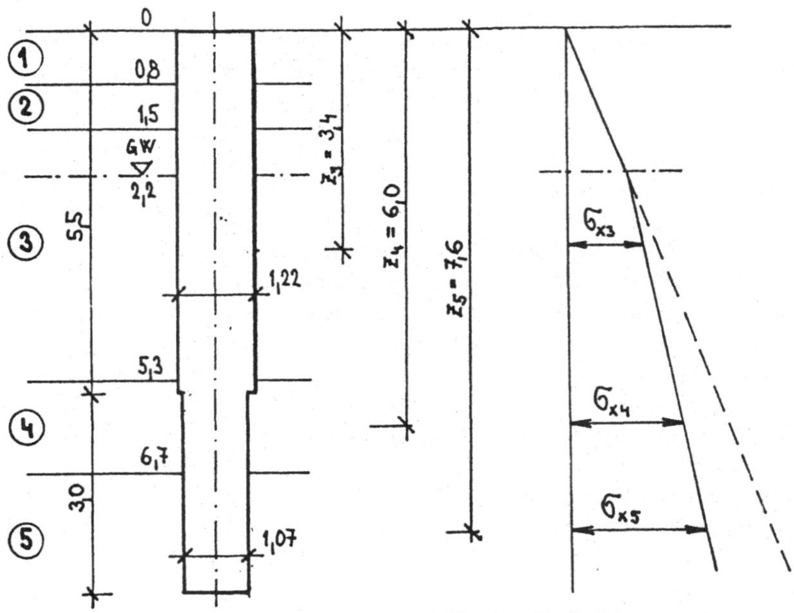

Figure 9 - Dimensions and stresses assumed in the pile design example.

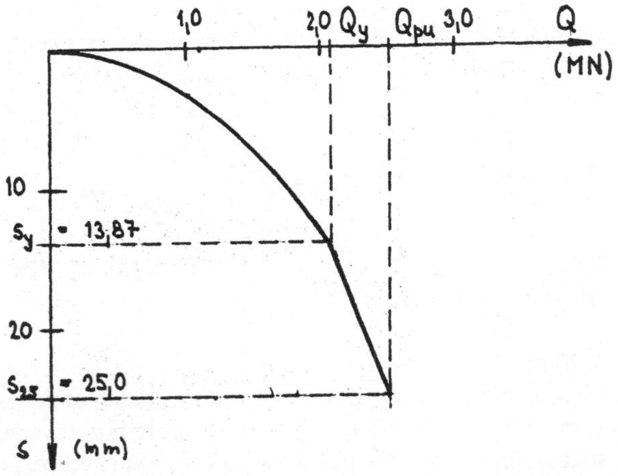

Figure 10 - Ultimate load-settlement curve - the example.

layer **4** : $\varphi_{d4} = 24/1.4 = 17.14°$, $C_{d4} = 10/2 = 5$ kPa, $\sigma_{or4} = 21 \times 6.0 - 10 \times 3.8 = 88.0$ kPa,
$Z_4 = 6.0$ m < 10 m→ $K_2 = 1.0$, $\sigma_{x4} = 88.0$ kPa, $\gamma_{r2} = 1.0$,
$f_{g4} = 88.0.\text{tg}\,(17.14/1.5) + 5/1.0 = 22.79$ kPa
layer **5**: $\varphi_{d5} = 30/1.4 = 21.42°$, $C_{d5} = 20/2 = 10$ kPa,
$\sigma_{or5} = 21.5 \times 7.6 - 10 \times 5.4 = 109.4$ kPa,
$Z_5 = 7.6$ m < 10.0 m→ $K_2 = 1.0$, $\sigma_{x5} = 109.4$ kPa, $\gamma_{r2} = 1.0$,
$f_{s5} = 109.4.\text{tg}\,(21.42/1.5) + 10/1.0 = 37.85$ kPa
$U_{fd} = \pi.(1.22 \times 3.8 \times 19.96 + 1.07 \times 1.4 \times 22.79 + 1.07 \times 1.8 \times 37.85) = 626.66$ kN

$U_{vd} = 2\,021.16 + 626.66 = 2\,647.82$ kN > Vd = 2 250 kN.

6.2 *Serviceability limit state design (part 5.6)*

Computer code VP has been used, fig. 10 shows the result. Settlement at the ultimate skin friction $s_y = 13.87$ mm, $Q_y = 2099.09$ kN, $Q_{pu} = 2530.5$ kN; for the working load V = 1800 kN the settlement $s = 13.87\,(1800/2099)^2 = 10.2$ mm.

7 QUALITY CONTROL AND MONITORING

Loading tests of piles and their evaluation follow exactly the standard ČSN 731002. A dead weight up to 3000 kN is used, for higher loads anchorage is applied.

With dynamic pile tests vibrations of the pile are recorded after a horizontal or vertical impact of a ram. This non-destructive method enables :

- to discover serious deficiencies;
- to reveal any pile discontinuity due to the soil invasion into the pile shaft, concreting interruption, non satisfactory connection of various concrete batches, water filtration through concrete at the ground water level, incomplete filling of the boring by the concrete mixture near the reinforcement, insufficient cleaning of the boring bottom etc.;
- to reveal, for driven piles, the destruction due to the driving procedure or the damage of the reinforcement-concrete interaction;
- to measure the stiffness of the pile-soil embedment.

Simultaneously one may measure the concrete strength using Schmidt hammer. Ultrasonic test may inform about the homogeneity of the concrete in the pile head.

For the pile continuity, concrete homogeneity and pile length, the reflection and impedance methods may be used (vibration of the pile body or reflection of the dynamic wave are recorded using accelerometers).

Another method for checking piles is to use an inclinometer. By its inclination through the pile body, the depth, direction and rate of various underground movements may be identified. The apparatus passes through a PVC casing (installed permanently in the pile) provided with a groove. The accuracy of horizontal displacement measurements reaches ± 2 mm for 30 m depth.

This method is suitable for measuring of the inclination of underground trench (barrette) walls and piled walls.

To check the integrity, concrete homogeneity and quality of the underground foundation elements an ultrasonic ("transparency") method may be used. Several gauge tubes are installed in advance into the foundation element and the ultrasonic impulses of the transmitter in one tube are recorded by the receiver in the other one.

Acknowledgement
Following parts of the report were authored by: J. Krásný - 1, P. Juranka - 2, J. Masopust 3 to 6, V. Pachta - 7, J. Feda - coordinating and editing head of the team. Dr. T. Orr kindly reviewed this report.

8 REFERENCES

Bažant, Z. 1979. *Methods of foundation engineering*. Elsevier, Amsterdam, 615 pp.

Bažant, Z. & J. Masopust 1978. Velkoprůměrové piloty (in Czech). *Inž. stavby* 8: 376-390.

Bažant, Z. & J. Masopust 1981. Drilled pier design based on load-settlement curve. *Proc. X ICSMFE,* Stockholm, 2: 615-618.

ČSN 731001. 1987: *Základová půda pod plošnými základy* (in Czech). Praha, 75 pp.

ČSN 731002. 1989.: *Pilotové základy* (in Czech). Praha, 25 pp.

Desai, C. S. 1974. Numerical design analysis for piles in sands. J. *Geot. Eng. Div.* ASCE, 100, 6:661-678.

Ellison, R. D., E. D'Appolonia & G. R. Thiers 1971. Load-deformation mechanism for bored piles. J. *Soil Mech. Found. Div.* ASCE, 97, 4: 661-678.

Feda, J. 1977. *Interakce piloty a základové půdy* (in Czech). Studie ČSAV No. 9, Academia, 156 pp.

Masopust, J. 1978. *Únosnost vrtaných pilířů* (in Czech). Kand. dis. práce, ČVUT Praha, 224 pp.

Masopust, J. 1980. Volba modulu deformace při výpočtu velkoprůměrových pilot (in Czech). *Sb. Konf. "Výpočet únosnosti pilotových základů ", K.* Vary, ČSVTS: 33-41.

Masopust, J. 1981. Návrhové parametry pro výpočet velkoprůměrových pilot (in Czech). *Sb. konf "Zakladanie stavieb"*, V. Tatry, ČSVTS: 165-169.

Masopust, J. 1994. *Vrtané' piloty* (in Czech). Nakl. Čeněk a Ježek, Praha, 263 pp.

Masopust, J. and P. Muehl 1990. Velkoprůměrové vrtané piloty (in Czech). SNTL Praha, 316 pp.

Pochman, R. et al. 1989. *Komentář k ČSN 731002* (in Czech). Praha, 75 pp.

Poulos, H. G. 1972. Load-settlement prediction for piles and piers. J. *Soil Mech. Found. Div.* ASCE, 98: 615 640.

Design of Axially Loaded Piles – European Practice, De Cock & Legrand (eds)
© 1997 Balkema, Rotterdam, ISBN 90 5410 873 8

Pile foundation – Danish design methods and piling practice

Rikard Skov
cp test a / s, Vejle

0 INTRODUCTION

Piles have been used all over Europe for many years to carry the loads of various structures through soft soil layers to soil strata with high bearing capacity. However looking on the national codes today there is a wide variation from country to country regarding pile types, design - and control methods and safety philosophy.

With the new open borders within Europe there is a tendency that building contractors in the future will consider Europe as one big local market. Efforts have already been made to establish common rules. For the pile foundation industry the philosophy and principles have been outline in the proposal to Eurocode EC7.

The work has been in progress for many years and experience shows that it is difficult job to harmonise design methods and traditions for pile foundations in the involved countries. However when the topic is discussed amongst engineers , it is an advantage to have at least a brief knowledge how piling is carried out in other countries. ERTC3 has taken the initiative to a seminar where each country is invited to contribute with a short description of how pile foundation is put into practice in their respective areas. The enclosed contribution is intended to outline and describe the Danish piling practice. Without going to much in details it describes how piling is handled in the Danish Code. Two foundation cases are enclosed as practical excamples.

1 REGIONAL GEOLOGY

Denmark is situated close to the Scandinavian mountain area and the substratum is built up of sediments originated in that area. Over time movement has taken place, soil layers have been eroded and the landscape as we see it today has been formed over at least 3 glacial periods, of which the last one only covered a part of Denmark. The tertiary clay and glacial deposit can be strongly preconsolidated. Normal consolidated clay occurs in the northern part of Denmark.

Considering pile foundation the soil layer supporting the piles can typically be:

Limestone
Tertiary layer
Quaternary layer

Figure 1 shows a map of the surface just below the quaternary layers in Denmark.

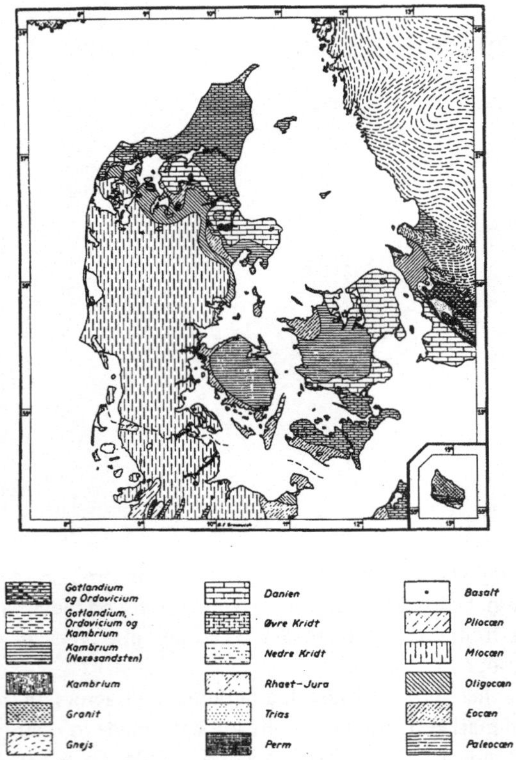

Figure 1

2 COMMON PRACTICE FOR SOIL INVESTIGATION

For Structures with pile foundations, boring are usually performed to a depth which ensures knowledge of soil conditions well below the level of pile base, and to an extent sufficient to permit a design evaluation of bearing capacity and negativ skin friction.

For clay and silt the undrained shear strength c_u is usually determined by measuring the vane strength c_v. Rules are given how to come from c_v to c_u for different soil conditions.

In sand and gravel the angle of friction is obtaind from test on undisturbed samples.

In situ test as SPT and especially CPT are also used more frequently .

3 PILING TECHNOLOGY: COMMON PILE TYPES, EXECUTION METHOD.

During the latest 25 years the prefabricated mild steel reinforced concrete pile has been the most used in Scandinavia . On a rough estimate this pile type make out a total of 95 - 98% of all installed piles in Denmark.

Pile fabrication as well as pile driving have undergone industrialisation - and standardisation process in the same period which secure a high and uniform quality with an economy competitive to other pile systems. Piles in clay are often driven to a specified toe level base on a static calculation while piles with the toe in sand often are stopped on a driving criteria derived from the Danish Driving Formula. Over the last 17 years pile capacities in both types of soil are more often verified by dynamic load test which is obvious to use in connection with driven piles and secure a reliable foundation. 25 years ago bored piles were more common to use and some large structures are standing on bored piles. The pile type was brought in disrepute when

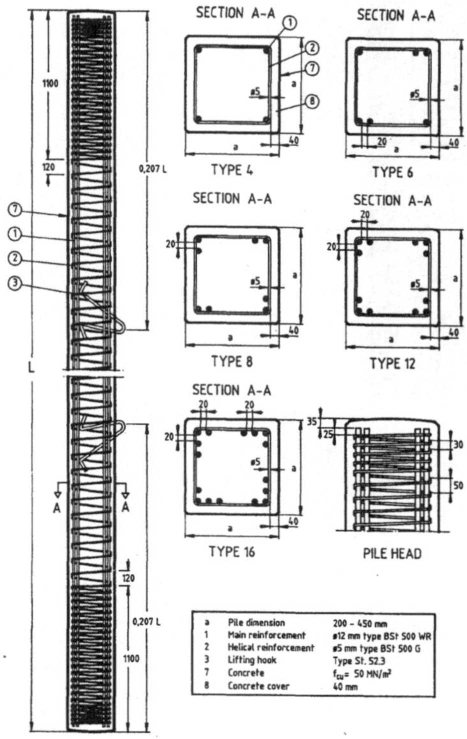

Figure 2.

a large bridge falled down the day before it was opened due to defective quality control. During the last few year bored piles have been used less frequently

In some larger bridge- or harbour project steel piles up to a diameter of 800 mm have been applied.

Underpinning of older building is often carried out using jacked segmental concrete piles or driven steel pile segment of diameter 120 - 150 mm. The piles are usually driven by hydraulic operated mini driving rigs or normal drop hammers typically 300 - 400 kg.

In average the total Danish piling market is 300.000 metre each year.

3.1 Prefabricated concrete pile.

As standard, precast piles can be delivered with cross-sectional area ranging from 200x200 mm² to 450x450 mm² with increments of 50 mm.

The maximum pile length in a single element is 18 metre. Where longer piles are required one or more pile joints may be used.

Pile joints can be manufactured to take the same bending moment and tension forces as the precast pile itself. The longest concrete pile used in Denmark is about 50 metre.

Precast piles are mainly manufactured automatically. Reinforcement cages can be manufactured to any length with little waste on a welding robot using steel from coils. See figure 2.

Typical concrete strength is 45 Mpa. Steel area/concrete area is usually between 0.75 % and 2.10 %, adjusted according to actual pile length.

With the right soil conditions a safe working load of 10 Mpa ore more on driven precast concrete piles is often obtained.

Figure 3.

Figure 4.

3.2 Pile driving

The most common equipment for driving piles is a hydraulic pilling rig (Banut, Uddcomb/Hitachi, Junttan). Hammer weight typically range from 40kN to 80 kN. Leaders are usually of length sufficient to drive a 18 m pile in one peace.

Environmental considerations are also important. Many pilling rig have a drill attached to one side of the leader in case it is necessary to predrill through upper hard soil layers to reduce vibrations on neighbouring structures. During pile installation vibration monitoring is often carried out on neighbour houses to ensure that regulations are observed.

During recent years experiments have been carried out to reduce noise level. On some occasions a reduction up to 12dB has been achieved. See figure 4.

4 THE DANISH CODE, DS415

The theory of plasticity was used very early in geotechnical design in Denmark. The theories of Brinch Hansen are also well known in other countries.

The first edition of the Danish Code of Practice for foundation engineering date back to 1965. This code was the first construction code at all based on limit state design with a pertain system of partial coefficient Latterly all Danish construction codes have been based on this principle.

The present code is the 3rd edition (1984). It is remarkable that the size is only 70 A5-pages and it contain real code text as well as a guidance for the user.

The code is a legal- and technical document which define general demand on function -and safety principles. It is characteristic that there are not many specific demands to procedure, numbers of test and quality of parameters. In this respect it is assumed that the user of the code has sufficient technical knowledge to be able to make relevant judgment himself. The reason for this is possibly that Danish engineers are relatively uniformly educated and in addition there is a practise for how things are done which is not commited to the code text.

It is characteristic when dealing with pile foundation that the code is almost only specific to driven precast piles.

The code itself is arranged after the different phases in a project.

5 DESIGN METHODS IN THE CODE

5.1 Foundation classe

As a beginning distinction is made between three foundation classes.

-low foundation class
-normal foundation class
-high foundation class

The foundation class determine the minimum requirement for extent and quality of the geotechnical analysis, design calculations and workmanship control . When determining the foundation class the following are taking into consideration the nature and size of the structure, loading conditions, soil conditions and conditions of adjacent structures. However there is also an economical aspect in the choice of foundation class. E. g. the partial coefficients in the low foundation class are 25% higher than in the normal foundation class and it might be economically viable to enlarge the extent of the preliminary investigation to carry out the project in the normal foundation class. The same advantage is not valid by selecting high foundation class in preference to normal foundation class. Only safety or technical considerations apply in this case.

5.2 Safety classes.

Depending on consequences of failure, the structure is placed in one of the following safety classes:

-low safety class
-normal safety class
-high safety class

Only the two last classes are applied in connection with pile foundation. The chosen class determine the partial coefficients to be used in the project. In the high safety class the partial coefficients are 10% higher than in normal safety class.

5.3 Preliminary investigations.

In *low foundation* class investigations should aim at establishing :

- the dept and extent of highly compressible layers within the area of the pile foundations, and the magnitude of negative skin friction.

In *normal foundation* class boring must be carried out to a level below pile toe to establish:

- soil strata
- strength- and deformations parameters for compressible layers
- strength - and deformations parameters for layers supporting the pile.

In clay and silt undrained vane test is often carried out. For sand and gravel the internal friction angle must be determined from laboratory test. Important parameters are density, grain size distribution, grain size and grain shape, void ratio stress level. Density can be determined indirectly by sounding methods. However which sounding methods should be use is not specified. The code allows engineering judgment.

In *high foundation* class additional investigations are introduced.

5.4 Safety.

Foundation structures should be examined both in the ultimate limit state and the serviceability limit state.

Generally, safety can be taken into consideration by one of the three methods:

- actual actions are used, material strengths are limited by division with a safety factor.
- actions are multiplied by a factor and characteristic material strengths are used.
- actions are multiplied by a partial factor and characteristic material strengths are divided by a partial factor.

The partial coefficient method is based on the last principle. The method make it possible by choosing factors to take into consideration how safe the actions or material parameters are determined. It can be good economy to spent extra time and money to determine the parameters with greater accuracy and hence be able to use lower partial coefficients in the design without reducing the overall safety.

Design actions are found by multiplying the load by a partial coefficients related to failure - or deformation analysis.

Design material parameters are in the same way determined by division with a partial coefficient related to failure- or deformation analysis.

The calculations must demonstrate at least equilibrium between design actions and design material parameters in failure - and deformation analysis respectively.

Partial coefficients used in normal foundation class:

Actions:		
Dead load:	1.0	
Moveable load:	1.3	
Pile capacity:	Safety class	
	Normal	High
Without test loading on site	2.0	2.2
With test loading on site	1.6	1.75
Pile subjected to test loading	1.4	1.55

5.5 Design

5.5.1Ultimate limit state analysis:

Actions:

It is considered important that in addition to actions transmitted from the structure above, a pile may also be influenced by actions due to consolidation of compressible deposits by surface actions, fill, lowering of the ground water table, etc.

The magnitude of the negative skin friction may be static calculated or derived from tests. Negative skin friction may be reduced by bitumen coating the pile surface.

Internal pile capacity:

The ultimate limit state of the pile material should be examined for the design pile action in compliance with the relevant structural codes.

External pile capacity:

The bearing capacity in the soil can be confirmed by one of the following methods:

Static load test
Dynamic load test
Driving formula
Static calculation

The Danish Driving Formula is commonly used in friction soil, see also example 2. For piles in cohesive soil static calculations are often used.

Vertical load-bearing capacity.

In the analysis of the vertical load-bearing capacity all soil strata are assumed to be supporting without taking into consideration deformation. For a vertical compression pile this results in the criterion

$$V_d < \frac{Q_p + Q_m + Q_{neg}}{\gamma_b}$$

where

V_d is the design pile action

Q_p is the characteristic pile base resistance

Q_m and Q_{neg} are the characteristic shaft resistance, respectively, in the soil strata where it is always positive, and in the soil strata where it may be negative, and

γ_b is the partial coefficient

Static calculation.:

The load-bearing capacity of a single or cylindrical pile with the base located in clay can be determined by:

$$Q = Q_p + \Sigma Q_m$$

where

$Q_p =$ $9c_uA_p$ in cohesive soil

$$Q_m = \begin{cases} N_m q'_m A_m & \text{in non cohesive soil} \\ m r c_u A_m & \text{in cohesive soil} \end{cases}$$

N_m = 0.6 for compression piles in all foundation classes

N_m = 0.2 for tension piles in low and normal foundation classes

m~ $\begin{cases} 0.1 \text{ for wood} \\ 0.8 - 1.0 \text{ for concrete, depending on surface structure} \\ 0.7 \text{ for steel} \end{cases}$

For driven piles in firm glacial till twice the specified base resistance ($Q_p \sim 18c_uA_pp$) can be assumed empirically.
Where no precise determination is carried out, the regeneration factor r may be assumed to be ~ 0.4, except in case of calculation of negative skin friction where r should be assumed ~1.0.

The following is mentioned about bored piles in the guidance part of the code: The load bearing capacity of bored cast in situ piles may be substantial less than that of corresponding driven piles. Provided that the boring process was efficient, so that the concrete may be cast against intact soil strata, the base resistance may be determined as for footing foundations, taking into account that use of design load- bearing capacity exceeding 1000 kN/m2 is only permitted in high foundation class. For bored piles cast in situ, a shaft resistance greater than 30% of the shaft resistance of the corresponding driven pile should not be assumed without a more detailed analysis. If drilling mud is used for stabilisation prior to casting, the shaft resistance may be even smaller.

5.5.2Serviceability limit state analysis.

For smaller pile foundations the serviceability limit state analysis may be performed by using the following design criterion, provided there is no highly compressible deposits below the pile bases. In compliance with this criterion.

$$V_d + Q_{neg} = \frac{Q_f}{\sqrt{\gamma_b}}$$

where

V_d is the design pile action with the partial coefficient γ_f corresponding to $\gamma_f = \sqrt{\gamma_b}$

Q_{neg} is the negative skin friction with the partial coefficient 1.0

Q_f is equal to $Q_m + Q_p$, which is the part of the characteristic load bearing capacity of the pile provided by the layers below the compressible deposits, and

γ_b is taken from the table in section 5.2

5.6 Workmanship and control

A pile driving contract normally commence with a test driving in which the entire pile driving process is recorded for a few test piles driven on different locations on the site. The test piles can be considered as a full scale sounding test to confirm the depth to the load carrying strata assumed in the borehole investigations. For the remaining piles it is usually sufficient to record the driving resistance during the final phase of driving.

6 EXAMPLES DEMONSTRATING THE METHODES

6.1 Example 1, piles in clay

Normal safety class.

Action: Failure analysis $Vd = 1 \cdot 200 + 1.3 \cdot 250 = 525$ kN/m
Deformation analysis $Vd = 1 \cdot 200 + 1.0 \cdot 250 = 450$ kN/m

Static calculation:
Pile toe assumed at level -7.5 ~ 3.3 meter in glacial till.

Skin friction:

Sand layer:

$Q_m = N_m \cdot q_m \cdot A_m$

$= 0.6\,(1 \cdot 18 + 0.25 \cdot 10) \cdot 4 \cdot 0.25 \cdot 1.5$

$= 18.5$ kN

Clay:

$Q_m = m \cdot r \cdot c_u \cdot A_m$

$= 0.9 \cdot 0.4 \cdot 52 \cdot 4 \cdot 0.25 \cdot 2.9$

$= 54.3$ kN

Glacial till.:

$Q_m = m \cdot r \cdot c_u \cdot A_m$

$= 0.9 \cdot 0.4 \cdot 420 \cdot 4 \cdot 0.2 \cdot 3.3$
$= 499$ kN

Toe resistance:

$Q_p = 18 \cdot c_u \cdot A_p$

$= 18 \cdot 420 \cdot 0.25 \cdot 0.25$

$= 472$ kN

Failure analysis:

Total capacity for the pile

$Q = Q_p + \Sigma Q_m$

$= 472 + (18.5 + 54.3 + 499) = 1044$ kN

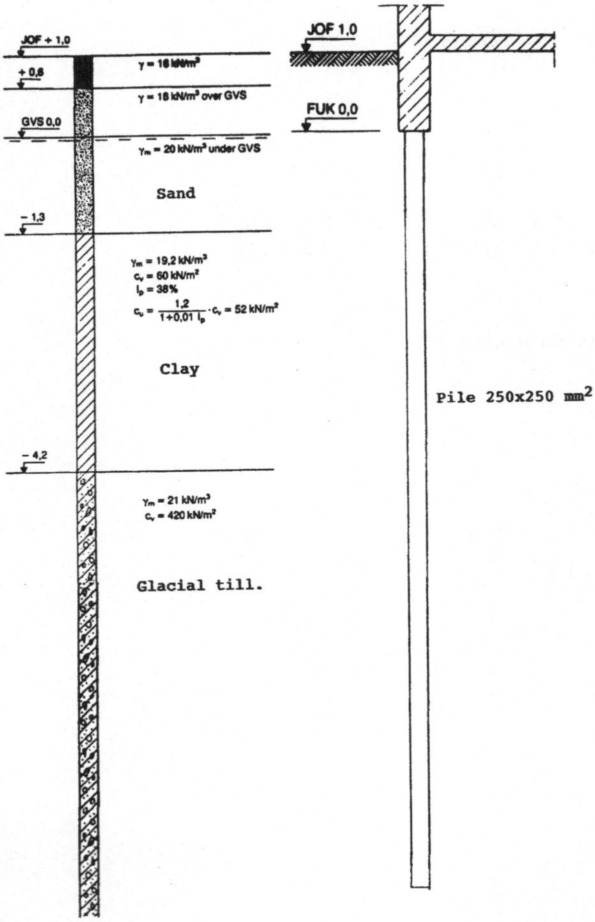

$$V_d \cdot a \;\le\; \frac{Q}{\gamma}$$

$$525 \cdot a \le \frac{1044}{2} \rightarrow a < 0.992 \text{ m}$$

Distance between piles in metre: a

Deformation analysis

$$(V_d + Q_{neg}) \cdot a \;\le\; \frac{Q_1}{\gamma}(450 + 0) \quad \cdot a$$

$$\le \frac{1074}{\sqrt{2}} \rightarrow a < 1.69 \text{ m}$$

Piles should be 250x250 cm² concrete piles. Driven length 8 meter. Toe level: -7.5

Distance between piles a = 1.00 metre

6.2 Example 2, Piles with the pile toe in sand.

Normal founding class.

Industrial hall:Floor direct on the ground. 4 piles 250x250 mm^2 supporting each column. Load on the floor 10 kN/m^2, 50% of this influence settlement. Original surface loaded by sand fill.

Actions on piles from column:

Failure analysis:
$\frac{1}{4}(1 \cdot 350 + 1.3 \cdot 850) = 339$ kN

Deormation analysis:
$\frac{1}{4}(1 \cdot 350 + 1 \cdot 850) = 300$ kN

Failure analysis:
$Q > V_d \cdot \gamma = 339 \cdot 2 = 678$ kN

Deformation analysis:

Negative load on the pile, minimum of:

1:Calculated resistance in soil strata above foundation soil strata
2:Settlement producing action Settlement producing action

1:Skin friction, static calculation
Sand layer:

$Q_m = N_m \cdot q'_m \cdot A_m$
Above water table
$= 0.6\,(5+0.2 \cdot 24+1.5 \cdot 18.5+0.9 \cdot 18) \cdot 1.8$
$= 0.6 \cdot 53 \cdot 8 \cdot 1.8$
$= 58.1$ kN

Below water table
$= 0.6\,(53.8+0.9 \cdot 18+0.6(20.10)) \cdot 1.2$
$= 54.1$ kN

Clay:
$Q_m = m \cdot r \cdot c_u \cdot A_m$
$= 1.0 \cdot 1.0 \cdot 70.3 \cdot 4 \cdot 0.25 \cdot 2.5$ = 176 kN

Negative load from skin friction 289 kN

2: Settlement producing actions after the piles have been installed:

Changes in load during building.

Relief:	$0.4 \cdot 16$	=	-6.4 kN/m^2
Sand fill:	$1.5 \cdot 18.5$	=	27.8 kN/m^2
Floor:	$0.2 \cdot 24$	=	4.8 kN/m^2
Floor load		=	5.0 kN/m^2
	Δq	=	31.2 kN/m^2

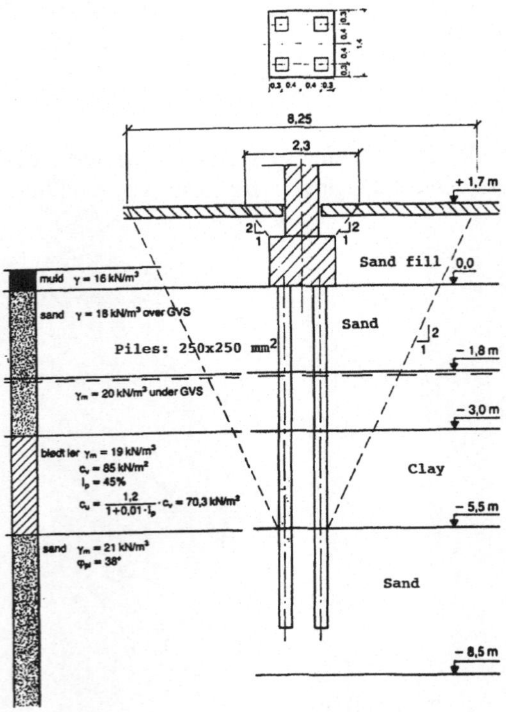

Load each pile $Q = \frac{1}{4}(31.2\,(8.25^2 \cdot 2.3^2) = 490$ kN

Q_{neg} = min. of 289 kN or 490 kN → Q_{neg} = 289 kN

$$V_d + Q_{neg} < \frac{Q}{\sqrt{\gamma}}$$

$$300+289 < \frac{Q}{\sqrt{\gamma}} \rightarrow Q > 833 \text{ kN}$$

Settlement analysis determine the capacity.

For piles with the toe in sand, the Danish Formula is used on the site to prove resistance.

$$Q_d = \frac{\eta \cdot h \cdot G}{s \cdot 0.5\, s_0} \quad \text{where } s_0 \cdot \sqrt{\frac{2 \cdot \eta \cdot h \cdot Glp}{AE}}$$

With a hydraulic hammer with the weight G = 50 kN, efficiency $\eta = 0.9$ and dropheight h = 0.60 meter:

$$s_0 = \sqrt{\frac{2 \cdot 0.9 \cdot 0.6 \cdot 50 \cdot 8}{0.25^2 \cdot 20 \cdot 10^C}} = 0.0186 \text{ m}$$

$$s > \frac{0.9 \cdot 0.6 \cdot 50}{833} - \frac{1}{2} \cdot 0.0186 = 0.0138 \text{ m}$$

or 15 blows/20 cm which will be the driving criteria given to the site.

7 PILES IN THE EURO CODE

The ideas of limit state design as it is also included in the new EC7, will not cause difficulties for Danish engineers. Official or unofficial limit state design has been used in Denmark in geotechnic for more than 40 years and extensive practical experiences have been gained which have shown that the system works satisfactory. For many Danish engineers it is a wish that future codes also will allow space for engineering judgement and not reduce the possibility to develop new methods in piling practice.

8 REFERENCES

1: Dansk Ingeniørforenings norm for fundering DS415, 3. Udgave 1984.
2: DGI-Bulletin No. 36, translation into English of Code of Practise for foundation engineers.
3: DGI-Bulletin No. 35, Copenhagen 1983, Knud Mortensen. Is Limit State Design a Judgement Killer?
4: DGI-Bulletin No. 1, J. Brink Hansen: Limit State design and safety factors in soil mechanics.
5: Kompendium i Pælefundering, 2. Udgave 1982, Hans Denver, Per Lagoni
6: Træk af Danmarks geologi, Gunnar Larsen, Danish Geotechnical Society
7: Teknisk Ståbi, Bent Hall, 15. Udg. 1986, Teknisk Forlag.

Design of Axially Loaded Piles – European Practice, De Cock & Legrand (eds)
© 1997 Balkema, Rotterdam, ISBN 90 5410 873 8

The bearing capacity of a single pile – Experience in Estonia

M. Mets
GIB Ltd, Tallinn

ABSTRACT: The present work gives a survey of experience how to fix the bearing capacity of a single pile in Estonia.

Estonia is located on the Russian platform attached side by side to the Fenno-Scandian shield where Cambrian clays, Ordovician and Silurian limestone and Devonian sand - and limestone and clay are opening. The rocks of the Paleozoic era are covered with Quaternary sediments: moraine, fluvioglacial, limnioglacial, marine, alluvial and marsh sediments. The thickness of Quaternary sediments is big in pre-glacial primeval valleys. Marine and limnoglacial sediments are represented by weak clayey soils and sands and here and there by sands containing organic matter, alluvial and marsh sediments by peat, mud and weak clayey soils. There is much moraine, a relatively weak one.

In Estonia the pile foundations are widely used in the areas of weak clayey soils where they are driven through weak clayey soils and sands containing organic matter into dense sand moraine. In Southern Estonia, on the area of distribution of a relatively weak moraine friction piles drilled into the weak moraine or supporting piles driven through weak moraine into strong moraine are used. In the areas of distribution of weak clayey soils, in the places where they are covered with a sufficiently thick medium dense sand layer short piles have been used, driven into the sand.

The picture about the piles' use area in Estonia can be received in the following classification :

1. Cohesion piles - CP. Piles that are all over in weak clayey soil and the basic part of their bearing capacity proceeds from the pile's side friction.
2. Friction piles - FP. Driven entirely into the sand or moraine.
3. Hold piles - HP. Piles that have been driven through weak soil into dense sand or moraine.
4. Point bearing piles - PP. Driven through Quaternary sediments into the over-consolidated clay or lime and sandstone.

In the geotechnical investigations drilling together with testing is used, whereby the best result by testing is given by vibro-drilling and rotation drilling. Many undisturbed samples are taken, but to a great extent only the definitions of W_n; W_p and W are used. In Estonia very good correlation have been worked out by W_n and W_L for the evaluation of mechanical properties (*Cu*; *Mo*; *OCR* etc.)

Penetration by WPT; CPT and DPT is widely used. By using CPT and using correlation the bearing capacity of the pile is fixed, while better results will be achieved by penetrometre, with the help of which the summary friction of rods will be measured. The bearing capacity of edge piles can also be exactly measured. The results are considerably worse while using the friction muff.

By CPT mechanical properties of soils (φ; c; E; C_u; etc.) can be fixed. Especially in sands and moraines, where undisturbed sampling is practically impossible. For fixing the mechanical properties correlation are used, based on the correlation between plate load tests and CPT. Taking into account a great number of such parallel tests (over 300) with different test plates, these correlation should be seriously considered.

DPT gives good information about piles' counter possibilities and its results correlation with hammer's power enables to give a rather exact evaluation of rational countersinking depth of piles.

WPT is used as a specify of the geological section and if there are no CPT data, then also as the estimator of sand soil properties.

At fixing the strength of weak clayey soils vane tests are widely used.

With the help of vane tests the creep threshold C_{uy} of the maximum shear strength C_{uf} and the residue shear strength C_{ur} are fixed.

For fixing the piles' bearing capacity many dynamic piles tests have been used in Estonia and also the pile load tests. The number of loaded piles is nearly 2000. There are good correlation between the dynamic testing and the pile load test and the dynamic method, due to the before-mentioned, has widely been used in Estonia.

For evaluating the bearing capacity of the pile model piles from steel pipes with a diameter 89; 127; 146; 209 mm are used. Usually the model piles are dynamically and statically tested. Dynamic testing gives a picture about pile bearing capacity and the depth of countersinking. The comparison of pile load tests of model piles and of industrial piles has shown that for the transmission from one to other there are no unique different in a different geological section. Good results can be achieved, when the special side friction of the model pile and special resistance of the toe are fixed. These numbers are in good correlation with similar numbers of industrial piles in the same geological section. At the use of model piles with tenso-system is simpler and the test is cheaper. It is quite probable in the future they would be widely used. The use of model piles has enabled to estimate the differences in side friction while pulling out and pressing in.

During the Soviet period only concrete piles were used. Mainly the concrete piles with a cross-section of 30×30 and 40×40 for driving were used. For linking the piles a jointed bolt extension was used that has justified itself in the building of more than 2000 houses. Using a rigid joint extension gave bad results and while using the rigid extensions, many piles were broken. The relevant investigations proved the advantages of the jointed extension. The lack of other technologies made the geotechnical work out the methods near the dwellings and using the relevant means, the results were not good, but acceptable.

At present Fundex and Vibrex technologies are widely used and on the basis of these technologies the piles up to 30 metres have been made justifying themselves.

In the Estonian market the technologies of Fundex and Vibrex have turned out to be more competitive than pile drivers and we believe in their future.

The use of 2..3 m wedge piles should be pointed out. Here we have a wedge-formed pile, which thickness is 30 cm, the point is 8 cm and its head's width is 60…80 cm.

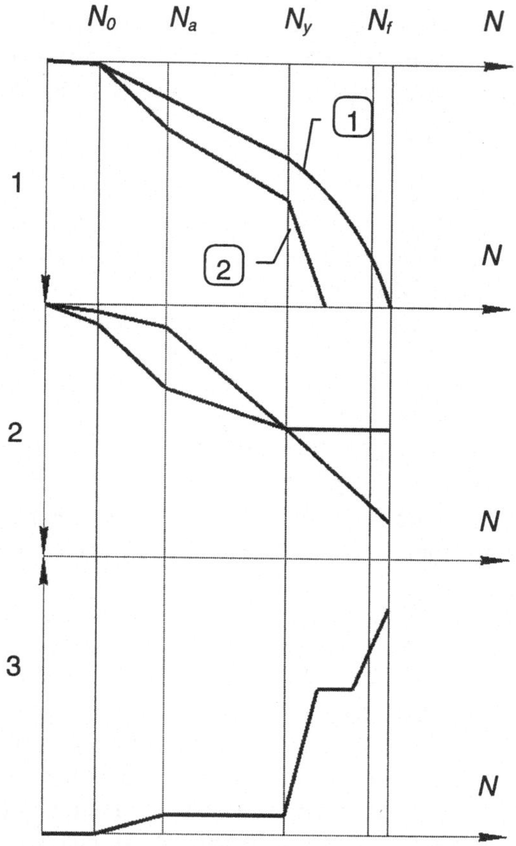

Figure 1 - Characteristic Points on curve $S = f(N)$
1. $S = f(N)$ dependence between load and settlement.
2. Division load between pile shaft and toe.
3. $b = f(N)$.
1 - $S = f(N)$ load test date.
2 - $S = f(N)$ calculated final settlement.

Such piles have been used under 200 buildings and they are about 2 times cheaper than usual foundations. 80 wedge piles are loaded and there is a good correlation between the wedge pile's bearing capacity and CPT.

For a long time the Soviet pile standards (SNiP) have been used in Estonia, but using them our own experience was also taken into account and it enabled to avoid many mistakes. We had got the project of our own standards, but in Soviet conditions we could not confirm it. Grounding on the last-mentioned, on Eurostandard and SNiP, our own Estonian standards are worked out now.

Designing the pile foundations the pile load test has been the basis for a long time, also the pile dynamic test and CPT. In the most cases the permitted load for the pile is above the proportionality (or creep) limit. It guarantees the work of the pile base in the linear zone. About the pile tests used for determining it will be the following story.

Basing on pile tests, CPT and here and there on soil properties, thousands of pile foundations have been designed in Estonia and it must be said that rather successfully.

From among the buildings built on piles about 70 buildings are under settlement observation and that is a good feedback with the made. Ones the analysis of buildings has shown that if the load on the pile $N<N_o$ - from the side friction - the settlements of buildings are 1…3 cm and will be finished in 1…3 years. In case $N_a< N < N_y$, the settlements are 3..6 cm and they last 6…12 years.

Pile load test. The dependence $S = f(N)$ is received by the use of pile load test.

The dependence $S = f(N)$ or the piles's behaviour is in concordance with the growth of loading. A great number of experimentations with tensiopiles and pile-sounds (60 piles in all) enabled to evaluate the pile's behaviour in connection with the growth of the load and to stress the characteristic points which are important for evaluating the bearing capacity of piles and the settlement of pile foundations (see fig.1) These points are especially important because of their characterization of the changes of the division of strength between the pile's toe and pile's side [1]. To evaluate these points on common pile tests one may use the methodics of EPT [2] worked out by the author. According to the methodics the dependence between the settlement of the pile and time is estimated by the formula $S = a(t/t_0)^b$ (S- the settlement of the pile at the moment interesting for us, a initial settlement at the moment t_0 and b- the factor characterizing the speed of the settlement.

By means of the mentioned dependence the settlement of the pile could be fixed at the moment we are interested in, or evaluate (fix) the moment when the pile's settlement is practically finished.

In the figure 1 there are curves, the 1st one characterizes the settlement of the pile in the course of the test, the 2nd one - during the long period interesting for us. In the second part of the figure there is a typical dependence when the load is divided between the toe and the side of the pile, in the third part - b changed together with the growth of the load - that- as one can see in the figure- enables more exactly to fix the below observated characteristic points.

N_0 - characterizes the load which practically lacks the settlements in time- $b = 0$ and they are connected with elastic deformations of the pile and the deformations of the extensions of the pile. As the relative exactness of measurements in this sector is smaller, in this part the division of strength between the toe and the side of the pile is not clear.

N_a - this load comes rather clearly out almost in all the tests and here the change in the division of strength between the saft and the toe of the pile takes place. The majority of tests (95%) gave the result that N_a was equal to the maximum bearing capacity of the pile's saft at the pile's sond test.

N_a -ni - the pile's side receives the majority of the load and the smaller part (15-30%) from the added load hits the toe of the pile. But surpassing N_a the picture changes and the majority of the load is starting to hit the pile's toe up to the load N_y. The factor b between the loads N_a and N_0 is growing linearily. The settlement of the pile's head N_a depends on many factors (the length of the pile, the geological cutting, the type of the pile etc.), but usually (in case of piles with diameter 10...40 cm) the settlement is 1.5...3.5 mm and 80% of the piles have the settlement of 2... 2.7 mm.

Our investigations and the settlement observations of the buildings built on piles have shown the great importance of the characteristic point N_y - the proportionality limit. The dependence between S and N is practically rectilinear up to N_y and especially in the calculated curve of the settlements 2. The factor b between the loads N_a and N_y is

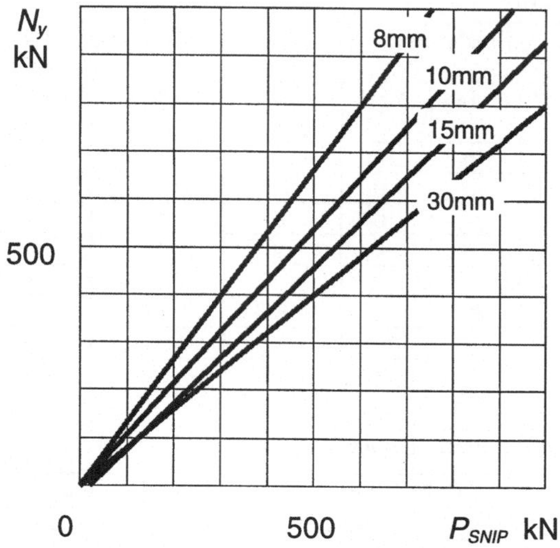

Figure 2

constant. After N_y the pile's toe and the nonlinear dependence between the settlement of the pile and the load added to it will begin.

In the figure 2 the dependence between the load fixed in the course of pile test at settlements S 8; 10; 15; 30 mm and the pile's proportionality limits, whereby the load relevant to the given settlement has been divided 1.2.

At the observed tests (nearly 300) (concrete piles with diameter 30×30 cm) good correlative dependence between the load fixed by the settlement (N_8; N_{10} … N_{30}) and the proportionality limit (in the figure) were received. It could be seen in the figure that not depending on the length of the pile and the geological section, around the pile at settlements 8 and 10- mm the fixed load is smaller than the proportionality limit N_y, at settlements 15 and 30 mm bigger N_y.

The determination of pile bearing capacity by calculations is complicated due to the fact that practically we have to solve two tasks: to fix the bearing capacity of a pile's side and its one too. While determining the general bearing capacity of a pile, it is necessary to additionally take into account the total effect of both factors which in different conditions will be completely different. Considering the above, following both of them will be treated separately and their total effect will be also dealt with.

The formation of side friction was investigated by pile tests made by special pile sounds and tenso-concrete piles. The objective of the investigations was to find out the influence of engineering geological conditions and loading manner and pile parameters on the formation of side friction of a pile.

The piles loaded in different engineering-geological conditions showed that side friction of a pile does not practically depend on a pile's length, material and diameter (table 1).

The lack of influence of a pile's diameter on a special side friction could be explained by the fact that the properties of the weak soil surrounding the pile do not practically change by driving, side friction of a pile depends only on the soil strength.

Table 1 - Special side friction of piles kPa in weak soils

	Metal piles					Concrete piles	
Pile's length in m and soil	**∅ 89**	**∅ 127**	**∅ 146**	**∅ 219**	**Number of tests**	**30×30 cm**	**Number of tests**
4 m (Varved clay of Pärnu)	12	10	10	10	20	10	2
6m (Varved clay of Pärnu)	10	10	10	10	18	10	11
12 m (Varved clay of Pärnu)				10	2	8	4
4,5 m (Soft silts of Tallinn)	9	8	8.5	9	14	8	2
12m (Soft silts of Tallinn)						8	8
5 m (Soft moraine of Valga)	22	21	20	22	8	23	2
12 m (Soft moraine of Valga)				20	1	20	3

Table 2 - Dependence of pile's side friction on loading speed

Duration of loadingstage	180	1200	7200	Standard subsidence (growing of settlement below 0.2 mm in 2h)
Side friction τ_k , Pa	9.5	10.3	11.2	11.1

The lack of influence of a pile's length could be explained by non-changing of depth of creep threshold fixed by undrain shear strength. The lack of influence of a pile's material can be explained by the formation of a surrounding "shirt" in the course of driving-fixing the size of side friction of a pile. The side friction of a pile depends upon the loading speed of the pile (see table 2).

On a rapid loading the redistribution between the pile's side and its tip could be not be completed, the load under the pile's tip could quicker reach its maximum value and the base of the pile will weaken.

The difference in special side friction by rapid and slow loading amounts to soil's rigid cohesion in its value.

By rapid settlement side friction diminishes also on a limit load and usually leads to the breaking of the pile's basement. In this case the diminishing of side friction is also equal to rigid cohesion.

In weak soil the special side friction does not depend on a resting-time pause of the pile as the made tests have shown (table 3). The lack of increase in a pile's side friction in weak soils during their stopping-time could be explained by the fact that soil properties on the pile's side do not practically change in the course of driving due to the lack of rigid connections in weak soils and to the relatively small sensitiveness of the investigated weak soils.

In clayey soils of plastic and hard consistency the pile's resting-time has a strong influence on the pile's special side friction due to which it increases many times over. The special side friction of piles driven into moraine before the limestone shore of Tallinn was after the driving 0.02 MPa and during a month it increased up to 0.1 MPa. The special side friction of the piles driven into Cambrian blue clays immediately after the driving was 0.01 MPa, after a week's resting-time it was 0.05 MPa and after a month's resting-time it was 0.1 MPa. The growth of the pile's special side friction in the time in plastic soils and in soils of hard consistency can be explained by a relatively large rigid cohesion of the last-mentioned and by the fact that while driving into these soils the soil structure is completely destroyed. In the course of time the broken

Table 3 - Dependence of pile's special side friction

Name of the soil	Resting time				
	4 hours	1 week	2 weeks	3 weeks	7 months
Soft silts of Tallinn	8.3	8.0	8.5	8.9	-
Valga moraine	2.2	2.5	-	2.5	-
Pärnu varved clay	10.6	10.7	10.6	-	10.7

connections will be restored, a new structure will break out and pile's side friction will be considerably increased.

The comparison of special side friction measured in the course of pile tests to the creep threshold fixed at undrain shear test showed that in case of weak soils they practically coincide (figure 4). In varved clays, in marine silts and in sands containing organic matter, the coincidence of τ_k and τ_r is very good and usually the difference does not exceed 10 %. The deviation is greater in case of marine clays, because their properties are very much changeable in the geological section. Much worse is the pile's special side friction characterised by threshold in plastic moraine, due to a certain violence of structure in preparing samples and perhaps due to the soil improvement on the pile's side in the course of driving.

The soils which shear strength on the creep threshold is characterised by the angle of internal friction and cohesion, the first one influences the pile's special side friction up to a certain critical depth that could be considerably well fixed by static penetration, because from this depth on the special cone resistance (that so far has increased) in this soil is constant and does not increase together with the increase of depth.

More complicated is the determination of piles' special side friction in dens sands, hard clays and hard moraines. On investigating the soils mentioned it came out that the pile's special side friction is here and there greater than the soil's maximum shear strength. Usually the soil shear strength is fixed in the samples of disturbed structure and which density is equal with natural density of the soil, but which structure connections do not respond to the soil of undisturbed structure.

The testing of sand soils of undisturbed structure shows that dens sands at the creep threshold are characterised by a great cohesion that, together with the hooking depending upon vertical pressure, bring about a great side friction - 0.06...0.15 MPa. For hard-plastic clays it is difficult to reach the actual creep threshold values due to unevenness of overconsolidated cracked clays.

The analysis of pile tests shows that the pile's special side friction is considerably less dependent upon soil's consistency as compared to soil's clay containing and the geological history of the surface. The consistency of Pärnu varved clays and of marine clays spread in Tallinn is almost the same, but due to the older age, the creep threshold of Pärnu varved clays is greater and therefore their special side friction is also greater (fig. 3, groups I and II). The same can also be said about marine and glacio-lakey silts spread in Tallinn (fig. 3, groups III and IV), whereby here in the last ones the piles' special side friction is nearly three times greater than in the first ones. In case of clay soils of the same age the special side friction depends on clay containing. The marine sediments of Tallinn (fig. 3, groups II and III) having the same consistency, have a greater pile's side friction in silts than in clays. Piles' special side friction depends upon engineering-geological section of the pile's base. During the tests in Tallinn, in Kadaka

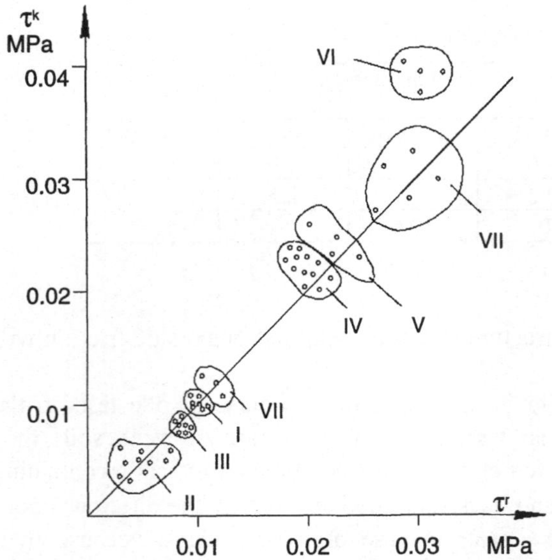

Figure 3 - Dependence of special side friction τ_k and creep threshold τ_r I-varved clay, II-marine clay, III-marine silt, IV-glacio-lakey silt, V-yielding moraine, VI-plastic moraine, VII-sands containing organic matter, VIII-sands of medium strength

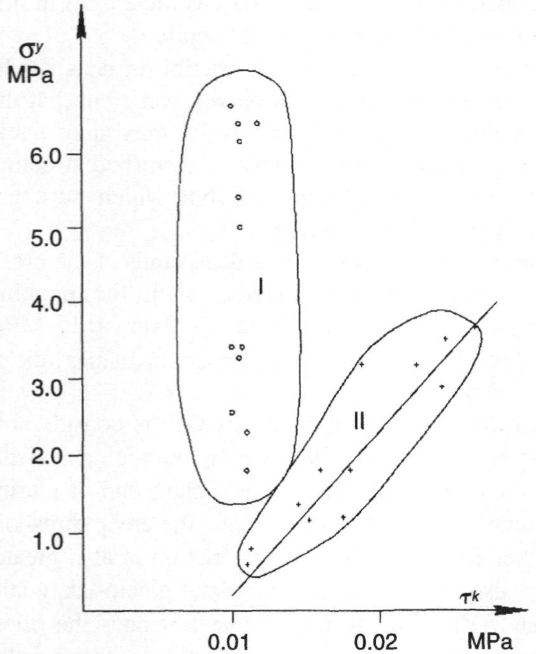

Figure 4 - τ_k=f(σ_y).

tee, the piles driven into the depth of 4.8 m of marine silt were loaded. Their special side friction was 9 kPa, not depending on the depth, diameter or resting-time. After the loading the piles were countersunk into the glacio-lakial silt located 20...50 cm deeper, where by the bearing capacity of the pile's tip increased nearly 2...3 times. All that was accompanied by the increase of pile's special side friction up to about 16 kPa. That refers to different working conditions of the pile, depending upon the fact whether the pile's tip has been driven into weak or hard soils. In the last case the origin of creep processes at the pile's side is made more difficult due to the smaller settlement of the pile conditioned by the greater bearing capacity of the tip and the value of pile's special side friction becomes greater.

In figure 4 the dependence between the pile's special side friction τ_k and the tension characterising the special pressure σ_y of the pile's tip at the proportionality limit N_y in Pärnu have been described. The 1st group involves the compound piles with polt joint, driven through varved clay moraine or through the sands of a glacier river. The special side friction of these piles does not depend on the bearing capacity of the tip and is equal to creep threshold of varved clays. The second group is formed of the piles, consisting of one link and driven through varved clays. In their case the dependence of side friction upon the tip's bearing capacity is evident. The special side friction of only these piles which tips are in varved clays responds to the creep threshold.

The special side friction of other piles is growing linearly to the growth of the bearing capacity of the pile's top. The piles with non-rigid extension have creep possibilities in the pile's extension and due to that the special side friction of a pile cannot grow together with the growth of the bearing capacity.

If the weak soils in the pile's base are located under the soils of greater bearing capacity, the pile's special side friction depends on the creep threshold of the deeper weaker soil. In the centre of Tallinn there is a complex of yielding clayey soils at the 10...12 m depth under the layer of marine sands which thickness amounts to 20...25 m. The special side friction in marine sands of the piles with the length of 8 m was 16 kPa. After driving the piles into the depth of 16 m (in sands 12 m and 4 m in weak clayey soils), the special side friction of the pile decreased to 10 kPa, characterising the creep threshold of these clayey soils. In Pärnu the special side friction of piles in marine sands was 20...30 kPa; after driving the tips of piles into the yielding varved clays by 300...350 mm, the special side friction of piles diminishes up to 10 kPa, responding to the creep threshold of varved clays. In weak soils the creep occurs at smaller shear tensions, bringing about the origin of creep deformations in the sands located above and the diminishing of a pile's side friction.

The data about side friction on pushing the piles inside and pulling them out is given in the figure 5. The graph shows that the side friction of τ_{kss} pushing in is usually bigger than that of pulling out τ_{kvt}, whereby the interval is as big as the side friction. The only exceptions are these piles (marked with crosses), driven through sands into weak soils. In their case the side friction of polluting out τ_{krt} is bigger than the driving-inside friction.

The dependence between side friction on the grounds of correlation calculations is $\tau_{nvt}=0.8\tau_{kss}^{0.5}$

As the maximum side friction of a pile does not mean the exceeding of maximum shear strength of a soil, but the formation of creep processes on the side surface of the pile (depending upon the shear strength of the pile's base and on the bearing capacity of the pile's tip), it is more correct to consider the pile's side friction the maximum

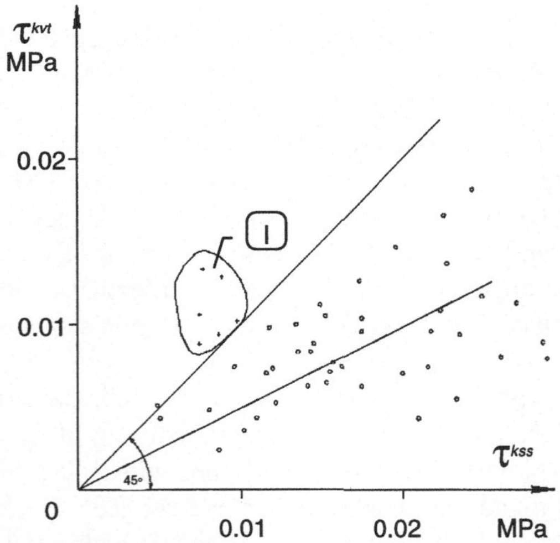

Figure 5 - τ_{kvt}=f(τ_{kss})

strength that the pile is able to bear to the surface through its side. After the origin of creep processes on the pile's side, the whole load will be carried to the tip of the pile.

Pile toe. The pile tests made in different engineering-geological conditions showed that the bearing capacity of the pile's tip at the loads N_y and N_u practically does not depend upon the loading speed. The strength of the investigated soils does not diminish and they do not become very much denser under the pile's tip. Depending upon the filtration modulus of the soil, the settlement characterising the filtration consolidation will occur either in the course of driving soft silts and moraine or there will be no densening together with the pile's driving or the pile will be driven on the account of pressing the soil; side or upwards (soft clay and loam). At the pile's following loading test the pile's settlement will be caused by skeleton's creep on silts in the dense prism formed under the pile's tip. On weak clays the process is influenced by the void pressure after the ramming and the time of loading is usually too short to bring about a remarkable consolidation densening. Pressing out at greater loads does not involve the soil densening under the tip.

The made tests showed that the bearing capacity of the pile's tip is influenced by the depth of pile driving at which the influence of the depth depends upon the angle of internal friction of soil under the pile's tip and for determining it G.Meyerhoff's calculation scheme is the best.

In weak soils, as the pile tests carried out in Pärnu figure 6 showed, the depth does not have any impact on the bearing capacity of the pile's tip immediately after the driving and due to the maximum values of poor pressure, under the pile's tip - $N_y = N_u$. In the course of the rest following, N_y reaches its maximum value in 2...4 hours, at which the depth of dependence will be fixed by γ_h. The growth of N_u during the rest will be considerably longer and will end with complete vanishing of poor pressure in the pile's base cropped up during the driving), which, depending on the value of filtration modulus, will be reached after a rather long resting-time. After the vanishing of poor pressure, the depth's influence on the ultimate load in varved clays depends on the value of the angle of internal friction of the soil on the unconsolidated-drained shear. The

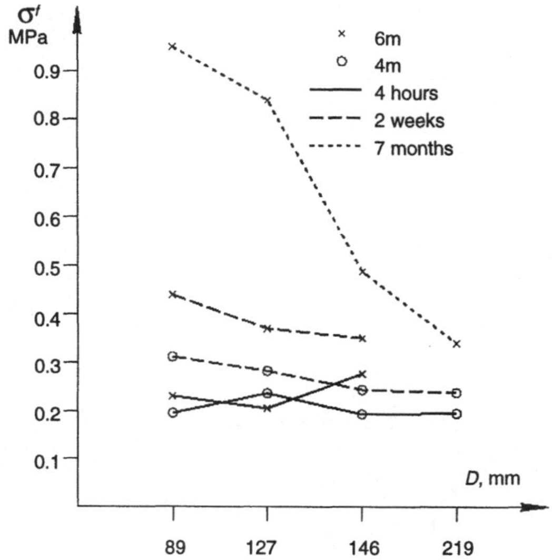

Figure 6 - σ_f = f(pile D and resting time)

influence of the pile's diameter on the bearing capacity of the pile's tip depends on the soil's filtration properties. At the tests made in marine and glacio-lakial silts the bearing capacity of the pile's tip did not depend on the pile's diameter and the resting-time and N_u and N_y had reached their maximum value in 4 hours already. At the tests made at Valga polygon in the yielding moraine, the poor pressure vanished in 3…7 days and after that N_y and N_u of the piles of different diameter were equal. Further on N_u reached the maximum. During the resting-time N_y did not change. The influence of the resting-time and diameter on the bearing capacity of the pile's tip in varved clays is characterised in figure 6. Immediately after the driving and after the 4-hour stopping time the special load of the pile's tip at $N_3(\delta_y)$ is equal to the pile's tip's special load at $N_5(\delta_f)$. Afterwards when the resting time is increasing, δ_y will remain practically constant and the growth of δ_f will begin. It is rather slow and depends on the pile's diameter, proceeding more rapidly in case of a smaller diameter. At these tests only a pile with a diameter of 89 mm reached the maximum δ_f value after the stopping time of 7 months. But the other piles were relatively far from this result, whereby the poor pressure of the pile with a diameter of 219 mm, had vanished only in 40 %.

In comparing the made pile tests with calculation formula e, it came out that both of them δ_y and δ_f will be fixed by G. Meyerhoff's formula. In calculating δ_y the shear parameters should be the bases, characterising shear proportionality limit and for fixing δ_f - the shear parameters characterising the maximum shear strength (see table 4).

Driving piles into silts and moraine, under the pile's tip a densened area of rather wide spread will arise. In figure 7 one can see a densened area under the pile's tip, formed while driving of 5 m long piles into the soft glacio-lakial silt, that in the natural condition is flowing and has moisture of 25 %. After the driving the soil under the tip of the pile was in a super compact condition (W_n=10%). The thickened zone amount to the 1 m depth and between moisture and depth there is practically a linear dependence, such thickened zone should considerably increase the bearing capacity of a pile, but as the calculations show, the bearing capacity of a pile is not determined by the strength of a

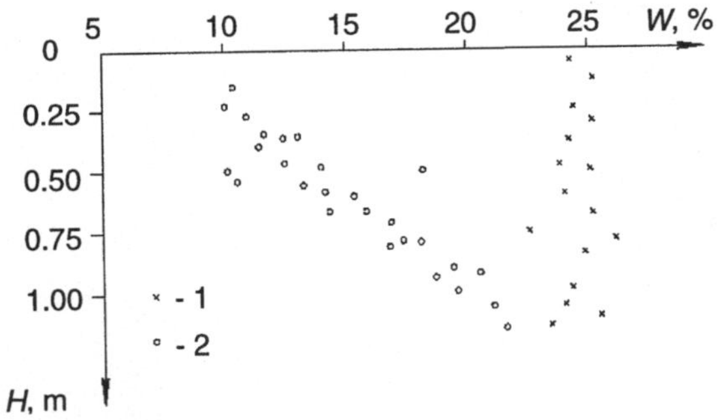

Figure 7 - Change W_n under pile tip

Table 4 - Calculation of different soils σ_y and σ_f by G. Meyerhoff's formula

Name of the soil	σ_y, MPa		σ_f, MPa		Number of tests
	test	calculation	test	calculation	
Varved clay	0.22	0.20	1.00	0.96	36
Marine clay	0.12	0.10	0.75	1.20	3
Marine loam	0.40	0.45	1.25	2.15	4
Marine silt	0.70	0.65	1.50	8.00	8
Glacio-lakey silt	1.80	1.70	3.80	8.00	7
Yielding moraine	2.00	1.90	6.00	9.10	5
Plastic moraine	6.00	5.50	12.00	14.00	3
Hard moraine	8.00	7.50	16.00	17.00	1
Cambrian clay	12.00	13.50	22.00	24.00	6
Sand containing org. matter	1.40	1.20	3.00	7.25	5
Medium thick sand	5.00	5.50	9.50	10.00	3
Thick sand	9.00	8.50	15.00	16.00	2

thickened material, but by the strength of a natural material. Therefore the thickened cushion is a connecting link transmitting the load and not much increasing the bearing capacity of a pile's tip. But it diminishes the deformation of the soil under the tip of the pile, because the active area of the pile is in the pile itself and further up to the proportionality limit - only dry unit is represented here.

As it can be seen in the table the calculation coincides with the test on the creep threshold much better than on driving. It is so in case of silts, moraines and in case of sands containing organic matter. In fixing the maximum shear strength of the named soils by drained-unconsolidated shear, τ_{max} is usually overestimated. In the course of testing silt and sand become thicker and by the maximum shear strength the soil's internal friction does not characterise the natural soil, but the soil mass formed in the course of the test. It does not exist at the creep threshold and therefore the shear parameters fixed at the creep threshold will suit better.

The comparison of marine and glacio-lakey silts shows that the first ones according to the pile test data, are nearly two times weaker that is also in accordance with the data of settlement observations and penetration tests. Testing them at the maximum shear strength they have similar parameters and using them in calculation schemes we overestimate the bearing capacity of both of them. The same picture characterises also

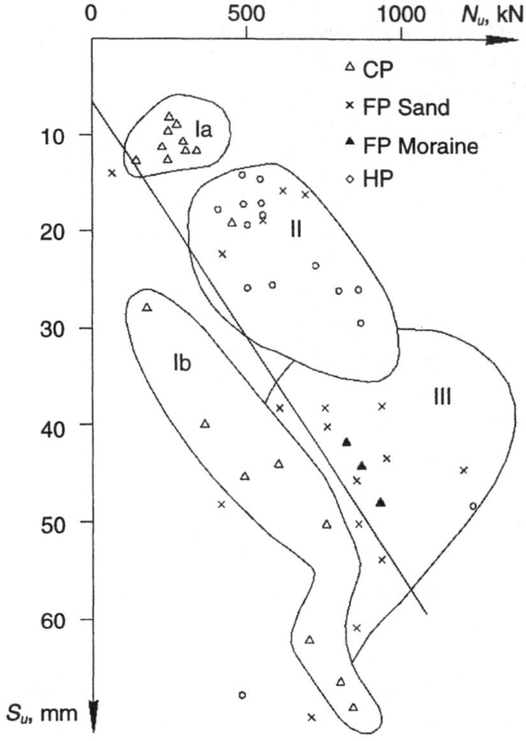

Figure 8 - S_u = f(N_u)

the compression tests that usually underestimate marine silt and overestimate the compressibility of glacio-lakey silts.

The table shows that it is not right to take as the basis of fixing bearing capacity of the pile's tip the consistency of the soil. In marine clays, varved clays and marine silts of the same consistency, the bearing capacity of the pile's tip is totally different.

The same picture characterises glacio-lakey silts and moraine and Cambrian clays and hard moraine. The consistency indicator of the named pairs is similar, but the bearing capacity of piles is entirely different. On the basis of the aforementioned we may draw the conclusion that the geological history of clayey soils and their clay containing have much more influence on their bearing capacity than consistency does and not considering them, one may receive an incorrect result.

At the ultimate load N_u the settlement of the pile's head S_u depends figure 8 on the geological section of the pile's base, on the geotechnical properties of the soil and on the pile's length, but as it can be seen in the figure 11, there is a certain dependence between S_u and N_u (η=0.55) and generally together with the growth of N_u, the settlement at this load also increases. Here we can more clearly see the influence of the geological structure of the pile's base on the named sizes and according to them four groups can be distinguished in the figure. The group I_a is formed by the CP piles driven into the yielding clays and silts. They reach their ultimate condition at the settlement of 10...15 mm. The group I_b is also formed by CP piles, but they have been driven into the soft silts; their settlement is considerably bigger - and the dependence between N_u and S_4

could be clearly seen inside the group. The group II is formed by hold piles, which settlement at N_u is 15...35 mm and increases simultaneously with the growth of the load. In the group III there are piles of internal friction, driven into moraine and sands and they are characterised by a very big settlement - 35...64 mm at N_u.

Taking into consideration the fact that the groups Ib and III involve the soils which are characterised by a great angle of internal friction at τ_{max}, it may be concluded that for the complete realisation of the power of internal friction at such soils a considerably bigger settlement is needed and only then it is N_u accompanied by the driving of the pile point's base. Loam and clays belonging to the group I have a small internal friction and the N_u of piles' bases will arrive at considerably smaller deformations. As to the CP piles belonging to the group II the situation is somewhat complicated, because the pile's side is in the weak soil and its point in the stronger soil and perhaps due to this the settlement necessary for the base's breaking is smaller than at the piles of internal friction. The additional influencing factor may be the fact that the hold piles have been given into the strong layer only with their points, weak soils are straight above the pressing-out prisms and therefore the trajectory of pressing-out prisms is shorter.

In most of the pile standard the bearing capacity of a pile is found from the load N_u by the factor of safety and for estimating its size the connection between N_u and N_y was compared with pile tests made in different engineering-geological conditions. The data of the comparison is given in the figure 9 and it shows that there is a good correlative dependence between the named sizes N_y=0.6 N_u (η=0.75).

The first group I is formed by the piles driven into clays and loams, characterised by a relatively small interval between the named quantities (N_y=0.8N_u). In the second group there are the connection piles together driven into silts; in their case the difference between N_y and N_u is considerably greater. The different behaviour of piles in clays and silts can be explained by the differences of shear strength at the creep threshold and at the maximum shear strength that is small in clays and big in silts. In silts where the growth of shear strength at the maximum shear strength is based on internal friction for realising it (as it could be seen earlier) bigger deformations are needed - causing a greater growth of N_u. The 3 rd group is formed (in the figure) by hold and internal friction piles.

It may be concluded from the above that the bearing capacity of the pile is best characterised by a proportionality limit that sets apart two different stages of the pile's behaviour: at a smaller load the soil becomes thicker at a bigger load the soil pressing from under the pile's point will start. The factor *b* characterising the temporality of settlements makes a jump in this point. The exceeding of proportionality limit is always characterised by a considerably bigger final settlement. At the pile's proportionality limit the pile's side friction (τ_k = τ_r) has completely been realised and the pile's point behaves itself linearly in accordance with the regularities of the deforming environment. Up to this load for describing the dependence between deformations and the load it is possible to use elasticity theory. At bigger loads the use of elasticity theory is not justified and for describing the piles' settlement it is necessary to use empirical methods taking into account the temporality of piles' settlement.

The above does not mean that N_y is the maximum permitted load for a pile. The proportionality limit may be exceeded but the exceeding must be equal for all the piles under the building in construction with a view that all the settlements proceeding from it and fixed by some empirical methods, would be more or less equal. The most dangerous is the condition of a building when a part of the piles has been loaded by the load

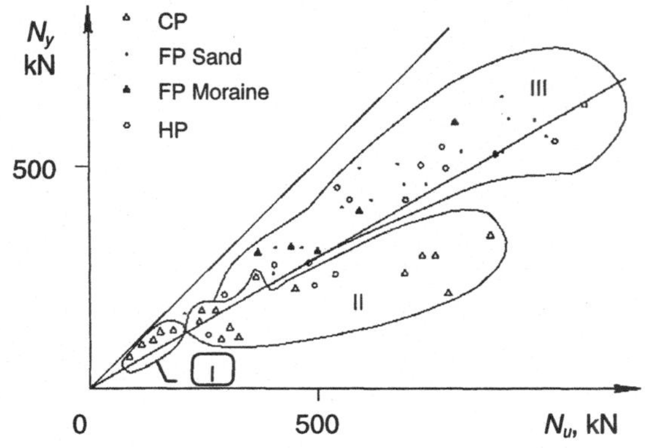

Figure 9 - $S_y = f(N_u)$

smaller than the proportionality limit and a part of the piles by the load bigger than that one which due to the increase in the intensity of settlement causes a different settlement of the building's foundations.

Dynamic Testing of Piles

The dynamic method of determining the bearing capacity of piles is one of the oldest. Until the present day more than 200 formulas have been worked out to this end. All of them can be divided into empirical and theoretical ones. The last mentioned take into account the conditions of energy balance necessary for driving. Some authors eliminate from balance equation the energy going to constant deformations of soils, others - the energy spent to elastic deformations. The third ones do not account any of them. Nowadays such formulas are widely used take all the components of balance equation into account.

N.M.Gersevanov's investigations were an essential contribution in this issue, being the first taking into consideration the pile's behaviour in the soil. N.M.Gersevanov noted at the beginning of this century already that by dynamic methods it would be possible to fix only the load of the pile's proportionality limit N_y (elasticity limit). The investigators posterior to him, among them the authors of standard documents valid in the Soviet Union, do not often consider this position. It is paradoxical that the formula of Gersevanov himself should be used up till now on the grounds of standard documents used for fixing the piles' ultimate load.

In Estonia the possibilities of fixing the dynamics of pile's proportionality limit by formulas were investigated. Special attention was paid to Gersevanov's formula as the most wide-spread in Estonia.

For analyses the test data of 70 concrete piles and 50 steel pipe piles, loaded 1...3 month after the driving were used.

Gersevanov's formula for fixing N_{D6} can be used at equivalent - set of 2...15 mm, in case of smaller equivalents the formula overestimates, but in case of bigger equivalents it underestimates the actual value (fig. 10).

On the basis of previous investigation results it can be concluded that on piles' static test loading with the loads exceeding side friction and proportionality limit, the accompanying settlements will be in the interval of 1.5...2.5 mm and 5...15 mm. Accordingly at equivalents of below 2 mm - the side friction is not exceeded and

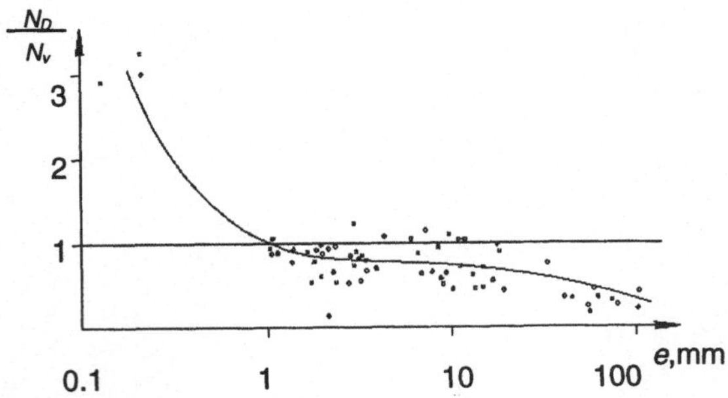

Figure 10 - Dependence N_D/N_y and set - *e*

therefore it takes more energy to drive the pile, as compared to bigger equivalents, causing the overestimation of the bearing capacity of the pile. At the equivalents 15…20 mm bigger the structure of the soil will break, the pressing out if the equivalent will increase.

According to the previous investigation results for fixing N_D (according to driving data) Gate's formula is the best (fig. 11).

If to fix the factor k represented in the formula, taking into account only the equivalent, we get correlation factor 0.9, but if to take into account the energy of blows and the relation between the weights of the pile and the hammer, the values of N_D fixed at the pile's static test loading and calculated by Gate's formula are similar in case of concrete as well as steel pipe piles. If the equivalent of concrete piles $e < 5$mm, then $k = 3$, but if $e > 5$ mm, then $k = 2$. At steel pipe piles in the interval of equivalents 5…40 mm $k = 1.5$, at equivalents 2…5 mm $k = 3.0$ and at equivalents 1…2 mm $k = 4$.

Dynamics formulas are not suitable for evaluating the bearing capacity of piles in the following engineering-geological conditions:

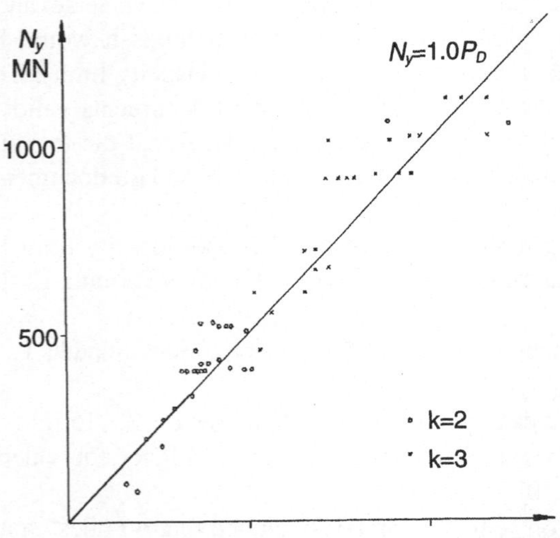

Figure 11 - The dependence of loads N_y fixed at static test loading and N_D (proportionality limit) calculated by Gate's formula

1. In the geological section there are Paleozoic clays or moraine before the limestone shore of hard consistency. Under the influence of driving shear strength of such soils will be decreased in 70...90 % and fixed dynamics N_D=250...500 MN. In 1...3 month after the driving the shear strength will be restored in a scale of 50...80 % and N_D, fixed at static test loading, will exceed 3...4 times the one fixed immediately after the driving.
2. There is weak silt in the geological section, whereby the equivalent is below 3 mm or more than 15 mm,. In case of equivalents 3 mm smaller, in the thickened zone under the pile the pressure is negative, causing a considerable increase in the bearing capacity. At equivalents more than 15 mm, the structure of silt will break and the base will become weaker (τ=0). Therefore, in this case the blowing energy must be such one that the equivalent will stay in the interval of 7...15 mm.
3. In driving into the sand, gravel containing colloid organic matter and other soils sensible to dynamic influence. The soil structure will be partially broken and shear strength will diminish. N_D calculated by the equivalents received in such conditions, is considerably smaller than the one, fixed at piles' static test loading.
4. In driving into fine sand, containing pressure water. The hydrodynamic pressure of this water causes a considerable lessening of the equivalent and thus the values of N_D, we shall get, are considerably larger than the actual ones.

Using CPT for the evaluation of the bearing capacity of the pile, has shown that the use of the equipment enabling to fix the saft friction of the whole sound, gives better results. The use of the friction muff is not justified in Estonia. Using CPT rather good correlations between the special resistance of the pile's toe (by N_y) and q_c have been gained. The correlations between the relevant special resistance of N_u and q_c are worse.

Recommended dependences are the following :

$$q_y = 0.2\, q_c \quad (r = 0{,}9),$$
$$t_c = 0.8\, q_a \quad (r = 0.85),$$

where q_y - the special resistance of the pile's toe by N_y.

t_a - the pile's special side friction,

q_a - the special side friction of the CPT,

q_c - the cone's special resistance.

To evaluate the bearing capacity of piles in Estonia mainly the pile's proportionality limit fixed by the pile load tests- is used and it has been dealt with as a permitted load. Such an approach as a whole may be dealt with as a wight and on about 30 objects [2], where the pile's bearing capacity has been fixed on the grounds of N_y's value, in the course of settlement observations the fixed settlements have remained within the limits of 2...5 cm. This approach was in the contradiction with the requirements of SNiP II-17-77 p. 6.7, according to which the piles had to be loaded with bigger loads than N_y. But it would have been accompanied with considerably greater and prolonged settlements.As one of the examples let us analyze the settlements of the two buildings given in the figure 12.

The first one is the 24-storeyed building founded on the 8m- piles. The piles were driven through the weak clay soil into the medium dense sand, under which there were loose sands 10m and under them 2 metres soft clayey silts that covered moraine. The N_y of the pile was on the grounds of tests 450...500 kN and N_f 700 kN and they were loaded

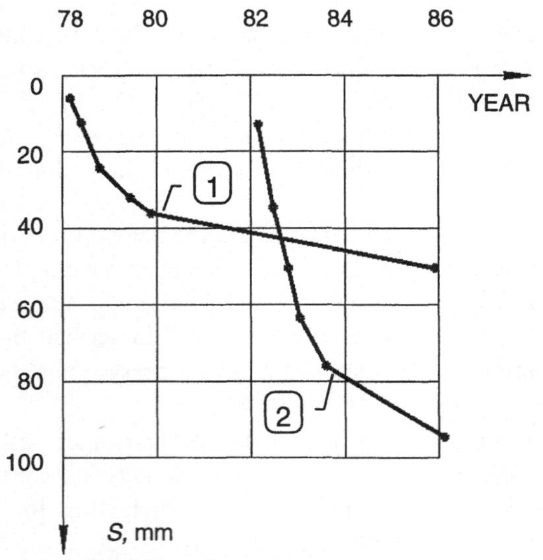

Figure 12 - Two houses settlements
1. $N < N_y$.
2. $N > N_y$.

with the load of 400 kN. The settlement of the building - was calculated by the author - was 6cm (written in the pile test report) and as it can be seen in the figure -it is close to the actual settlements.

Under the second building there are 12m-piles which were driven through the filling, sand, weak clay soil into the sand with medium density. The pile's N_y in the sand was N_y = 450 kN, N_u =700 kN and the piles were loaded with N = 60 kN. The prognosticated settlement was 14cm that could be most likely.

As it can be seen from this example, in principle the piles could possibly be loaded with loads surpassing N_y, but the settlement analyze must be concurrent with them and it must be based on the pile load tests.

Resting upon such an analyze the pile foundation of the Tallinn Town Hall was projected. A part of the piles was loaded with the load surpassing N_y and the prognosticated settlement was 15 cm and the other part with the load that was under N_y and their prognosticated settlement was 5 cm. The building was ready in 1980 and the settlement observations have been the proofs of the presented prognosis (the difference - 25%) [4].

REFERENCES

Mets, M. 1977. Vaia kandevõime. *Ehitus-geoloogilia kogumik IV.* Valgus. Tallinn, 161-175.

Mets, M. 1968. Vaivundamendi vajumisest. *Ehitus ja arhitektuur nr. 1.* Tallinn, 19-21.

Killar, E., Mets, M., Vares, J. 1977. Vaia kande-võime määramine rammimisandmeil. *Ehitus-geoloogia kogumik IV.* Valgus. Tallinn, 176-182.

Mets, M. 1991. Iseloomulike punktide meetod. *Ehitusgeoloogia kogumik V.* Valgus. Tallinn, 42- 62.

Design of Axially Loaded Piles – European Practice, De Cock & Legrand (eds)
© 1997 Balkema, Rotterdam, ISBN 90 5410 873 8

Design of axially loaded piles – Finnish practice

J. Heinonen & J. Hartikainen
Tampere University of Technology Geotechnical Laboratory
A. Kiiskilä
VR-Rata Ltd, Helsinki

1 REGIONAL GEOLOGY

The soil deposits in Finland were mainly formed during the last glaciation or thereafter as a result of various geological processes. The material either derives from the bedrock (mineral soils such as till, gravel, sand and clay) and plant remains (organic deposits such as peat), or is made up of precipitates of compounds dissolved in water (lime gyttja). Moraine and till deposits cover over 50% of the total surficial deposits in Finland. 15% are formed of peat deposits, 13% of rocky terrain and 10% of clay and fluvial deposits. The distribution of Quaternary deposits is uneven, depending on land forms, the elevation of an area above sea level, regional and local characteristies of the activity of the continental ice sheet and its meltwaters, the rate of uplift and climatic factors.

Basal till is the predominant sediment type in Finland, and it often acts as a substratum for other sediments. Basal till covers the bedrock, being the thickest in depressions. The average thickness of the sedimentary deposits in southern and western Finland is on the order of magnitude of about 10 meters. The thickness of sedimentary silty deposits, met with in interior of the country, varies generally from five to ten meters. Even in valleys, the thickness of sedimentary deposits can vary considerably: a ridge of bedrock or moraine can rise to the surface in the middle of a valley.

Because the densely populated areas are situated mostly in the low-lying coastal areas of southern and south-western Finland, the marine and lacustrine deposits, silt to clay, have great importance from the engineering point of view. In central Finland, which was submerged for a relative short time, the deposits are small and shallow. In southern and south-western Finland, as well as in Ostrobotnia, the deposits are coherent, extensive and up to 60 m thick.

The geologic stratigraphy is geotechnically problematic in Finland. Under the fine-grained and soft sediments there is usually a very stiff layer of moraine. The layer between moraine and clay can be thin consisting of silt, sand and gravel, or the layer can be absent. Very often the layers beneath the clay and silt contain stones and boulders. (Halkola, H. &Törnqvist, J. 1995)

2 COMMON PRACTISE FOR SOIL INVESTIGATION

2.1 Principle

The purpose of the soil investigation is to determine the surface relief and soil layers at the construction site, properties of the soil layers, ground water balances in the layers and the location of the rock face. Moreover, the structure of the bedrock is determined in order to make a reliable plan for the foundations and other bottom structures of a building and to facilitate safe foundation engineering that does not harm the environment.

The extent and quality of the soil investigation are determined by the quality of the soil, loads on the soil, existent and forthcoming constructions, and the environment. The extent and quality of the soil investigation are assessed by a geotechnical designer who also observes the development and the results of the investigation. Moreover, he must supplement the investigation plan if needed. The investigator is responsible for the competence of the investigation tools and investigation methods as well as the reliability of the results.

2.2 Sounding

Soundings are performed according to the instructions published by the Finnish Geotechnical Society (Sounding Guide I - V). If a Finnish sounding method instruction is not available the soil is investigated using a generally agreed method that is known to work well. The most common sounding methods that are in use or can be adapted in Finland are presented in Figure 1.

Soil investigation must be extended below the soil layer which is defined to be an objective level for the pile points. This facilitates the use of end-bearing piles, friction piles and cohesion piles that all have a different working principle. Soil investigation must be performed using such methods that a working principle and design values of the piles and pile groups can be reliably defined.

Weight sounding test is generally used to determine foundation conditions of light and medium weight structures and buildings. The cone penetration test is used in the clay regions of the country, such as Southern and South-western Finland, and Ostrobotnia.

Grain size distributions of the soil layers are determined in laboratories using disturbed samples

The main use of the sounding method ●

Secondary use of the sounding method, the method has low precission for the matter to be determined ○

Sounding methods \ The matter to be determined	Location of the rock surface	Location of dense bottom layer	Boundaries of the soil layers having different density	Approximate strenght of the soil layers	Exact strenght of the soil layers	Approximate density of the soil layers	Soil type	Determination of the lenght of the driven pile
Weight sounding test	○	●	●	○		●	●	○
Dynamic probing	○	●	○	○		●	○	●
Cone penetration test		○	●	●		●	●	○
Vane shear test					●			
Percussion drilling	○	●					○	○
Percussion drilling (air pressure drilling)	●	○						○

Figure 1. The most common sounding methods and their recommended usage (TPO-83).

or based on observations made during sounding. Strength properties and limit of consistency of the cohesive soil are investigated using vane shear tests.

Bottom conditions of heavy structures and buildings, and machine bases are mainly investigated using dynamic probing. If the piles are designed to be supported by the rock the exact height of the rock face is measured by percussion drilling.

The granularity of a soil layer is studied and the classification properties of soil layers are determined using disturbed samples and observations made during sounding. Strength properties are determined using field and laboratory tests. It is often necessary to take undisturbed samples to assure sufficient tests in a laboratory.

2.3 Soil Investigation Using Different Types of Piles

2.3.1 End Bearing Piles

When end bearing pile foundations are used, soundings reach the rock face or a solid bottom layer. General structures of the penetrable layers and the bottom layer are determined. The location and relief of the rock face are investigated, especially when cohesion layers reach the rock face or the rock face is declining and the non-cohesive soil layer on the rock face is loose. Furthermore, if the non-cohesive layer or moraine layer on the declining rock face is thin, location and relief of the rock face should be determined.

If the diameter of the piles is smaller than 300 mm and the maximum ultimate load of the piles is about 1.5 MN, dynamic probing should be made on different sections of the construction site in addition to weight sounding test. When the ultimate load of the pile is more than 1.5 MN it is necessary to carry out dynamic probing or other suitable soundings to determine location of the dense bottom layer. Determination of the thickness and structure of the bottom layer may require the use of more powerful sounding methods than dynamic probing. The location of the rock face must then be reliably determined by percussion drillings.

The rock face should be investigated using percussion drillings if end bearing piles touch the rock and have a rock shoe or are made of reinforced concrete or steel. The location of the rock face should be assured using percussion drillings, especially when the soil investigation is made for large capacity, rock-supported end bearing piles. In some special cases it is necessary to investigate the structure of the rock using rock sample drillings.

2.3.2 Friction Piles

When friction piles are used, the boundaries between soil layers are investigated as well as the properties (such as granularity and constructional density) of the layers penetrated by the piles. Especially the layers which are designed to be functional should be investigated. If sufficient results on the loading tests of the piles and test pilings are not available in design, bearing capacities should be determined, especially in demanding cases. Test pilings and loading tests should be done in similar ground conditions Otherwise instructions given in section 2.3.1 for end bearing piles will be applied.

The cone penetration test can be used with friction piles. When investigation methods are selected and the investigation is performed, special attention must be paid to the cohesive layers that may be between non-cohesive layers.

2.3.3 Cohesion Piles

When cohesion piles are used, not only general ground conditions at the construction site but also strength and deformation properties of the cohesive layers are determined. The properties must be investigated over the entire cohesive layer in order to determine the ultimate bearing capacity and settlements of the cohesion piles as a function of the pile lengths.

Deformation properties of the cohesive layer are investigated in order to determine the expected settlements of the piles and the extra loads on the piles caused by the cohesive layer. Typically the undrained shear strength in Finnish cohesive soil deposits varies 10 to 20 kPa in range, the most common value is ca 10 kPa.

The use of cohesion piles is very rare in many areas in Finland. Due to the geotechnical circumstances, the cohesion piles can only be used in Southern and South-western Finland and in Ostrobotnia. The use of the cohesion piles cannot be recommended in any circumstances and their use is mainly limited to the foundations of lightweight buildings. For example in the design of high capacity piles the resistance of cohesive soil layers is only allowed to be utilized in the design of short term loads.

2.4 Special Requirements for the Soil Investigation

The soil investigation report must include the following information in addition to the requirements presented in the general geological description of the construction site, ENV 1997-1:

- the ground level at all investigation points using one of the known leveling systems
- locations and properties of soft, loose and disturbance sensitive soil layers
- information about in-fills which have large voids. Mass deflects may therefore exist in the layers when the piles are concreted
- locations and thicknesses of all bearing layers
- information about large stones, boulders or other large objects which make piling more difficult and which may require penetration or removal tools
- locations, extents and thicknesses of the soil layers which are sensitive to water filtration or to the increase in pore water pressure. Compacting and vibrating effects caused by the piling work are reasons for the water filtration and to the pressure increase.
- information regarding the levels of perched ground water and ground water, range of the levels and information about artesian ground water
- information about transmissive layers
- study on the aggressivity of the soil and ground water if it is probable that the aggressivity is exceptionally high and can therefore have a harmful effect on concrete or steel parts of the piles
- depth and decline of the rock face
- fractures and weakness substages observed in the rock
- information about waste fills and contaminated soil layers

With the aid of the soil investigation it must be able to assure that below the supporting level of the piles there exists no soft soil layers. The piles can penetrate to the soft layer and therefore lose their bearing capacity or the soft layer can compress and cause settlement of the piles.

3 PILING TECHNOLOGY

3.1 Pile Types

Pile types can be divided into three groups according to their geotechnical working principle:

1. end bearing piles
2. friction piles
3. cohesion piles

This type of division is common in Finland. However, high capacity piles are divided into two main groups based on geotechnical design.

1. displacement piles
2. replacement piles

Either end bearing or friction piles should be selected according to the geotechnical action.

The end bearing pile is the most common type used in Finland. The geotechnical circumstances in Finland are suitable for relatively short piles that are supported by hard bottom layer or bedrock. Since the depth of the bearing stratum is only 10 - 20 m in many areas, the use of friction and cohesion piles is often both technically and economically non-profitable.

The second most common pile type is the friction pile which is sometimes used at the esker margins and regions where thick and loose non-cohesive soil layers exist. Friction piles are mainly used in the foundation of relatively light weight residential and commercial buildings.

The pile types can also be determined based on the embedding and fabrication methods of the piles. According to this division the pile types used in Finland are:

1. prefabricated driven piles
2. driven piles cast in place
3. driven piles with base enlargement
4. bored piles cast in place

3.2 Piling Technology

3.2.1 Prefabricated Driven Piles

Prefabricated driven piles are made of reinforced concrete or steel. Reinforced concrete piles are the most commonly used pile type in the foundations of light and medium weight structures and buildings. Piles are typically embedded using a hydraulic hammer where the weight of the hammer is 30 - 50 kN and the height of fall is 0.2 - 0.5 m. The operation load of the piles is about 500 - 1500 kN. The most commonly used cross-sections of the reinforced concrete piles are 250 x 250 mm^2 and 300 x 300 mm^2. Steel piles used in Finland are x- and G-piles, and RR-piles the diameter of which is less than 300 mm. RR-piles are the most commonly used steel piles.

High capacity piles which are used in the foundations of heavy structures, such as bridges and wharfs, are generally steel pipe piles. Their most common diameters are in the 500 - 1200 mm range, the thicknesses of the pipe walls are 8 - 16 mm. The operation load of this type of piles is about 3 - 9 MN and therefore the piles are mainly supported by the bedrock. The piles are almost always equipped with a rock shoe to protect the pile against breakage and to facilitate the attachment of the pile point to the rock.

The high capacity piles are embedded using high energy pile driving devices such as hydraulic hammers or Franki-pile driving devices which drive the pile from the base. The rated blow energy of the Franki devices is 60 - 250 kNm or more. Moreover, diesel hammers have been used and some experience in the use of IHC-hydrohammers have been gathered. However, high capacity steel pipe piles are mainly embedded using hydraulic hammers.

3.2.2 Driven Piles Cast in Place

In Finland this type of piles are mainly Vibrex-piles but they are not particular commonly used. The diameter of the pile pipe is typically about 500 mm and the operation load is about 2 MN. Vibrex-piles are normally embedded using similar pile driving devices as the high capacity steel pipe piles.

3.2.3 Driven Piles with Base Enlargement

Franki-pile, Franki-pipe pile, Franki-mixte pile and onion (as bawang) pile are the piles of this group which are fabricated in Finland. The diameter of the rod of a Franki-pile is about 500 - 600 mm and the operation load about 1.5 - 3.5 MN. Piles with base enlargement are fabricated using a Franki drop hammer which is used to drive the pile base. The weight of the hammer being 100 - 150 kN and the falling height as much as 10 m.

The diameter of the base enlargement of the onion pile is about 1.5 to 2 times the diameter of the pile rod corresponding the allowed capacity in same range as Franki-piles.

3.2.4 Bored piles cast in place

The use of bored piles is relatively common in Finland. The diameters of the piles are typically about 900 - 1500 mm and the maximum load in operation conditions is ca. 3 - 17 MN. The piles are fabricated using piling device which digs the soil inside the working pipe and simultaneously presses and grinds the pipe. The piles are bored underwater and therefore they are concreted using submerged concreting technique.

Obstacles, such as large stones and boulders, which hinder pile embedding are broken by chiselling or explosion. When the pile is designed to be supported by a declined rock face, the pile must be attached to the rock using an anchor pin in order to prevent the pile base from slipping.

4 RELEVANT NATIONAL DOCUMENTS WITH REGARD TO PILE DESIGN

4.1 General Documents

The regulations concerning foundation engineering in Finland are given in Volume 3 of the Finnish Code of Building Regulations. Recommendations given in the Finnish Foundation Engineering Standard (RIL 121-1988) published by the Association of Finnish Civil Engineers are based on the Finnish Code of Building Regulations and represent good foundation techniques. The principles to be obeyed when foundations and other burdening structures are designed and constructed are presented in the code. RIL 121-1988 has been written such that geotechnical design of foundations can be made using the limit state method and partial factors. The guide uses the concepts of the limit state design but also facilitates geotechnical design using the global safety method or allowed stress method.

An alternative for the foundation engineering and pile design is to use the European Prestandards (so called Eurocodes) for the design of load-bearing structures. A National Application Document of the European Prestandard ENV 1997-1:1994 Eurocode 7: Geotechnical design was published in Finland 9.5.1996.

4.2 Documents Concerning Pile Design

For the design of the pile foundations, the Finnish Foundation Engineering Standard is supplemented by the Design Code for Driven Piles (LPO-87) and the Design Code for High Capacity Piles (RIL 212-1995, SPO-95). The Design Code for Driven Piles (LPO-87) is written for piling where rectangular piles having cross-section smaller than 350 x 350 mm^2 or spherical piles having diameter smaller than 500 mm are driven using a device equipped with a drop hammer or other striking bodies. The Design Code for High Capacity Piles RIL 212-1995 discusses the general principles of designing and fabrication of high capacity piles. The guide can be used for piles whose bearing capacity is 1.5 MN or more and a diameter of at least 300 mm. The Design Code for High Capacity Piles was published in 1995 and it is a temporary guide: the European standardisation of foundation engineering is ongoing but the work concerning high capacity piles has not been finished. When the Finnish guide was renewed, the goal was to fit the Finnish design and construction procedures to the European requirements.

In addition to the general instructions mentioned above, various organisations publish guides of their own. The Finnish National Road Administration uses the Foundation Engineering Standard for Bridge Design when designing bridge foundations and other structures supported by the soil. This guide supplements the recommendations given in the Finnish Foundation Engineering Standard RIL 121-1988. Chapter 10 - Pile Foundations - of the guide was renewed in 1996. A need for the renewal of the piling chapter raised when the Design Code for High Capacity Piles was finished in 1995. Steel pipe piles for bridge foundations are designed according to the Steel Pipe Piles guide by Finnish National Road Administration. Steel pipe piles for railway bridges are designed using the Steel Pipe Piles for Railway Bridges guide published by Finnish Railways.

5 DETAILED DESCRIPTION OF THE NATIONAL DESIGN METHODS

5.1 General Philosophy

The foundation structures must be designed such that they and the structures borne by them have sufficient tolerance against collapse, breakage, fractures and excess elastic or plastic deformations. The compression and side resistance capacity and tensile capacity must be checked in the service limit state and in the ultimate limit state.

The limit states in the foundation structures or in the constructions supported by the foundations structures must be checked in the service limit state and in the ultimate limit state. The possible limit states are caused by the displacements of the piles. In the ultimate limit state the pile loads must be multiplied by the appropriate partial factors. Moreover, the capacity of the pile must be divided by its partial factors. In the service limit state, the characteristic values of the pile loads must be used and the characteristic values of the capacity are used as a pile capacity.

5.2 Design on Basis of Static Loading Tests

Static loading test can be applied to test piles and foundation piles. The diameter of the test piles must be more than 50 % of the diameter of the actual piles. The material and the assembling method of the test pile must correspond to those of the pile to be designed.

When the ultimate load is determined using the static loading test, the test must be performed in most adverse (geotechnically) foundation conditions found at the construction site. Moreover, the load-tested pile must correspond to the pile to be designed and the loading test must be performed in similar circumstances as the designed pile will have.

The static loading test is recommended to be performed according to the "Axial Pile Loading Test, Suggested Method" published by the ISSMFE Subcommittee on Field and Laboratory Testing. The method has been published in the ASTM Geotechnical Design Testing Journal, June 1985, pp. 79 - 80.

The load used in the static loading test must correspond at least to the maximum load multiplied by a safety factor.

In the analysis of the test results, the base and the shaft resistances of the pile can be roughly approximated by a displacement of 0.05 D for the base resistance and 0.005 - 0.01 D for the shaft friction. Furthermore, the elastic deflection of the pile must also be taken into account.

5.3 Design on Basis of Ground Tests

5.3.1 General

When the geotechnical ultimate load is determined using the bearing capacity formula that is based on the strength properties of the soil or using empirical methods based on sounding resistance, the geotechnical bearing capacity R_c is the sum of base and the shaft resistance:

$$R_c = R_b + R_s$$

Where R_b = the design value for the base resistance; R_s = the design value for the shaft friction.

In the partial factor method, the characteristic values of the resistances are computed according to the ENV 1997-1 and its clarifications presented in Section 5.6.

The characteristic values for the base and the shaft resistance can be computed using the following equations:

$$R_{bk} = q_{bk} A_b$$

$$R_{sk} = \sum_{i=1}^{n} q_{sik} A_{si}$$

Where A_b = cross-sectional area of the pile base; A_{si} = area of the pile shaft in the ith soil layer; q_{bk} = characteristic value of the pile base resistance per unit area; q_{sik} = characteristic value of the pile shaft friction per unit area in the ith soil layer.

5.3.2 Ultimate load on basis of static bearing capacity formula

When the bearing capacity of the pile is estimated using the static bearing capacity formula, the internal friction angle of the soil must be determined on the pile shaft and in the base layer of the pile. The estimation can be made using triaxial tests made in a laboratory, in- situ methods or the sounding resistance.

The characteristic value of the pile base resistance per unit area is obtained in non-cohesive soil layer from:

$$q_{bk} = \frac{\sigma_v N_q}{\xi}$$

Where N_q = bearing capacity factor of the pile base, see Figure 2; σ_v = effective vertical stress at the level of the pile base; ξ = 1.6, see section 5.6.

The average friction angle in the base layer of the pile must be used in the selection of the bearing capacity factor N_q. The base layer of the pile includes the soil layer extending 5 D above and 3 D below the pile base (D is the diameter of the pile). If the friction angle is larger than 40° it must be determined in a geotechnical laboratory using triaxial tests or in-situ tests.

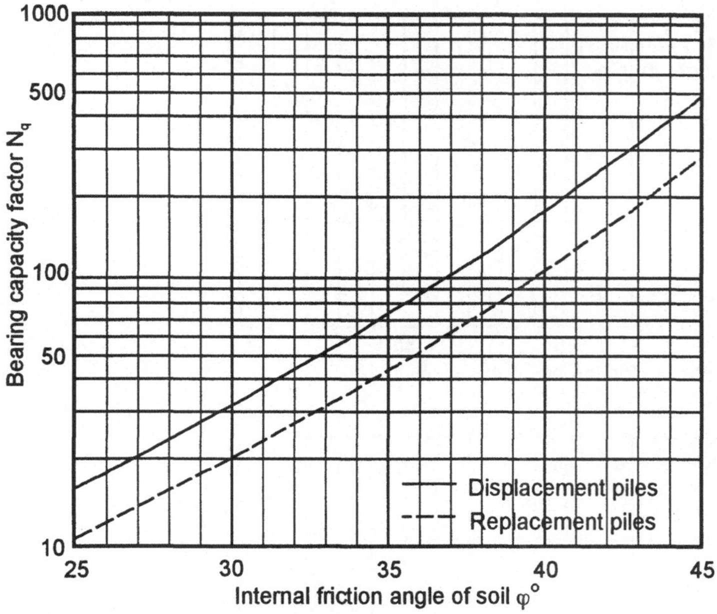

Figure 2. The bearing capacity factor N_q as a function of the internal friction angle of the soil.

The effective vertical stress σ_v at the level of the pile base can be computed by taking into account the effective load of the soil layers extending less than 10 D above the pile base.
The characteristic value of the pile shaft friction per unit area in a non-cohesive soil layer is obtained from the equation:

$$q_{sik} = \frac{\sigma'_v K_s \tan\phi_a}{\xi}$$

Where $K_s \tan f_a$ = shaft friction factor depending on the pile material, embedding method and internal friction angle (obtained either from Figure 3a or 3b); σ'_v = effective vertical stress on the pile shaft; ξ is 1.6, see section 5.6.

The effective vertical stress σ'_v on the pile shaft is determined by taking into account the load of soil layers extending less than 10 D above the examined level.

In the cohesive soil layers, the characteristic value of the pile shaft resistance is the adhesion between the pile and the soil. It can be estimated using the undrained shear strength s_u and the adhesion factor a:

$$s_{uik} = \frac{\alpha s_{ui}}{\xi}$$

Where a is the adhesion factor corresponding to the pile material (obtained from Figure 4); S_{ui} is the undrained shear strength in the ith soil layer; ξ is 1.6, see section 5.6

The development of the adhesion between the pile and the soil takes at least a month depending on the properties of the soil type.

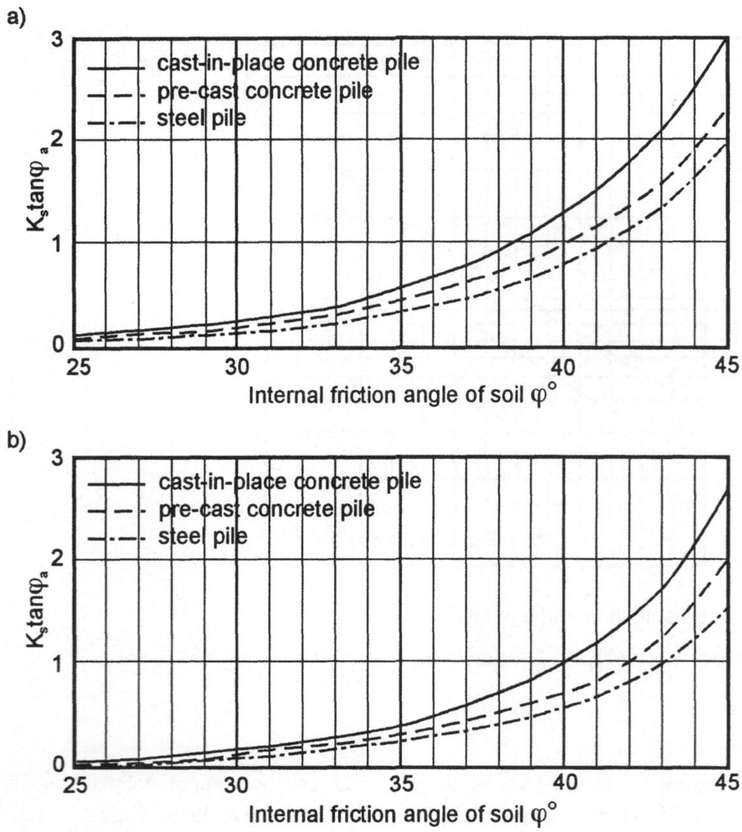

Figure 3. The shaft friction factor of the pile $K_s tanf_a$ for a) displacement piles and b) replacement piles.

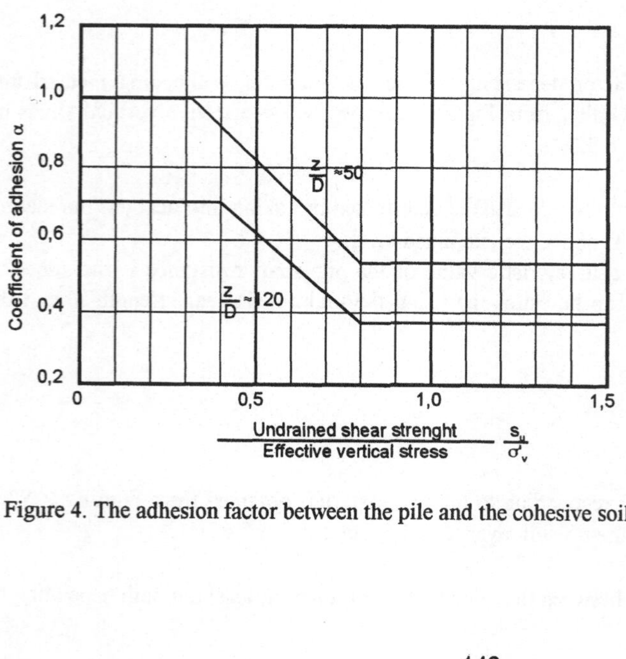

Figure 4. The adhesion factor between the pile and the cohesive soil.

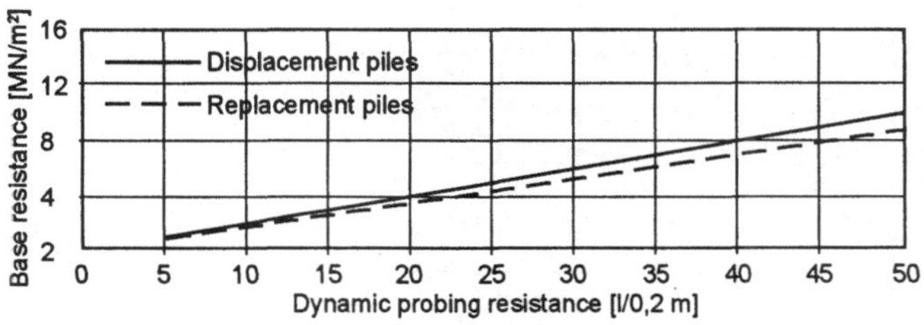

Figure 5. The base resistance q_b of the pile.

5.3.3 Ultimate load on basis of dynamic probing resistance

The ultimate load q_b per unit area of the pile base can be determined using the average sounding resistance in dynamic probing, see Figure 5.

The average sounding resistance can be determined in soil layers which extend 5 D above and 3 D below the pile base using dynamic probing resistance based determination of the pile base resistance.

The ultimate load per unit area on the pile shaft can be determined using the average dynamic probing resistance on the pile shaft (obtained using Figures 6a and 6b, depending on the pile shaft material and pile type).

The characteristic values for the base and shaft bearing capacities can be computed from the ultimate bearing capacity:

$$R_{bk} = \frac{R_b}{1,6}$$

$$R_{sk} = \frac{R_s}{1,6}$$

The shaft friction caused by the coarse-grained soil layer located above the cohesive layer must not be included in the bearing capacity due to possible negative shaft friction.

5.3.4 Plugged open-ended steel pipe pile

The base resistance of the plugged open-ended pile is estimated based on the plugging factor:

- in moraine

$h = 0.8$ when $z/D = 10$

- in sand and gravel

$h = 0.8$ when $z/D = 15$

Where z is the embedding depth of the pile in the plugging soil layer; D is the diameter of the pile

The design base resistance of the plugged pile is obtained from the design base resistance of the closed-ended pile.

$$R_{pbk} = hR_{bk}$$

Where R_{bk} is the design base resistance of the closed-ended pile, whose diameter corresponds to that of the plugged pile.

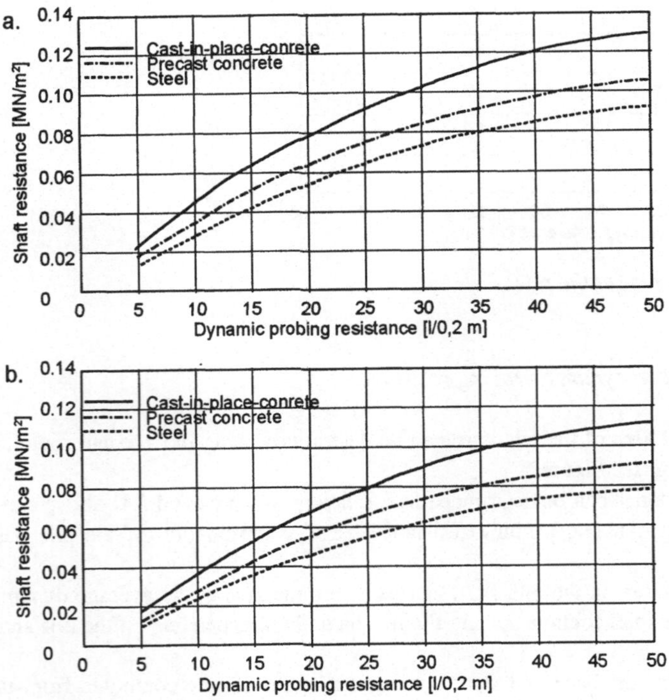

Figure 6. Shaft resistance q_s for a) displacement piles and b) replacement piles.

When the ratio z/D decreases, the factor h is linearly decreased. The pile can be considered as a plugged pile when the settlement of the soil inside the pile is more than half of the settlement of the pile in the plugging soil layer.

5.3.5 Driven pile supported by the bedrock

The rock shoe of the pile can be considered to be supported by the rock, when the pile driving observations and soil investigation confirm that the pile base has reached the bedrock. In uncertain cases, stress wave measurements are used to assure that the driven piles supported by the rock.

The point resistance of the pile equipped with rock shoe supported by the rock is obtained from the unconfined compression strength using the following equation:

$$q_p = \frac{2q_{syl}}{\xi}$$

Where q_{syl} is the unconfined compression strength.

5.3.6 Bored pile supported by the bedrock

Before casting the pile, it must be assured that bored pile has reached the bedrock using chiseling performed at the bottom of the pile excavation. Before casting, weathered or poorly supporting surface of the rock must be removed from the bottom of the pile excavation. When needed, percussion drilling can be used to assure that the bored pile is reached the rock surface.

In demanding piling, the possibility to check and improve the rock contact must be assured by injecting the base of the bored pile after concreting.

The principle for design a bored pile is the same as mentioned in the preliminary CEN document Execution of Special Geotechnical work: Bored Pile. The base resistance of a bored pile supported by the bedrock can be designed based on the strength of the pile material if the contact with the rock is confirmed. The exception to this principle are design works carried out according to the design code of practise of various national organisations such as example the Finnish National Road Administration.

5.4 Design on Basis of Driving Formula

The final set during end blows computed solely using the dynamic pile driving formula can only be used to confirm the bearing capacities of the reinforced concrete and steel piles whose capacity is less than 1.2 MN. For the high capacity driven piles, the dynamic pile driving formula can only be used in the preliminary estimation of the ultimate load. The ultimate load of these piles must always be confirmed by dynamic or static test loading.

Two examples of dynamic pile driving formulae applicable to the preliminary estimation of the ultimate capacity of the high capacity driven piles are Hiley's and Gates' formulae:

Hiley's formula

$$P_u = \frac{e_f E}{s + \frac{1}{2}c} \frac{W_h + n^2 W_p}{W_h + W_p}$$

Where E is the impact energy [kNm]; e_f is the efficiency factor of driving; n for hard wooden cushion block = 0.5, for hard plastic cushion n = 0.8 and for steel die without pile cap n = 1.0; s is the permanent settlement of the pile [mm]; c is the elasticity [mm]; W_p is the weight of the pile [kN]; W_h is the weight of the hammer.

Gates' formula

$$P_u = 96(2{,}4 - \log s)\sqrt{e_f W_h H)}$$

Where e_f is the efficiency factor of driving; W_h is the weight of the hammer [kN]; H is the falling height of the hammer [m]; s is the measured settlement per drive of the pile [m].

When the concentrated bearing stress of a reinforced concrete pile is less than 7 MPa in the operation state, the sufficient geotechnical ultimate load of an end bearing pile can be confirmed

Table 1 Instructions to final set the reinforced concrete piles using hydraulic or other free falling hammers.

The maximum stress in operation state	The size of the pile	The largest allowable settlement [mm]/10 blows				The falling height of the hammer [m], when the length of the pile is					
		2 t	3 t	4 t	5 t	5 m	10 m	15 m	25 m	35 m	50 m
< 9 MPa	250x250	-	7	10	13	0,25	0,27	0,29	0,31	0,35	0,40
	300x300	-	-	7	10	0,25	0,27	0,29	0,32	0,35	0,40
< 7 MPa	250x250	8	15	20	25	0,24	0,26	0,28	0,31	0,34	0,39
	300x300	(5)	10	15	20	0,24	0,26	0,28	0,31	0,34	0,39

using final set condition based on dynamic bearing capacity formulae, and experience. If the concentrated bearing stress of the reinforced concrete pile is less than 9 MPa in the operation state, the validity of the end blow condition must be confirmed by static or dynamic loading test.

The end blows are made, when the pile base reaches the solid subgrade and the settlement of the pile clearly decreases. The end blows typically include 3 -5 series á 10 drives and they are made on the basis of recommendations on the hammer and the falling heights for piles of different lengths (see Table 1). The number of end blow series depends on circumstances. If the settlement of the pile decreases slowly more series are made and if the settlement decreases fast less end blow series are needed.

5.5 Design on Basis of Wave Equation Analysis

In demanding building projects, the geotechnical bearing capacity of the driven piles must be confirmed using stress wave measurements. It is recommended to confirm the geotechnical bearing capacity of the driven piles in other building projects, too - and at least in uncertain and suspicious cases - by stress wave measurements. Stress waves should be measured for each different pile size, pile type and ground conditions. The number of piles to be measured should be at least 10 % of the total number of the piles.

The person making the stress wave measurements and the person analysing the measurement results must be familiar with the pile driving and especially with the behaviour of the piles in the driving stress. Moreover, they must have sufficient knowledge of the stress wave theory.

In bearing capacity computation based on stress wave measurements, the damping coefficient of the pile laying on a soil layer must be confirmed by making at least one stress wave theory based analysis. The analysis can be for example of SIGNAL MATCHING or CAPWAP type but it must be based on measurements made at the construction site. There is no need to make the above mentioned analyses for the piles laying on the rock.

In suspicious cases and in the cases where the base resistance and shaft friction clearly depend on the construction section, it is recommended to make several analyses to assure that the damping coefficient and the geotechnical bearing capacity of the pile are sufficient.

The bearing capacity can be estimated using the CASE method. In the CASE method, the geotechnical ultimate bearing capacity is computed using results of the stress wave measurement:

$$R_t = \frac{F(t_1) + F(t_2)}{2} + \frac{MC}{2L}[v(t_1) - v(t_2)]$$

$$R_s = R_t - J_c[F(t_1) + zv(t_1) - R_t]$$

Where R_t is the total driving resistance; F is force; v is the particle velocity of the pile; t_1 is instant of time for the maximum force / maximum velocity; t_2 is $t_1 + 2\,L/c$; M is the mass of the pile; L is the distance between the measurement point of the stress wave and the pile base; c is the stress wave velocity in the pile; R_s is the static load of the pile; J_c is the damping coefficient; z is the impedance of the pile.

5.6 Prescribed Global Factors for Safety and Applied Partial Factoring

5.6.1 Finnish Foundation Engineering Standard RIL 121-1988

The Finnish Foundation Engineering Standard RIL 121-1988 publication has been written such that geotechnical design of foundations can be made using limit state, global safety or allowed stress methods. The standard provides minimum values for global safety factors to be used in pile design. The minimum values are presented in Table 2.

Table 2 Minimum values for global safety factors.

Target	Global Safety Factor
Static load tests	1.8
Dynamic load tests	2.0
Empirical or analytical calculation	2.2*

* When calculations are carried out at least two independent calculation method and the global safety factor is applied to the lowest value.

Table 3 Partial factors for actions.

Action	Partial Factor
Permanent action	1.2
Variable action	1.6

Table 4 Partial factors for ground properties.

Target of the factor	Partial Safety Factor
Tangent of angle of shear resistance (bearing capacity of pile)	1.25
Cohesion (bearing capacity of pile)	2.0

Table 5 Partial factors to derive a design value for the piles in the ultimate limit state.

Component Factors	Base Resistance γ_b	Shaft Resistance γ_s	Total Resistance γ_t
Driven piles	1.3	1.3	1.3
Bored piles	1.6	1.3	1.5

Table 6 Factors ξ to derive R_{ck}.

Number of loaded piles *)	Factor ξ				
	< 10%	≥ 10%	≥ 25%	≥ 50%	100%
Static load test					
a) ξ to the average load test result	1,35	1,30	1,25	1,20	1,10
b) ξ to the lowest load test result	1,25	1,20	1,15	1,10	1,00
Dynamic load test					
a) ξ to the average load test result	1,5	1,45	1,40	1,35	1,25
b) ξ to the lowest load test result	1,4	1,35	1,30	1,25	1,15

*) When the number of load-tested piles is two or more, the percentage values given in the table are computed for equal type of piles in equal ground conditions of the building site. When only one pile is load-tested, ξ is 1.5.

Piles can also be designed using the limit state method. In the service limit state all partial factors are 1.0. Partial factors used in the ultimate limit state are given in Table 3.
Partial factors for strength parameters of soil layers in the ultimate limit state are presented in Table 4.

5.6.2 The Design Code for High Capacity Piles RIL 212-1995

According to the Design Code for High Capacity Piles, the piles are designed using the limit state method. The design values for the bearing capacity are obtained from

$$R_{cd} = \frac{R_{bk}}{\gamma_b} + \frac{R_{sk}}{\gamma_s}$$

$$R_{ck} = \frac{R_{cm}}{\xi}$$

Table 5 presents the partial factors provided in the guide to derive design values for the geotechnical bearing capacity of the piles in the ultimate limit state.
When the geotechnical capacity of the pile is computed, the characteristic value of the capacity is obtained by dividing the theoretical ultimately load by factor ξ. Table 6 gives the values for the factor ξ, when static or dynamic loading tests are applied. Otherwise, the factor ξ is 1.6.
The Finnish National Application Document for the European Prestandard Eurocode 7 provides the same partial factors and ξ factors as the Design Code for High Capacity Piles RIL 212-1995.

5.7 Rules for Serviceability: Particular Rules for Pile Settlements

Table 7 presents the limit values for equal total settlements and angular displacements of the foundations given in the RIL 121-1988 publication (see also Figure 7).

The values given in Table 7 are only applicable in the design of house buildings. For bridges the general limits for the allowed total settlements are given in the National Application document. The allowed equal total settlement is 20 mm and the corresponding allowed difference in settlement between bridge footings is 10 mm. If these values exceed, the additional load effect caused by settlement shall be taken into account in the loading of the bridge.

Table 7 Limit values for even total settlements and angular displacements

Type of Structure	Limit Value for the Total Settlement s_{max}[mm]	Limit Value Range for the Angular Displacement δ/l Friction Soil	Cohesive Soil
Solid rigid structures	100...150	1/250...1/200	1/250...1/200
Statically determined structures	100...150	1/400...1/300	1/300...1/200
Statically not determined structures			
- Timber structures	100...150	1/400...1/300	1/300...1/200
- Steel structures	80...100	1/500...1/200	1/500...1/200
- Mansoried structures	40...80	1/1000...1/600	1/800...1/400
- Reinforced concrete structures	60...100	1/1000...1/500	1/700...1/350
- Reinforced concrete element structures	40...80	1/1200...1/700	1/1000...1/500
- Reinforced concrete framed structures	30...60	1/2000...1/1000	1/1500...1/700

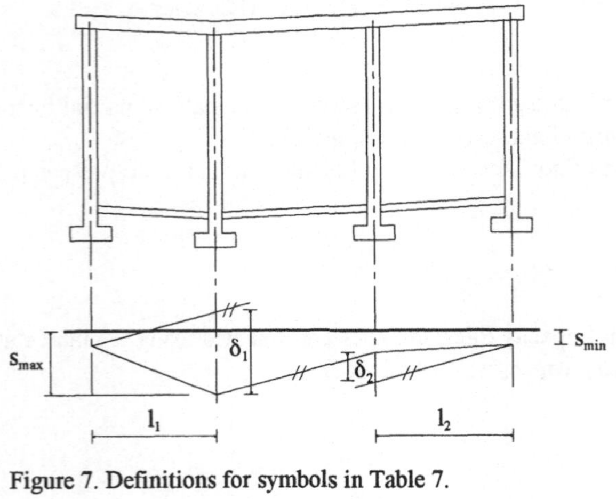

s = settlement
$\frac{\delta}{l}$ = deflection

Figure 7. Definitions for symbols in Table 7.

6 PARTICULAR EXAMPLES TO DEMONSTRATE THE METHODS

6.1 General

The example bridge has been founded using closed steel pipe piles. The geotechnical ultimate load of the pile is computed using a method based on dynamic probing. The method is presented in the Design Code for high Capacity Piles (RIL 212-1995). The piles are designed using the global safety method and the limit state method. The data of the example are based on a real bridge. The bearing capacity of the piles is confirmed using dynamic load tests.

Both dead and live loads are applied to the closed steel pipe pile. The characteristic values for the dead and live loads are 3000 kN and 1350 kN, respectively. The design load is computed using the global safety method and the limit state method.

The design load of the pile using the global safety method:

$G_{d1} = 3100 + 1350 = 4450$ kN

The design load of the pile using the limit state method:

$G_{d2} = 1{,}2 \cdot 3100 + 1{,}6 \cdot 1350 = 5880$ kN

The steel pipe pile has been embedded using Franki devices. The soil data, dimensions of the closed steel pipe pile and the dynamic probing resistance diagram are presented for the intermediate bearing of the bridge in figure 8.

6.2 Bearing Capacity on Basis of Dynamic Probing Resistance

The ultimate load of a pile using dynamic probing resistance is obtained from the equation:

$$R_{cm} = \int_0^z \pi d f_s ds + A_{pk} q_p$$

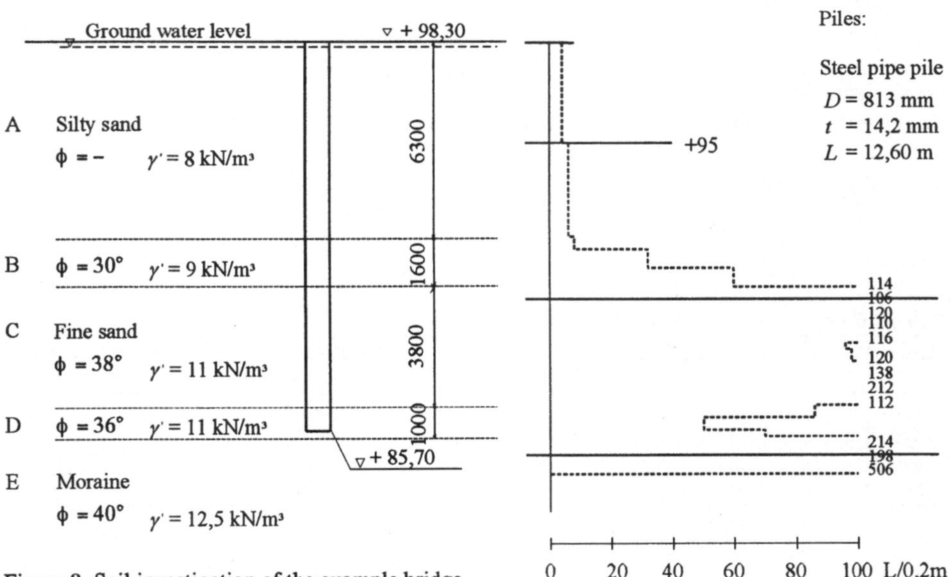

Figure 8. Soil investigation of the example bridge.

The average dynamic probing resistance at the pile base:

$5d = 4.065$ m
$3d = 2.439$ m

$$\frac{0{,}2\cdot70+0{,}4\cdot(50+86)+1{,}4\cdot157+0{,}6\cdot96+1{,}065\cdot115+1{,}2\cdot206+1{,}239\cdot506}{4{,}065+2{,}439}=206\ \mathrm{l/0{,}2m}$$

Base resistance:

$\Rightarrow q_p = 20$ MPa $= 20\ 000$ kN/m^2

The bearing capacity of a steel pipe pile base:

$$R_{bm} = A_{pk}q_p = \pi(\frac{0{,}813}{2})^2\cdot 20000 = 10382\ \text{kN}$$

Table 9 presents the computation of the shaft bearing capacity layer by layer (RIL 212-1995).

Total shaft bearing capacity of a steel pipe pile (RIL 212-1995):

$$R_{sm} = \int_0^z \pi d f_s ds = 327 + 1456 + 230 = 2013\ \text{kN}$$

Ultimate load of the pile (RIL 212-1995):

$R_{cm} = 10382 + 2013 = 12395$ kN

Design of a closed steel pipe pile using the global safety factor method:

$$F_d = \frac{R_{cm}}{G_{d1}} = \frac{12395}{4450} = 2{,}8 > 2{,}2$$

Design using the limit state method:
The characteristic value of the ultimate load of the pile:

$$R_{ck} = \frac{R_{cm}}{\xi} = \frac{12395}{1{,}6} = 7747\ \text{kN}$$

Table 9 Computation of shaft bearing capacity

Soil Layer	Height [m]	Driving Resistance [l/m]	Shaft Friction [kN/m^2]	Shaft Bearing Capacity [kN]
Layer A	6.3	4	-	-
Layer B	1.6	37	80	327
Layer C	3.8	127	150	1456
Layer D	1.0	68	100	230

The design bearing capacity of the pile:

$$R_{cd} = \frac{R_{ck}}{\gamma_t} = \frac{7747}{1,3} = 5959$$

Safety using the limit state method:

$$F_d = \frac{R_{cd}}{G_{d2}} = \frac{5959}{5880} = 1,01 > 1,0$$

6.3 Bearing Capacity on Basis of Wave Equation Analysis

Two piles in the intermediate bearing at the Kutujoki bridge have been load-tested. The number of steel pipe piles is four. The ultimate loads obtained from the dynamic load tests are 10 040 kN and 10 720 kN. The mean value of the load tests is 10 380 kN and the lowest value 10 040 kN. Design of a closed steel pipe pile using global safety factor method:

$$F_d = \frac{R_{cm}}{G_{d1}} = \frac{10040}{4450} = 2,3 > 2,0$$

Design using the limit state method:
The characteristic value of the ultimate load of the pile:

a) for the mean value of the load test results

$$R_{ck} = \frac{R_{cm}}{\xi} = \frac{10380}{1,35} = 7689\,\text{kN}$$

b) for the lowest value of the load test results

$$R_{ck} = \frac{R_{cm}}{\xi} = \frac{10040}{1,25} = 8032\,\text{kN}$$

The design bearing capacity of the pile:

$$R_{cd} = \frac{R_{ck}}{\gamma_t} = \frac{7689}{1,3} = 5915\,\text{kN}$$

Safety using the limit state method:

$$F_d = \frac{R_{cd}}{G_{d2}} = \frac{5915}{5880} = 1,01 > 1,0$$

7 QUALITY CONTROL AND MONITORING

7.1 Quality Control

The supervisor must be competent and experienced person, and he is responsible for the following:

- the work is done according to the instructions, qualitative requirements defined for the work,
- the work description or a proper working method,
- the piling plan and other essential working plans are properly made and inspected,

- controlling measurements are made and measurement record is maintained,
- the developer and/or the designer are informed of exceptional situations and conditions observed at the construction site, as well as of all the other matters affecting the working process,
- specialists, such as welders, must fulfill the competence requirements.

7.2 Monitoring

Piling must be supervised by measurements and all appropriate information must be written in driving records (see Annex 1 - 2). The driving record must include

A) Piling site and piles
- piling site, types of piles, sizes and depths, steel types and the types of the pile tips when needed

B) Soil information
- soil type, layer boundaries, level of the ground water, possible level of the perched ground water and pressure level of the artesian ground water
- layers difficult to penetrate
- quality of the soil or rock at the bottom of the pile

C) Boring / digging / drilling
- equipment and tools
- assembly of the working pipe
- boring phases
- the use of bentonit supporting the pile excavation
- water level inside the working pipe
- fabrication and inspection of the base enlargements
- cleaning the bottom of the pile

D) Pile driving
- driving devices
- supports for the pile during driving
- driving phases
- pile extension
- final drives, measurement of the bearing capacity and the failure level of the pile

E) Reinforcement
- inspection of the cleanliness of the pile and cleaning when needed
- type of reinforcement, dimensions, composition and length
- assembling depth of reinforcement and bonds, supporting and centering

F) Concreting
- underwater concreting, centering in dry circumstances
- ready-mixed concrete or concrete mixed in place
- granularity, mix proportion and consistency of the dry aggregate
- casting, consumption of concrete, cast duration, rising rate and the final level
- hoisting the working pipe
- hoisting the casting pipe

G) Injecting

- injecting of the rock contact: number of injecting pipes, working order, composition and properties of the grout, time, speed, pressure

The supervisor must be informed of all exceptional situations and conditions affecting the working process and the results of the work.

The measurement record of the controlling measurements must be written immediately. The records must be stored at the working site until the work is completed.

The piling records must be delivered to the builder's supervisor and/or designer according to agreement.

After piling, an implementation drawing must be made to specify locations of the piles, pile sizes, foundation level and accessories used in the piles.

Measurement plan, measurement results and other piling documents must be stored as agreed in the contract agreement or regulations concerning contract work.

REFERENCES

Association of Finnish Civil Engineers RIL. RIL 121, Pohjarakennusohjeet 1988. Helsinki 1988. ISBN 951-758-178-5. ISSN 0356-9403. p. 92 (In Finnish.)

Association of Finnish Civil Engineers RIL & Finnish Geotechnical Society SGY. RIL 212-1995, SPO 1995, Suurpaalutusohje 1995. Lahti 1995. Suomen Rakennusinsinöörien Liitto RIL r.y. ISBN 951-758-342-7. ISSN 0356-9403. p. 141 (In Finnish.)

Broms, B. & Holeyman, A. The Bawang Pile - a New Type of Pile. *Franki International Technology Report 5/86.* Belgium.

EN 1536:1996. Execution of special geotechnical work: Bored piles. Draft March 1996.

Finnish Geotechnical Society SGY & Rakentajain Kustannus Ltd. Lyöntipaalutusohjeet LPO-87. Vaasa 1991. p. 135 (In Finnish.)

Finnish Geotechnical Society SGY & Rakentajain Kustannus Ltd. Talonrakennuksen pohjatutkimus-ohjeet TPO-83. Vammala 1983. p. 94 (In Finnish.)

Finnish National Road Administration. Pohjarakennusohjeet sillansuunnittelussa. Helsinki 1991. ISBN 951-47-2727-4. TIEL 2172068. p. 69 (In Finnish.)

Finnish National Road Administration. Pohjarakennusohjeet sillansuunnittelussa Luku 10. Perustaminen paaluille, tarkistus. Helsinki 1996. ISBN 951-726-278-7. TIEL 2170010. p. 16 (In Finnish.)

Finnish National Road Administration. Teräsputkipaalut (Steel Pipe Piles). Helsinki 1993. ISBN 951-47-6857-4. TIEL 2173448. 77 s.

Halkola, H. & Törnqvist, J. National Report, Finland. *Proceedings CPT'95. Volume 1.* pp. 63-66.

Hartikainen, J. & Saari, M., & Zadroga, B. Bearing capacity of open-ended and closed-ended steel pipe piles. *Application of Stress-Wave Theory to Piles, F.B.J. Barends (ed.) 1992.* Rotterdam: Balkema. ISBN 90 54100826. pp. 571-576.

Hartikainen, J. Road and Railway Bridge Foundations of Steel Pipe Piles. *XII Nordic Geotechnical Conference 1996.* Reykjavik: Island. pp. 225-232.

Heinonen, J. & Hartikainen, J. Steel pipe pile foundations for large structures in difficult soil conditions. *5th International Conference and Exhibition on Piling and Deep Foundations. 1994.* Brygges: Belgium. p. 5

Ministry of the Environment. National Application Document, for Prestandard SFS-ENV 1997-1:1994 Eurocode 7: Geotechnical design, Part 1: General rules. 1996. 34 s.

Ministry of the Environment. Suomen rakentamismääräyskokoelma 1992; B3 Pohjarakennus (Finnish Code of Building Regulations 1992, B3 Foundation Engineering). (In Finnish.)

VR-Rata Ltd. Teräsputkipaalut rautatiesilloissa. p. 21 (internal report in Finnish.).

General instruction concerning pile design:

Suomen rakentamismääräyskokoelma B3 (Finnish Code of Building Regulations B3)

RIL 121 Pohjarakennusohjeet (Finnish Foundation Engineering Standard)
ISBN 951-758-178-5. ISSN 0356-9403 (In Finnish.)
RIL 212 Suurpaalutusohje 1995 (Design Code for High Capacity Piles)
ISBN 951-758-342-7. ISSN 0356-9403. (In Finnish.)
Lyöntipaalutusohjeet LPO-87 (Design Code for Driven Piles) (In Finnish.)

Design codes of practice published by the Finnish National Road Administration:

Teräsputkipaalut (Steel pipe piles)
ISBN 951-47-6857-4. TIEL 2173448
Pohjarakennusohjeet sillansuunnittelussa
ISBN 951-726-278-7. TIEL 2170010 (In Finnish.)
Pohjarakennusohjeet sillansuunnittelussa, Luku 10 Perustaminen paaluille, tarkistus
ISBN 951-726-278-7. TIEL 2170010 (In Finnish.)

Design codes of practice published by VR-Rata Ltd.:

Teräsputkipaalut rautatiesilloissa (In Finnish.)
Siltapaalujen laadunvarmistus iskuaaltomittauksilla (In Finnish.)

DRIVING RECORD FOR STEEL PIPE PILE **ANNEX 1/1**

COMMON PART

Contactor ______________________ Pile type and piling method ______________________

Site ______________________ Piling equipment ______________________

______________________ Drawing no ______________________

Pile group ______________________ Pile numbers ______________________

Pile data		
Shaft		
• concrete		☐
• composite structure		☐
Geotechnical function		
• end bearing pile		☐
• friction pile		☐
• pile group with tension piles		☐
Pipe		
• external diameter [mm]		
• wall thickness [mm]		
• steel quality		
• manufacturer		
• delivery document		
Pile toe		
• rock shoe		☐
• base plate shoe		☐
• strengthened toe (open pile)		☐
Joint		
Tolerances for pipe		
• external diameter		
• linearity		
• circularity		
• right angle accuracy of head		
• wall thickness		
• length		
• irregularities of head		

Pipe driving	
Piling rig	
Pipe driving	
• free fall hammer	☐
• hydraulic hammer	☐
• diesel hammer	☐
• vibrating hammer	☐
• some other	☐

Reinforcement		
Drawing no		
Longitudinal bars [mm]	ϕ	mm
Transverse bars [mm]	ϕ	mm
Type of spacers		
spacing [mm]		

Concrete	
Class of concrete	C
Consistency	
Prefabricated concrete	☐
Concrete mixed in situ	☐
• cement quality (deliverer)	
• quantity of cement [kg/m³]	
• aggregate (maximum size) [mm]	
Water-cement ratio w	
Concrete admixtures	
• % of cement weight	
Workability time [h]	

Concreting			
Submerged conditions	☐		
Dry conditions	☐		
Method of concreting			
Tremie pipe	☐	ϕ	m
Pumping hose	☐	ϕ	m
Different placing method	☐		
• description			

Comments/Observations

☐ Mark the proper square

DRIVING RECORD FOR STEEL PIPE PILE

ANNEX 1/2

SEPARATE PILE, PAGE 1/2

Pile number ____________________ Compressed pile ☐ Tension pile ☐

Pile data	Designed	As-built	Difference
Location	X	ΔX	mm
	Y	ΔY	mm
Level of head			mm
Level of toe			mm
Length of pile [m]			mm
Inclination of pile [mm/m]			mm/m
Direction of pile	°	°	°
• survey line			

Hammer and cushion		
	Hammer	Cushion
Material		
Mass [kg]		
Section [mm²]		
Length [m]		

Reinforcement		
Longitudinal bars	ϕ	mm
Transverse bars	ϕ	mm
Length of longitudinal bars		
Head level of longitudinal bars		
• before concreting		
• after concreting		
Lower end of longitudinal bars		

Concreting	
Date and time of placing	
Water table in pipe before commencement of placing	
Ambient temperature [°C]	
Concrete temperature [°C]	
Codes of specimens	
Concrete volume [m³]	
• teoretical	
• as-built	
Cleaning of pile bottom/check	
Deliverer	

Monitoring of pile capacity
Stress wave measuring

Date and signature
Foreman for piling
Supervisor of client

DRIVING RECORD FOR STEEL PIPE PILE

SEPARATE PILE, PAGE 2/2

Pile no ______________________

Pile driving and concrete placing (events for driving and concrete placing shall be recorded in proper columns in time order)

Working level ____________ Ground surfice level ____________

Date	Time	Distance from working level to lower head of pipe [m]	Height of fall [m]	Number of blows	Settlement [mm/blow]	Elasticity [mm]	Distance from working level to casting level [m]	Concrete volume [m³]	Comments

DRIVING RECORD FOR FRANKI-PILE

COMMON PART

ANNEX 2/1

Contractor ______________________ Pile type and piling method ______________________

Site ______________________ Pilingg equipement ______________________

______________________ Drawing no ______________________

Pile group ______________________ Pile numbers ______________________

Pile data		
Shaft • concrete		☐
• composite structure		☐
• pre-fabricated element		☐
Geotechnical function		
• end bearing pile		☐
• friction pile		☐
• pile group with tension piles		☐
Pipe or pre-fabricated element		
• external diameter [mm]		
• wall thickness [mm]		
• quality of concrete/steel		
• manufacturer		
• delivery document		
Enlarged base • normal		☐
• onion		☐
Joint		
Tolerances of pipe or element		
• length of external diameter [mm]		
• linearity		
• circularity		
• right angle accuracy of head		
• wall thickness		
• irregularities of head		

Reinforcement		
Drawing no		
Longitudinal bars [mm]	ϕ	mm
Transverse bars [mm]	ϕ	mm
Spacers, type		
spacing [mm]		

Concrete	Shaft	Enlarged base
Class of concrete	C	
Consistency		
Concrete of enlarged base		☐
Prefabricated concrete	☐	
Concrete in situ	☐	
• cement quality (deliverer)		
• quantity of cement [kg/m³]		
• aggregate (maximum size) [mm]		
Water-cement ratio w		
Concrete admixtures		
• % of cement weight		
Workability time [h]		

Pipe driving
Pile rig

Method of shaft concreting		
Pumping hose	☐ ϕ	m
Different placing method	☐	
• description		

Comments /Observations

☐ Mark the proper square

DRIVING RECORD FOR FRANKI-PILE

SEPARATE PILE, PAGE 1/2

Pile number ____________ Compressed pile ☐ Tension pile ☐

Pile data	Designed	As-built	Difference
Location	X	ΔX	mm
	Y	ΔY	mm
Level of additional placing			
Level of head			mm
Level of base			mm
Length of pile [m]			mm
Inclination of pile [mm/m]			mm/m
Direction of pile	°	°	°
• survey line			

Hammer	
Material	
Mass [kg]	
Section [mm²]	
Length [m]	

Reinforcement	
Longitudinal bars	pieces ϕ mm
Transverse bars	k , ϕ mm
Length of longitudinal bars	
Head level of longitudinal bars	
• before concreting	
• after concreting	
Changes	

Concreting	Antura	Varsi
Date and time of placing		
Ambient temperature [°C]		
Concrete temperature [°C]		
Codes of specimens		
Concrete volume [m³]		
• teoretical		
• as-built		
Cleaning of pile bottom/check		
Deliverer		

Concreting of enlarged base Working period

Plug T [cm] and batch A [l]	Rise of casing [cm]	Number of blows	Heigth of fall [m]	Plug at the end of period [cm]	Observations/ acceptance of foreman
T					
A					
A					
A					
A					
A					
A					
A					
A					
A					

Monitoring of pile capacity
Stress wave measuring

Date and signature
Foreman for piling
Supervisor of client

DRIVING RECORD FOR FRANKI-PILE

ANNEX 2/3

SEPARATE PILE, PAGE 2/2

Pile no ______________________

Pile driving and concrete placing (events for driving and concrete placing shall be recorded in proper columns in time order)

Working level ___________ Ground surfice level ___________

Date	Time	Distance from wor-king level to lower head of pipe [m]	Height of fall [m]	Number of blows	Settlement [mm/blow]	Elasticity [mm]	Casting level [m]	Concrete volume [m³]	Comments, heigth of plug when striking last blows etc

Design of Axially Loaded Piles – European Practice, De Cock & Legrand (eds)
© 1997 Balkema, Rotterdam, ISBN 90 5410 873 8

Design of axially loaded piles – French practice

M. Bustamante
Laboratoire Central des Ponts et Chaussées (LCPC), Paris
R. Frank
CERMES (ENPC-LCPC), Cité Descartes, Champs-sur-Marne

ABSTRACT : After reviewing the complex and varied geology of France, the soil investigation commonly performed for pile projects is described. The types of piles and the execution methods used in the country are also given. The design rules for axially loaded piles, recommended by the French code for foundations '*Fascicule 62*-V' (1993), based on Ménard pressuremeter test (PMT) and on cone penetration test (CPT) results are presented in detail (including formulae and charts). The relevant factors of safety are given, both for ultimate and for serviceability states. The t-z method for obtaining the axial load-axial displacement behaviour of a pile, from PMT results, is described. Finally, the national practice for integrity testing and for load testing of piles is given.

1 REGIONAL GEOLOGY

France covers an almost hexagonal territory, approximately 1000 km long in any direction, with a total surface of about 550 000 km^2. The country comprises five main types of geological and geomorphological units (Figure 1, Frank and Magnan 1996) :
- three main chains of hercynian mountains: the Armorican Massif, the Vosges-Ardennes Massif and the Central Massif, which was partially covered by volcanic rocks during the tertiary and early quaternary periods,
- two large tertiary massifs (Pyrenees and Alps), which culminate at 3404m and at 4807m respectively,
- two large sedimentary basins (Paris Basin and Aquitaine Basin) covered with secondary formation,
- two main «sillons» corresponding to the Rhône-Saône valley and to the Rhine valley in Alsace, filled with sediments from the neighbouring mountains,
- numerous river valleys and coastal plains, mostly covered with quaternary sediments, including soft clays, sands and gravels, peat, etc.

The tertiary and quaternary deposits are the most important grounds for civil engineering structures foundations. Arnould (1968) describes the main types of soils that can be found in France:
- eolian sands frequent in the form of dunes on the Mediterranean, Atlantic and Channel coasts. The eolian Landes sand covers thousands of square kilometres of land along the Atlantic coast, north of the Pyrenees, with thicknesses up to 20 meters;
- marine sands found both on recent shore deposits and in tertiary deposits such as the Beauchamp and Fontainebleau horizons in the Paris Basin;
- fluvial sand and gravel deposits frequent in the main river valleys, often encountered in the form of terraces;

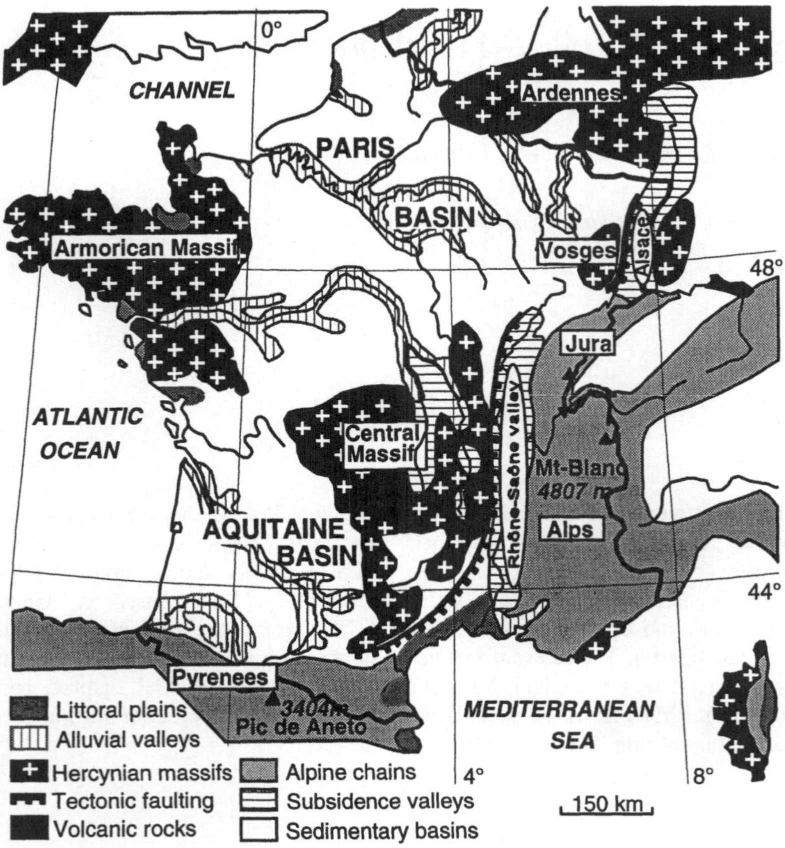

Figure 1. Main geological units of France.

- surface layers of residual (eluvial) sands, existing in all the regions, described as feldspathic rocks on geological maps (mainly in the hercynian mountains);
- silty and loessial soils of various origins in all parts of France: for example brown eluvial silts produced by the decalcification of chalk or limestone found in most parts of the Paris Basin, and loessial deposits found in layers up to 30-40 m in Alsace, up to 20 m near Lyon, up to several metres in the North of France, in the Paris Basin and Normandy;
- clayey overconsolidated stiff soils of marine, lacustrine or fluvial origin common in the sedimentary basins;
- clayey muds and soft organic clays found along the Mediterranean and Atlantic coasts of France and in the valleys of the main rivers. The thickness of these soft deposits is currently of the order of 5 to 10 m, with 20-30 m or even more than 50 m in the Adour valley and on the Côte d'Azur;
- more or less hard marls common in most parts of the country in secondary and tertiary deposits;
- glacial and fluvio-glacial deposits less frequent in France than in the northern Europe, since the inlandsis stopped at some distance of the country. Yet, ice covered the Alps and the Pyrenees and left some typical marks in the landscape and in the soil properties, such as lines of morainic deposits, glacial and fluvio-glacial sands and gravels, heavy overconsolidated by the ice-cover. The whole Rhône-Alpes region is therefore covered with stiff soils, which are suitable for the construction of buildings and structures, but sometimes quite hard to penetrate;
- peat is relatively rare in France. Except for the large Normandy Carentan marsh and some

smaller peat bogs, peat can mostly be found in the estuarine deposits of the main rivers or in same places north of the Seine river.

The above description shows that the bedrocks and surface layers of France are extremely rich and varied. Due to this variety a significant part of the piling experience had to be derived from foundations in complex materials such as weathered rock, hard materials often coupled with a great spatial variability, with the presence of boulders in medium-grained to coarse-grained matrixes, karstic and dissolutions of gypsum or limestones, etc.

At last a special attention should be paid to the groundwater regime in all the regions of France, since it can be as varied as the grounds. Artesian water pressures are frequently encountered all over the country. Close to the seafronts, the nature of the water as well as the range of the tidal variations (especially along the Atlantic coast) require careful investigations to ensure they will not affect the piles during construction works. In sloping and mountainous grounds the groundwater regime, frequently erratic, require hydraulic gradient measurement since significant flow can affect pile integrity.

2 COMMON PRACTICE FOR SOIL INVESTIGATION

The French practice of soil investigation reflects necessarily the richness and the complexity of the soil conditions all over the country. As a basic approach it is accepted that the programme of ground investigation should be flexible and based on various existing techniques, field as well as laboratory (Table 1).

The choice of the investigation techniques depends both on the type of ground and on the type of geotechnical structure and the general approach is similar to the one followed in most countries. As a preliminary, an attempt is always made to use simple and low cost investigation techniques, such as test pits dug by a backhoe for shallow depths or auger borings for deeper investigations. These methods permit to retrieve representative but usually severely disturbed samples.

The investigation of deeper formations and recovery of 'disturbed', 'slightly disturbed' or 'undisturbed' samples are systematically required when deep foundations are planned. In these cases samples are taken from boreholes sunk using powerful and more or less sophisticated drilling machines. Deep boreholes are needed in the soil investigation of any serious piling project. The boreholes can be sunk using :

- rotary-core drilling with air or water flush;
- rotary tricone drilling with air or water flush;
- double- or triple-tube core barrels;
- percussion drills, down-the-hole hammers, etc.

The well known piston sampling, limited to shallow sampling, is used for 'research' class sampling, in association with special problems encountered in the research field.

Since the early eighties, classical boreholes techniques have been supplemented by destructive drilling performed with acquisition systems (Lütz or Enpasol), recording the drilling parameters such as : oil pressure, torque, weight on bit, etc. This technique has become very popular in France. When adequately used, it can provide instantaneously geotechnical information necessary to confirm or amend the project.

In addition to classical boreholes aimed at recovering remoulded or undisturbed samples for laboratory examination and investigations (shear and compressibility, chemical properties of soil and water), the range of used field tests includes:

- pressuremeter tests (with Ménard or self-boring pressuremeters),
- cone penetration (CPT) or, more rarely, piezocone tests,
- standard penetration tests (SPT),
- dilatometer,
- plate bearing tests,
- vane shear tests,
- in situ permeability tests.

Table 1. Usual site investigation techniques in France for deep foundation projects

Type of structures Type of soil	Small & medium structures	Large size structures (bridges, tall buildings, water-fronts, etc)
Organic soils and soft clays (1)	CPT Pressuremeter (PMT)	CPT Pressuremeter (PMT or self-bored) Vane tests Laboratory tests
Clayey soils (firm to hard)	Auger CPT Pressuremeter SPT (sometimes)	CPT Pressuremeter Laboratory tests (idend., shearing)
Sandy soils (loose to very dense)	Auger (identification) CPT Pressuremeter (slotted tube) SPT (sometimes)	CPT Pressuremeter Laboratory tests (gradation)
Gravels (medium to very dense)	Auger (identification) CPT (static or static-dynamic) Pressuremeter (slotted tube) SPT (sometimes)	Pressuremeter (slotted tube) CPT (static or static-dynamic) Laboratory tests (gradation)
Marl, chalk, gypsum	Auger (identification) CPT Pressuremeter tests (slotted tube)	CPT Pressuremeter (high pressure tube, slotted tube) Laboratory tests (ident., UCS) Deep sounding and geophysical testing for cavity detection
Rocks (weathered to fractured)	Auger (for decomposed material) CPT (static-dynamic) Pressuremeter (high pressure probe, slotted tube)	Rotary-core drilling CPT (static-dynamic) Pressuremeter (high pressure probe, slotted tube) Dilatometer (rarely) Destructive sounding and geophysical testing Laboratory tests (ident., UCS)

(1) for upper part of the pile

For the design of deep foundations the two more popular field tests, are by far the CPT and the Ménard pressuremeter test (PMT), carried usually together.

The presence of hard formations however, very often encountered all over the French territory limits the versatility of the CPT. Bustamante and Gianeselli (1996) report that, for a total of 147 sites (located in France and 11 other countries) where pile load tests were to be carried out, CPT tests could not be performed to full depth on 88 sites, i.e 60% of the total number (see Table 2). SPT tests are carried out more rarely, when investigating soils for small to medium structures. The other in situ tests as self-boring pressuremeter, vane tests or dilatometer tests are used occasionally.

Where appropriate and for major projects (dams, very large bridges or industrial plants, tunnels, harbours facilities, etc.) or for peculiar geological conditions, geophysical methods may be used to supplement the field or laboratory tests mentioned above. The main recognised techniques are :

- electrical resistivity,
- seismic refraction or reflection,

Table 2 Field and laboratory tests feasability after Bustamante & Gianeselli (1996)

Test	carried out to full design length (1)	incomplete test (2)	not carried out (3)	irrelevant (4)
PMT pressuremeter (p_l)	112	0	35	0
CPT (q_c)	40	51	19	37
laboratory tests (C, φ)	17	41	47	42

(1) including the full length of pile + additional meters below the pile point
(2) due to premature refusal for CPT; not possible sampling for laboratory tests; soil inadequacy for SPT
(3) feasible but not planned when the investigation campaign was decided
(4) considered from the beginning as inadequate with respect to soil nature or compacity

- magnetic method,

and, more recently,

- radar and tomographic techniques.

3 PILING TECHNOLOGY : COMMON PILE TYPES, EXECUTION METHODS

The term 'piling' is often employed to indicate, generally speaking, all types of deep foundations, i.e. foundation units that provide support for a structure by tip resistance and by shaft resistance in the soil in which they are placed.

The range of deep foundations used in France nowadays is particularly varied (more than 20 basic techniques available) and it should be noted that the design methods try to reflect this fact. Deep foundation can be pre-manufactured or cast-in-place. They can be driven, jacked, jetted, screwed, bored or excavated. All these installation techniques can be supplemented by grouting, a technique to which contractors and designers often have recourse to in France. Piles with a cylindrical section are the most commonly used deep foundations, but H-section steel piles and sheet piles, as well as bored bearing elements (barrettes) of any shape are also frequently used. The constitutive materials are : concrete, grout, mortar, steel, or combinations thereof. The timber piles have only a historical meaning and they are very rarely used, to support negligible loads only.

The classification for the design methods, presently used in France, is :

preformed piles :

- *driven precast pile*
- *driven steel pile (box, H, tube)*
- *prestressed tubular pile*
- *driven covered pile*

rotary bored or excavated piles:

- *bored with no temporary casing pile*
- *bored with a casing*
- *flush bored pile (including barrettes)*
- *continuous flight auger*
- *cast screwed pile*

micropiles:

- *micropile type I*
- *micropile type II*
- *micropile type III*
- *micropile type IV*

dug-hole piles
jacked piles:
- *concrete jacked pile*
- *metal jacked pile*

driven cast-in-place piles:
- *driven by internal drop hammer piles*
- *driven cast-in-place pile with a loose shoe*

4 NATIONAL RELEVANT DOCUMENTS WITH REGARD TO PILE DESIGN

The code *Fascicule* 62-V : '*Règles Techniques de Conception et de Calcul des Fondations des Ouvrages de Génie Civil*' ('Technical Rules for the Design of Foundations of Civil Engineering Structures') was adopted by the French government in March 1993 (MELT 1993). It is fully applicable to the foundations of all public works of civil engineering type; it is not applicable to public buildings, which have to use the same codes as private works (called 'DTU' : *Documents Techniques Unifiés*). DTU 13.2 for deep foundations is, at present, being modified in order to incorporate the design rules of *Fascicule* 62-V.

5 DETAILED DESCRIPTION OF THE NATIONAL DESIGN METHODS

5.1 *General philosophy and experimental background*

French design methods are based on engineering principles of varying degree of sophistication. Generally the design should comply with the following requirements:
i) reasonable safety against bearing capacity failure of the ground,
ii) adequate margin against excessive pile settlements or horizontal movements which would impair the serviceability of the supported structure. Design procedures can be broadly divided into 4 categories:
a) empirical methods called usually the 'static' methods, based on correlations with soil in situ tests results and pile full scale testing results,
b) pile driving formulae and wave equation analysis for driven piles,
c) other 'static' calculation methods based on laboratory tests and on simplifying soil and rock mechanics principles,
d) sophisticated analytical (or numerical) techniques used rarely when peculiar deep foundations problems have to be considered.
The 'static' methods from in situ tests are by far the most popular. The importance given to them comes from the complexity of the French geotechnical context found all over the territory and the versatility offered by the Ménard pressuremeter and CPT tests used in practice always together. Their experimental value or 'representativity' is enhanced by the data provided by numerous full scale pile load tests (more than 400 at present time), often carried out on instrumented shafts. It is all this experimental background which is behind the *Fascicule* 62-V design rules.

5.2 *Definitions*

The design of axially loaded piles in *Fascicule* 62-V is based on the following characteristic loads: for compression piles Q_c : creep load
and Q_u : limit load
for tension piles Q_{tc} : tension creep load
and Q_{tu} : tension limit load.

They are determined by representative static load tests (see below, under §5.3) or, if no test is available, on the basis of Ménard test results (PMT) or cone penetration test results (CPT) (see below, under § 5.4).

It is common practice in France, since the 1960's, to use the creep load Q_c to characterise the axial behaviour of piles. ISSMFE (1985) defines the creep load as "a critical experimental load beyond which the rate of settlement under constant load takes place with a notably increased increment". In case of a static load test performed using the maintained load procedure, it is easily determined by a simple graphic construction (see, for instance : AFNOR 1991).

5.3 *Design on basis of static load test*

For carrying out static load tests, *Fascicule* 62-V refers to the well established French practice, described in the standard of AFNOR (1991). French practice is based on the maintained load procedure, and is consistent with ISSMFE (1985). It points out the need to reach or to be able to draw conclusions on the limit load.

The limit load is said to be reached when the displacement of the head of the pile is equal to B/10, where B is the width or diameter of the pile.

When deriving Q_c , Q_u , Q_{tc} or Q_{tu} from static load test(s), a reduction is applied to the measured value(s). In the case of *Fascicule* 62-V, both for compression and tension :

- if only one single pile load test is performed :

$$Q = Q_m/1.2 \qquad (1)$$

where Q_m is the measured value ;

- if several load tests are performed :

$$Q_k = Q_{min} (Q_{min}/Q_{max})^{\xi'} \qquad (2)$$

where Q_{min} and Q_{max} are the minimum and maximum measured values ξ' is given in Table 3.

5.4 *Design on basis of in situ test results*

The compression and tension limit loads are calculated in the following way :

$$Q_u = Q_{pu} + Q_{su}$$
$$Q_{tu} = Q_{su}$$

where Q_{pu} is the point limit load and Q_{su} is the shaft friction limit load.

The compression and tension creep loads are calculated in the following way :

- non displacement piles (ND) : $Q_c = 0.5\, Q_{pu} + 0.7\, Q_{su}$
  $Q_{tu} = 0.7\, Q_{su}$
- displacement piles (D) : $Q_c = 0.7\, Q_{pu} + 0.7\, Q_{su} = 0.7\, Q_u$
  $Q_{tu} = 0.7\, Q_{su}$

Q_{pu} and Q_{su} are derived from :

$$Q_{pu} = q_u \cdot A_p \qquad (3)$$

$$Q_{su} = \sum_i q_{si} A_{si} \qquad (4)$$

Table 3. *Fascicule* 62-V : Values of factor ξ'.

Number of load tests	2	3	4	5
ξ'	0.55	0.20	0.07	0.00

where A_p is the area of the base, A_{si} the shaft area in layer i, q_u the point failure pressure and q_{si} the limit unit shaft friction.

Fascicule 62-V gives the rules for determining q_u and q_s from the results of Ménard (standard) pressuremeter tests (PMT) and from the results of cone penetration tests (CPT). Note that in the case of open-ended driven steel piles, H piles and sheet piles the specific calculations that must be used are also given according to Bustamante and Gianeselli proposals (1991). For tension piles, the limit unit shaft friction is assumed to be the same as for compression piles.

PMT method

The PMT method is the one originally proposed by Ménard in the 1960's. The tip bearing factors and the limit unit shaft friction values have been readjusted following numerous full scale static load tests on instrumented piles performed by the *Laboratoires des Ponts et Chaussées* LPCs (Bustamante and Gianeselli, 1981).

The tip failure pressure is obtained from the simple relation :

$$q_u = k_p \cdot p_{le}^* \qquad (5)$$

where p_{le}^* is the PMT equivalent net 'limit' pressure, p_l-p_o, around the base of the foundation and k_p is the tip bearing factor, which is a function of the type of soil and the type of pile, and is given by Table 4 (ND = non displacement pile; D = displacement pile).

The limit unit shaft friction q_s is determined using one of the curves of Figure 2, together with Table 5 which indicates that shaft friction depends not only on p_l, the type of soil and the type of pile, but also on the construction conditions of the pile.

CPT method

The correlations between q_u and q_s and the CPT cone resistance q_c, proposed by *Fascicule*

Table 4. *Fascicule* 62-V : Tip bearing factors k_p (PMT) and k_c (CPT)

Soil type			p_l (MPa)	q_c (MPa)	k_p (ND)	k_p (D)	k_c (ND)	k_c (D)
Clay	A	soft	< 0.7	< 3	1.1	1.4		
Silt	B	stiff	1.2 - 2	3 - 6	1.2	1.5	0.40	0.55
	C	hard(clay)	> 2.5	> 6	1.3	1.6		
Sand	A	loose	< 0.5	< 5	1	4.2		
Gravel	B	medium	1- 2	8 - 15	1.1	3.7	0.15	0.50
	C	dense	> 2.5	> 20	1.2	3.2		
	A	soft	< 0.7	< 5	1.1	1.6	0.20	0.30
Chalk	B	weathered	1 - 2.5	> 5	1.4	2.2	0.30	0.45
	C	dense	> 3	-	1.8	2.6	-	-
Marl	A	soft	1.5 - 4	-			-	-
Calcareous marl	B	dense	> 4.5	-	1.8	2.6	-	-
Rock	A	weathered (1)	2.5 - 4	-	1.1 to 1.8	1.8 to 3.2	-	-
	B	fragmented	> 4.5	-	-	-	-	-

(1) use the value of the most similar soil.

Table 5. *Fascicule* 62-V : Choice of limit unit shaft friction curve q_s (PMT method)

Soils	Clay & Silt			Sand & Gravel			Chalk			Marl		Rock
Type of pile	A	B	C	A	B	C	A	B	C	A	B	
Dry bored	Q_1	Q_1 Q_2(1)	Q_2 Q_3(1)	-			Q_1	Q_3	Q_4 Q_5(1)	Q_3	Q_4 Q_5(1)	Q_6
Bored with mud	Q_1	Q_1 Q_2(1)		Q_1	Q_2 Q_1(2)	Q_3 Q_2(2)	Q_1	Q_3	Q_4 Q_5(1)	Q_3	Q_4 Q_5(1)	Q_6
Bored with temporary casing	Q_1	Q_1 Q_2(3)		Ql	Q_2 Q_1(2)	Q_3 Q_2(2)	Q_1	Q_2	Q_3 Q_4(3)	Q_3	Q_4	-
Bored with perma-nent casing	Q_1			Q_1		Q_2	(4)			Q_2	Q_3	-
Piers (5)	Q_1	Q_2	Q_3	-			Q_1	Q_2	Q_3	Q_4	Q_5	Q_6
Steel driven closed-ended	Q_1	Q_2		Q_2		Q_3	(4)			Q_3	Q_4	Q_4
Driven concrete	Q_1	Q_2		Q_3			(4)			Q_3	Q_4	Q_4
Driven moulded	Q_1	Q_2		Q_2		Q_3	Q_1	Q_2	Q_3	Q_3	Q_4	-
Driven coated	Q_1	Q_2		Q_3		Q_4	(4)			Q_3	Q_4	-
Low pressure injected	Q_1	Q_2		Q_3			Q_2	Q_3	Q_4	Q_5		-
High pressure injected (6)	-	Q_4	Q_5	Q_5		Q_6	-	Q_5	Q_6	Q_6		Q_7(7)

(1) trimmed and grooved at the end of drilling
(2) for long piles (longer than 30 m)
(3) dry excavation, no rotation of casing
(4) in chalk, q_s can be very low for some types of piles ; a specific study is needed
(5) without permanent casing (rough pile walls)
(6) low rate injection and repeated grouting at selected depths
(7) (6) plus preliminary treatment of fissured or fractured masses and filling of cavities.

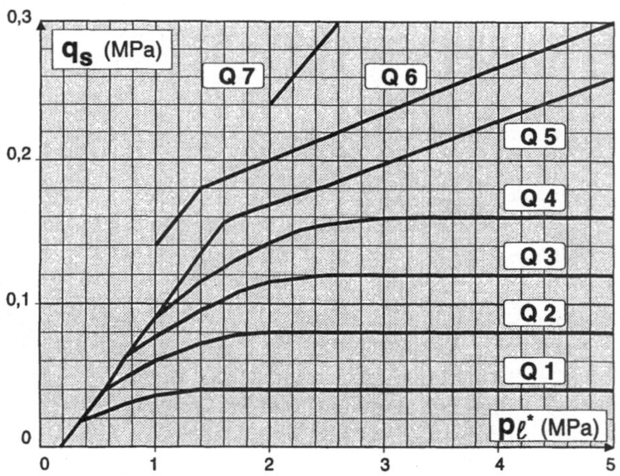

Figure 2. *Fascicule* 62-V : Limit unit shaft friction curves q_s.

Table 6. *Fascicule* 62-V : Limit unit shaft friction from CPT

Type of pile	Soils	Clay & Silt					Sand & Gravel			Chalk	
		A	B		C		A	B	C	A	B
Drilled	β	-	-	(1) 75	-	-	200	200	200	125	80
	q_{smax} (kPa)	15	40	80	40	80	-	-	120	40	120
Drilled removed casing	β	-	100	(2) 100	-	(2) 100	250	250	300	125	100
	q_{smax} (kPa)	15	40	60	40	80	-	40	120	40	80
Steel driven closed-ended	β	-	120		150		300	300	300	(3)	
	q_{smax} (kPa)	15	40		80		-	-	120		
Driven concrete	β	-	75		-		150	150	150	(3)	
	q_{smax} (kPa)	15	80		80		-	-	120		

(1) trimmed and grooved at the end of drilling
(2) dry excavation, no rotation of casing
(3) in chalk, q_s can be very low for some types of piles ; a specific study is needed.

62-V, also come from full scale pile tests performed by the LPCs (Bustamante et Gianeselli 1981 and 1982). The relation for the point failure pressure takes the form :

$$q_u = k_c \cdot q_{ce} \tag{6}$$

where q_{ce} is the CPT equivalent cone resistance around the base of the foundation ; k_c is the point bearing factor, which is a function of the type of soil and the type of pile, and is given in Table 4 (ND = non displacement pile ; D = displacement pile).

The limit unit shaft friction, q_s, is determined from the cone resistance, q_c, using :

$$q_s = \text{minimum value of } \{q_c/\beta \; ; q_{smax}\} \tag{7}$$

where β and q_{smax}, given in Table 6, depend on the type of soil and the type of pile.

5.5 *Design on basis of driving formula and wave equation analysis*

In France, many engineers continue to use driving formulae because they are easy to apply on site and they are still expected to give sufficiently correct estimates of the pile bearing capacity, independently of the fact that many analyses show that their reliability is poor (see, for instance, (Bustamante et al. 1985). They are commonly used for harbour, water-front, canal and off-shore constructions (for which driven steel profiles, i.e. H, tubular and sheet piles are used, because of the specific conditions of work). A few number of formulae are used, the most popular being the Dutch, the Crandall, the modified Engineering News Record and, more rarely, the Kummels formula. The common opinion is that there is not much to choose between them. For public works, when an important project comprises a large number of driven piles,

Table 7. *Fascicule* 62-V : K.tanδ values for negative friction assessment

Soils	Type of pile	Bored with casing	Bored	Driven
Peat	organic soil	0.10	0.15	0.20
Clay & Silt	soft	0.10	0.15	0.20
	stiff to hard	0.15	0.20	0.30
Sand & Gravel	very loose	0.35		
	loose	0.45		
	others	1.00		

the design must be based on the 'static' approach (from field test results, see under 5.1). Within a particular context of pile and soil types, the driving formulae are considered to yield an acceptable estimate of the bearing capacity, especially when they have been calibrated against results of static load test(s).

Dynamic load tests are also used in France, but no method is recommended in French national codes (neither 'Fascicule 62-V' nor 'DTU-13-2'). The methods based on wave propagation theory used are : Case, Capwap, TNO and Simbat. Whereas Case, Capwap and TNO analyses are mainly used for displacement piles, Simbat has been developed for bored piles.

It is recognised that these methods enable to obtain valuable information regarding :
- the energy delivered by the pile driving hammer,
- the maximum driving stresses (both tension and compression).

On the contrary, obtaining a correct evaluation of the static load-settlement response of the pile, or the location and extent of structural damage, is still being debated in France.

5.6 *Negative friction*

The recommendation of *Fascicule* 62-V, for negative friction (downdrag), follows the original work of Combarieu (1974). The method allows the determination of the maximum long term negative friction load, taking into account the hanging effect around the pile (which reduces the effective vertical stress σ'_v), as well as the existence of a neutral point (at which friction becomes positive). Precise rules for the assessment of negative friction on pile groups are also given. The method uses, in particular, the well known K.tanδ parameter, such that :

$$f'_n = K.\tan\delta.\sigma'_v \qquad (8)$$

where f'_n is the unit negative friction. The values of K.tanδ, obtained from full scale observations, are given in Table 7.

5.7 *Factors of safety at ultimate and serviceability limit states*

For ultimate limit states, according to *Fascicule* 62-V, the design axial load Q_d must be in between a maximum value (in compression) and a minimum value (in tension) :
- fundamental combinations : $-Q_{tu}/1.4 \leq Q_d \leq Q_u/1.4$ (9)
- accidental combinations : $-Q_{tu}/1.3^* \leq Q_d \leq Q_u/1.2$ (10)

(* for micropiles the minimum is : $-Q_{tu}/1.2$).

Note that *Fascicule* 62-V uses the same factor of safety γ_t (total material resistance factor) on the point and on shaft friction limit loads.

Fascicule 62-V uses the creep load Q_c to characterise the axial behaviour of piles under serviceability limit states. In a similar manner as for ultimate limit states, the design load Q_d must fulfil the following conditions (the frequent combinations of loads are not checked here) :

- rare combinations : $- Q_{tc} / 1.4^{*} \leq Q_d \leq Q_c / 1.1$ (11)
- quasi-permanent combinations : $0^{**} \leq Q_d \leq Q_c / 1.4$ (12)

(* for micropiles the minimum is $- Q_{tc}/1.1$; ** for micropiles the minimum is $- Q_{tc}/1.4$).

The bearing capacity of a group of piles must be examined by the two traditional approaches : by summing the individual bearing resistances and by considering a global block failure. When summing the individual resistances. *Fascicule* 62-V refers to the Converse Labarre formula or to other currently accepted, but crude, estimates.

Remark 1 : for fundamental ultimate limit states (persistent and transient situations), the partial load factors of *Fascicule* 62-V are very similar to case B of Eurocode 7 (ENV version of part 1, 1994). In particular : 1.35 on unfavourable permanent loads ; 1.0 on favourable permanent loads and 1.5 on the basic variable action (except for well known service loads, or loads having a specific character). The accidental combination of loads and the three combinations of loads for serviceability limit states are also very consistent with Eurocode 7 (for more details, see Frank 1994). Note that for overall stability, *Fascicule* 62-V uses a specific combination of loads, resembling Case C of Eurocode 7, recommended for soil nailing in France (Clouterre 1991).

Remark 2 : In *Fascicule* 62-V, negative friction is not cumulated with short duration variable axial loads : in all combinations only the larger design load of the two (Q_{nd} and Q_{Qd}) is added to the permanent and quasi-permanent axial loads Q_{Gd} when deriving the total design axial load :

$$Q_d = \max (Q_{nd} ; Q_{Qd}) + Q_{Gd} \tag{13}$$

Remark 3 : For the pile material, *Fascicule* 62-V refers to the relevant French standards (BAEL, limit state standard for reinforced concrete ; BPEL limit state standard for prestressed concrete; metallic construction standard). Some specific rules or adaptations for foundation materials are given in *Fascicule* 62-V.

5.8 *Prediction of load-displacement behaviour*

Fascicule 62-V also gives a method of determination of the load-displacement curve of a single pile under axial loading, based on the concept of t-z curves (curves linking the mobilised shaft friction at a given depth with the corresponding axial displacement). It suggests, in case a settlement estimate must be made, that the t-z curves and q-z_b curve (point pressure - point displacement curve) proposed by Frank and Zhao (1982) be used, as shown in Figure 3, with k_t

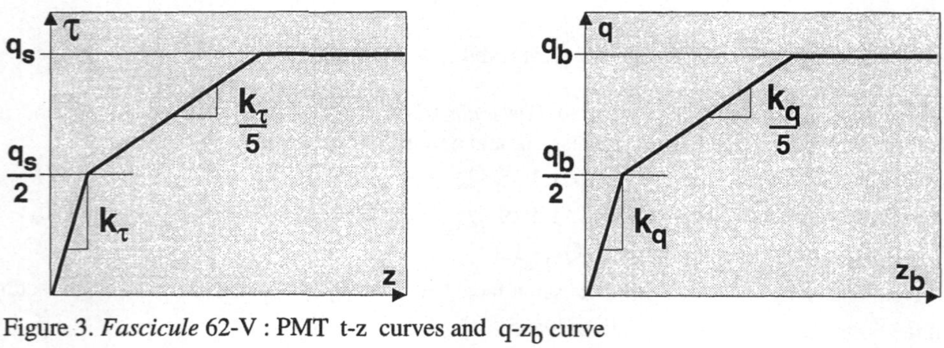

Figure 3. *Fascicule* 62-V : PMT t-z curves and q-z_b curve

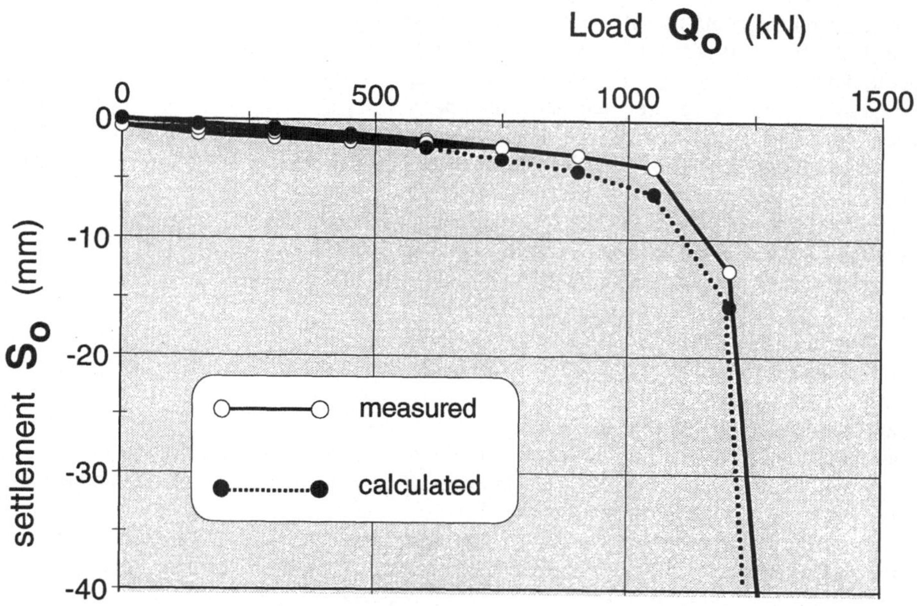

Figure 4: Comparison of measured and calculated load-settlement relationship for the Koekelare pile (Bustamante & Gianeselli 1993).

and k_q given as functions of the PMT pressuremeter modulus, E_M, and the diameter B of the pile: for fine grained soils

$k_t = 2.0\ E_M/B$ and $k_q = 11.0\ E_M/B$ (14)

for granular soils

$k_t = 0.8\ E_M/B$ and $k_q = 4.8\ E_M/B$ (15)

This method is expected to yield satisfactory results for maximum loads of 0.7 Q_c, i.e. maximum loads for quasi-permanent combinations. For larger loads, rate effects must be taken into account.

Figure 4 shows an application of this method for a cased skrew pile ϕ350/650 constructed in an Ypresian clay. The calculated load-settlement curve is compared to the experimental measured curve obtained from the full scale load test results

6 QUALITY CONTROL AND MONITORING

It is commonly accepted in France, both by public authorities and by the private sector that once the piles are completed, the quality controls can concern :

1) integrity tests (destructive or non-destructive),
2) performance under load (static or dynamic).

These two basic tests can be carried out separately or in conjunction, depending on the type of piles, the ground conditions, the importance and the size of the project, the anticipated construction defects and workmanship of the piling contractor.

6.1 *Integrity testing*

The most direct test of pile integrity is coring over the whole pile length or, in some cases, only over the bottom part. It is a costly method, but it is nearly systematically prescribed for bored piles of large diameter and heavily loaded. Coring at the tip of these piles, and more precisely over the last fifty centimetres of concrete and below the tip, over at least the same length, permits to determine the quality of the concrete (homogeneity or segregation and discontinuity) and the quality of the soil/concrete contact at the tip.

In special cases, the coring can be supplemented by a television camera observation. It permits in some conditions to view the sides and the bottom of a core hole on a television screen.

Depending on the importance of the defect detected in the shaft by non-destructive methods, coring of the pile shaft is sometimes imposed.

The most commonly used methods of non-destructive integrity testing in France are the following:

a) sonic-logging (or referred to as sonic coring),
b) vibration tests (referred to as impedance or transient dynamic response).
c) dynamic pile tests (PDA),
d) parallel sonic method (MSP).

Among these methods the first three ones are the most frequently used, the last one (MSP) being recommended in special cases (when the control of the actual length of pile embedment has to be checked for contentious reasons usually).

However, for major projects, the French administration (highway and bridges, railways, harbours, etc.) prescribes almost exclusively sonic coring : based on acoustic principles, one measures the propagation time of sonic transmission between two piezo-electric probes placed in access tubes (plastic or more usually steel) [cast in pile or diaphragm wall]. The impedance and dynamic methods are rather used by the private sector (housing, small and medium-sized industrial plants, etc.)

Some other methods, more popular in the seventies, can still be encountered for some particular problems and to supplement. These are :

a) echo method (usually limited to short piles, not exceeding 20 m),
b) gamma ray method (for detecting defects in bored pile and diaphragm walls and controlling proposed remedial works).

6.2 *Pile load tests*

For major or complex projects, it is commonly accepted that, above integrity tests, full scale load tests have to be carried out.

Pile load tests can have two distinct functions :

- to check whether the ground will support the loads transmitted by a pile of imposed size, length and type ;
- to check whether workmanship is satisfactory, the pile being randomly chosen.

When the pile load test is performed for checking the bearing capacity of the soil, the trial pile is usually not incorporated into the permanent foundation.The assessment of workmanship is always done on a pile incorporated into the finished work.

There are two main methods for carrying out pile tests : the Maintained Load (ML) and the Constant Rate of Penetration (CRP). Only the first one, ML, is recognised in France by the profession (public or private sectors). The ML method, fully described in the French standard (AFNOR 1991), is applied both when checking the bearing capacity and when estimating the work quality.

Since the early seventies, various governmental authorities have taken the opportunity of full load tests to instrument a large number of piles. The LPCs have developed their own extensometric system ('extensomètre amovible') which has become the standard equipment for axial static load tests (Bustamante and Jézéquel 1974). Over the last 25 years, a total of 230 deep foundations were equipped with this original device. The database set up with the measured values of unit shaft friction and tip resistance values has allowed to update (or even establish for some new types of piles), the design charts presently used in France (see PMT and CPT methods under 5).

Since the seventies, a lot has also been done for improving the data acquisition and monitoring systems for piles, for processing the results, etc.

7 REFERENCES

AFNOR 1991. *Essai statique de pieu isolé sous compression axiale. Norme Française NF P94-150*, . AFNOR : Paris. (in French).

Arnould M. 1968. Origine, formation et distribution des sols en France et en Europe occidentale. Chapter 1 in *Les fondations*, by G.A. Leonards (Ed.) et al., translated into French by a group of engineers of the Laboratoires des Ponts et Chaussées. Paris : Dunod.

Bustamante, M. & J.F. Jézéquel 1975. Mesure des élongations dans les pieux et tirants à l'aide d'extensomètres amovibles. *Travaux 489:* 48-52, *(in French).*

Bustamante, M. & L. Gianeselli 1981. Prévision de la capacité portante des pieux isolés sous charge verticale. Règles pressiométriques et pénétrométriques. *Bull. Liaison Labo. P. et Ch.* 113 : 83-108 (in French).

Bustamante, M. & L. Gianeselli 1982. Pile bearing capacity prediction by means of static penetrometer CPT. *2nd Europ. Symp. Penetration testing (ESOPT II), Amsterdam, Vol. 2* : 493-500. Rotterdam : Balkema.

Bustamante M., R. Frank, L. Gianeselli & A. Klos 1985. Vérification expérimentale des formules de battage. *Proc.11th Int. Conf. Soil Mech. Fdn Engng,* San Francisco (in French).

Bustamante, M. & L. Gianeselli 1991. Predicting the bearing capacity of sheet piles under vertical load. *4th Intern. Conference on Piling & Deep Foundations, Stresa (Italy), 7-12 April,TESPA, pp:6.*

Bustamante, M. & L. Gianeselli 1993. Design of auger displacement piles from in situ tests. *2nd Intern. Geotechnical Seminar: Deep Foundations on Bored and Auger Piles,Ghent, Belgium, 1-4 June: 21-34.* Rotterdam: Balkema.

Bustamante, M. & L. Gianeselli 1996. Private communication.

Clouterre 1991. *Recommendations Clouterre 1991. Soil Nailing Recommendations - 1991 For Designing, Calculating, Constructing and Inspecting Earth Support Systems Using Soil Nailing*. English Translation, July 1993, Presses de l'ENPC & U.S. Department of Transportation - Federal Highway Administration, Publication No. FHWA-SA-93-026.

Combarieu, 0. 1974. Effet d'accrochage et méthode d'évaluation du frottement négatif (Hanging effect and prediction method of negative friction. *Bull. Liaison Labo. P. et Ch.* 71 : 93-107 (in French).

Frank, R. & S.R. Zhao 1982. Estimation par les paramètres pressiométriques de l'enfoncement sous charge axiale de pieux forés dans des sols fins. *Bull. Liaison Labo. P. et Ch.* 119 : 17-24 (in French).

Frank, R. 1994. The new Eurocode and the new French code for the design of deep foundations. *Proc. FHWA Int. Conf. Design and Construction of Deep Foundations, Orlando, Florida, 6-8 December 1994, vol. 1* : 279-304. FHWA : Washington, D.C.

Frank R. & J.P. Magnan 1996. Cone penetration testing in France : National Report. *Proceedings International Symposium on Cone Penetration Testing (CPT'95), Linköping, 4-5 October 1995, vol. 3* : 147-156. Linköping : Swedish Geotechnical Society (Report 3:95).

ISSMFE Subcommittee on Field and Laboratory Testing 1985. Axial pile loading test - Part 1 : Static loading. *Geotechnical Testing J.* 8(2) : 79-90.

MELT 1993. *Règles techniques de conception et de calcul des fondations des ouvrages de génie civil, CCTG, Fascicule N°62 - Titre V.* Ministère de l'Equipement du Logement et des Transports : Paris.

Design of Axially Loaded Piles – European Practice, De Cock & Legrand (eds)
© 1997 Balkema, Rotterdam, ISBN 90 5410 873 8

Design of axially loaded piles and pile groups – German practice

R. Katzenbach & Chr. Moormann
Institute of Geotechnics, Darmstadt University of Technology

ABSTRACT: In this paper a survey report on the design methods for axially loaded single piles and pile groups in Germany is presented, giving an overview of present-day design methods applied in practice and of the factors affecting the bearing capacity of piles. Besides an detailed report concerning the actual German design methods for the different types of piles the application of these methods within the framework of Eurocode 7 by the corresponding German National Application Document is described. The design methods are exemplary illustrated by a static pile load test recently carried out for a bored pile in Berlin. Furtherone a few results of actual research work including numerical studies for the bearing behaviour of pile groups and of Combined Piled-Raft Foundations are presented demonstrating the influence of soil-structure interaction for the bearing behaviour of axially loaded pile groups.

1 INTRODUCTION

Today pile foundations have become the most important form of deep foundations. Especially in Germany during the last years the number of buildings founded on pile foundations and Combined Piled-Raft Foundations has increased significantly. Accompanied by the increasing shortage and costs of foundation sites in urban areas the present day tendency is to build high rise buildings with heavy loads and in the same time to construct large volume of underground constructions. The modern buildings involve the problem to transfer more and more growing up loads as well as more concentrated loads into the ground with limited settlements. Besides in urban area the adjacent high quality urban buildings and traffic lines demand for the foundation of new buildings an efficacious reduction of the deformations caused by the new constructed building. This development combined with economical reasons have initiated the realization of pile foundations although in cases in which in earlier times shallow foundations would have been chosen.

The technical progess of large bored piles and the continuous improvements of construction procedure and of the used piling equipment today have created the posibility to drill piles down into never before reached depth. An actual impressing example is the foundation of the highest office building in Europe, the 299 m high Commerzbank tower in Frankfurt am Main (Katzenbach et al. 1994; Wend & Seitz 1995) on 111 bored piles with lengths up to 45 m which transmit the load of the building through the relatively weak Frankfurt Clay into the rocky Frankfurt Limestone (Katzenbach et al. 1997). Quite often pile foundations are the only means of the civil engineer to realise a building project in an economically and technically optimized way with reduced settlements.

While the methods which are prescribed in the German codes and recommendations give a framework for the design of axially loaded piles the reliable design of pile groups or of Combined Piled-Raft Foundations demands more sofisticated design methods based on analytical calculations and numerical studies. In this context during the last decades until today a remarkable research work has been done in Germany.

2 PILE PRACTICE IN GERMANY

The geotechnical situation in Germany is characterized by very different conditions. In the nothern part of Germany usually clay and sand can be found, while in the middle and in the south cohesive and cohesionless soils (sands and gravels, marls) and rock with varying strength (chalck, Bundsandstein, Schiefer etc.) alternate. Concerning this the commonly used soil investigation methods vary in a quite wide range depending on the local geology.

As a result of these conditions the types of piles used in Germany are manifold (Schnell 1996). Displacement piles constructed by timber, precast or prestressed concrete or steel and cast-in-place types are used as well as replacement piles bored by continuous flight augers or rotary augers or by buckets with casing or with suspension support. Injection piles with relative small diameter and micro piles are used for example to anchor walls and to improve historical foundations. In Germany about 30 to 35 % of constructed piles may be displacement piles while 65 to 70 % may be replacement piles. The total length of piles constructed in Germany during one year can be assessed at circa 10.000 kilometres. Nevertheless reliable values are not available. For the construction of pile foundations in urban area mechanical constructed bored piles with large diameter become more and more common. In the last years the bearing capacity of piles could have been succesfully improved by shaft and base grouting (Nendza & Placzek 1988).

Considering the increasing requirements for pile foundations the control of the quality of constructed piles have become more relevant in the last years. So ultra sonic measurments or dynamic integrity tests are now quite often used methods to check the done pile work (Seitz 1992); here additional activities seem to be necassary.

3 NATIONAL CODES FOR THE DESIGN OF AXIALLY LOADED PILES IN GERMANY

3.1 *The German codes for the design of axially loaded piles*

For the determination of the bearing capacity of most commonly used types of piles the German code DIN 1054 (1976) gives fundamental regulations. DIN is the short form for „Deutsche Industrie-Norm“ (German Industry-standardisation, executed by German Institute for standardisation). The code DIN 1054 is the basic standard for geotechnical tasks in Germany. So while this code is the framework for pile design in Germany, the codes DIN 4014 (bored piles), DIN 4026 (driven piles) and DIN 4128 (injection piles with small diameter) include more detailed recommendations for the design and the construction procedure of the particular piles. Concerning to the recently born Eurocodes DIN 1054 will be future replaced by DIN V 1054-100 (1996) which is based upon the partial safety concept.

In Germany the actual national codes still base on the global safety concept. The determination of the safety factors is part of DIN 1054. Table 1 gives a summary of the actual global safety factors for designing piles in Germany concerning the ultimate limit state.

To demonstrate that the pile foundation will support the design load with adequate safety against bearing resistance failure, it has to be proved that the characteristic value of the ultimate bearing resistance R_{1k} divided by η is equal or greater than the characteristic value of the axial load F_{1k} determined on the basis of DIN 1055 (1963), "Design loads for buildings". Depending on the design situation three load cases have to be distinguished:

- load case 1: permanent actions and regular variable actions,
- load case 2: unregular variable actions and actions during construction,
- load case 3: unusually variable actions and accidential actions.

As documentated in table 1 the numerical value of the global safety factor η decreases with an increasing number of carried out static load tests.

DIN 4014 (1990) is the basic German code for bored piles (replacement piles). The code defines the fundamental rules for the construction procedure and for the design methods to determine the bearing behaviour of single bored piles. For pile groups DIN 1054 gives some recommendations. The code DIN 4014 is valid for single bored piles with a diameter from 0.3 m to 3.0 m and also for barrettes (see also DIN 4126). Additional DIN 4014 concerns to

Table 1. Global safety factors η for the design of piles defined by the German standards (DIN 1054).

type of pile	number of carried out static load tests under the same conditions	global safety factor η for load case 1	2	3
		shall be equal or greater than		
compression piles	1	2	1,75	1,5
	≥ 2	1,75	1,5	1,3
tension piles with an inclination ≤ 2 : 1 *)	1	2	2	1,75
	≥ 2	2	1,75	1,3
tension piles with an inclination of 1 : 1 *)	≥ 2	1,75	1,75	1,75
piles with alternating stresses (tension and compression)	≥ 2	2	2	1,75

*) For tension piles with an inclination between 2 : 1 to 1 : 1 the global safty factor has to be interpolated linearily depending on the angle of inclination.

bored diaphragms too. The cast-in-place piles regulated by DIN 4014 can be installed by using rotary augers, continuous flight augers or conventional grabs. For bored piles conrete with a strength of at least B25 (DIN 1045) has to be used in Germany.

The German code DIN 4026 (1975) deals with the construction procedure and the design of driven piles (displacement piles) constructed by wood, precast concrete, prestressed concrete or steel. The displacement piles regulated by DIN 4026 comprise precast piles such as solid-section piles or hollow-section piles with an open or closed end, which are driven or jacked into the ground and thus displace the soil. The tip of the piles can be improved by wings. Driven cast-in-place piles like the Franki pile and other special constructed driven piles are also dealed by DIN 4026.

For piles with a relative small cross-section area the German code DIN 4128 (1983) gives some special recommendations concerning design, construction procedure and permissible loads. The injection piles dealed with in DIN 4128 should have a diameter less than 30 cm. DIN 4128 distinguishes two types of injection piles both not prestressed, the in-situ concrete piles with a diameter greater than 15 cm and the composite piles with a diameter greater than 10 cm. In both cases the load transfer to the soil is realized by injection of concrete or cement suspension.

3.2 *Characteristics of the German design approaches for single pile behaviour*

In Germany the design of piles is usually based on one of the following approaches. The first possibility is to carry out one or more static load tests. The proceeding for carrying out a static load test is described by the German standard DIN 1054. Static load tests are considered to be the safest and most exact, but also the most expensive method to determine the bearing capacity and the load-settlement behaviour of a single pile. So this method is clearly recommended by the standards, especially by the standards DIN 1054 and DIN 4014. DIN 4128 demands static load tests for two injection piles, but at least for 3 % of the total amount of piles on a site to determine the bearing capacity for a pile foundation. On this way the German codes take into consideration that the soil surrounding a pile is influenced and disturbed by the construction procedure and that in this sense an analytical determination of the bearing behaviour is confronted by special problems and requirements.

In cases there are no results of static load tests available, under certain circumstances the load-settlement curve of a single pile can be calculated with the values of experience given by the German standards. This values of experience are recommended for certain types of piles by the German standard specifications. The most important condition for the use of these values given by DIN to determine the resistance of the base and of the shaft or for the design value of the pile resistance are "simple" ground situations. Such "simple" ground situations could be assumed, if the strength of soil can be classificated sufficiently by the cone penetration resistance q_{ck} for cohesionless soils or by the undrained shear strength c_{uk} for cohesive soils.

By the German standards dynamic load tests are usually not permitted to determine the bearing capacity of single piles. The standards recommend dynamic load tests only in cases, in which the validity of this method has been demonstrated by static load tests in comparable situations, that means same subsoil conditions and same type and dimension of pile. So dynamic load tests or special testing methods like the statnamic test could be used succesfully only for certain exceptional cases. An overview about the German practice of dynamic pile testing is given by Balthaus (Balthaus et al. 1985). Stress wave measurements have proved good mainly to check the quality and integrity of constructed piles (Klingmüller 1993).

Furthermore the German codes don´t offer any recommendations to determine the ultimate bearing resistance from ground test results by special calculation rules. Although there are such calculation rules established and used in practice to correlate between the results of field or laboratory ground tests and the bearing behaviour of piles they are not specified by DIN.

Pile driving formulae could only be used to assess the ultimate bearing capacity of individual compression piles in a foundation, if the piles are displacement piles in cohesionless soils and only, if the validity of the chosen driving formulae has been demonstrated for the local site situation and for the piles by previous experimental evidence of good performance or static load tests on the same type of pile, of similar cross-section and similar length and in the similar ground conditions. Nevertheless especially for prefabricated driven piles constructed in northern Germany driving formulae have been successfully used to control the bearing capacity since decades.

In the following the both most commonly used methods to design axially loaded piles will be presented more detailed: the determination by carrying out static pile load tests and the calculation with the values of experience offered by the German standards.

4 STATIC AXIAL PILE LOAD TESTING

As already mentioned the experimental determination of the characteristic load-displacement line of a single pile by carrying out a static load test is recommended by the German standards as the most reliable method to design axially loaded piles for the given geometrical and geotechnical situation. Especially static load tests have to be carried out for those cases, in which:

- the piles will be loaded with more weight as permitted by calculating the bearing capacity with the values of experience given by DIN 4014 or DIN 4126,
- the sufficient bearing soil in situ is not thick enough,
- doubts increase by constructing the piles under the given circumstances.

Static load testing of piles is dealt with by DIN 1054, DIN 4014 and DIN 4020 and the european codes ENV 1997-1 (Eurocode 7) and EN 1536. DIN 4020 (1990) contains the general principles, while DIN 1054, EC7 and DIN EN 1536 deal in more detail with test requirements and implementation. The Recommendations for Static Axial Pile Load Testing worked out by the German Society for Geotechniques (DGGT) · Working Group 2.1 will be published in autumn 1997 and will give explanations and additional recommendations considering also other national and international standards and recommendations. The object of these recommendations is the documentation of the actual technical level of static axial pile load testing taking into account to the current developments of German and European standardisation. The recommendations shall later be supplemented with rules for static horizontal and for dynamic pile testing.

In Germany static load tests of piles have to be carried out as maintained load tests. DIN 1054 describes the regulations, how to realise a static load test. By a maintained load test the load has to be increased step by step while watching the growing up settlements of the pile head. The load should be put up to the next step only, if the settlement stopped increasing.

As a result of the static load test the allowable working load of a pile R_{work} as a load that may be safely applied to a pile after reducing its characteristic ultimate bearing resistance is defined within the frame of the global safety concept as:

$$R_{work} \leq R_{1k} / \eta \quad (1)$$

By definition of DIN 1054 the ultimate bearing capacity R_{1k} of a pile is characterized by the point in the characteristic load-displacement line for which the slowly growing line changes into a fast growing up load-displacement curve (Figure 1).

Often the failure of a pile can not clearly be determined from the load-settlement curve. For this purpose a load corresponding to a specific settlement s_1 is defined. The respecitve settlement is given by DIN 1054 as:

s_1	$= 2.0$ cm	limit settlement for bored piles,
$s_{1,pl}$	$= 0.025 \cdot D_F$	limit plastic settlement for driven piles.

If the ultimate bearing capacity R_{1k} can't been reached during the static load test, the maximum tested load R_{max} has to be considered as R_{1k}. For the determination of $R_{work} \leq R_{1k} / \eta$ the global saftey factors η is given by DIN 1054 (Table 1).

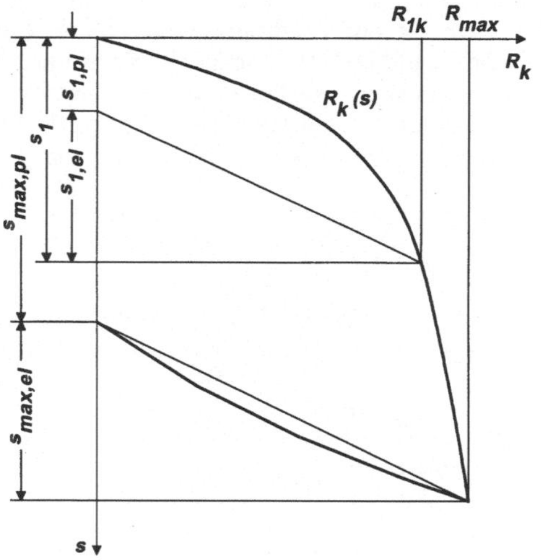

Figure 1. Definition of the ultimate bearing capacity R_{1k} of a single axially loaded pile by DIN 1054.

5 DETERMINATION OF THE VERTICAL LOAD BEARING CAPACITY BY CALCULATION WITH VALUES OF EXPERIENCE GIVEN BY GERMAN CODES

5.1 *General Remarks*

In cases there are no results of static load tests or experiences from tests under comparable conditions available the German standard specifications offer values based on longterm experiences for the determination of the bearing capacity or for the construction of load-settlement curves for single piles. Whether with this values of experience only the bearing capacity or also the load-settlement curve can be determined depends on the type of pile. In this context values of experience are values for the tip resistance $q_{bk}(s)$ and the ultimate skin friction q_{s1k} of a pile or also for the total pile resistance R_k. This values are the result of a wealth of static load tests and in this sense they represent the German experiences of designing piles in the last decades.

The values of experience are given by the standards for often used types of piles, namely for bored piles by DIN 4014, in the standard DIN 4026 for driven piles or certain other types of displacement piles and in the code DIN 4128 for injection piles with small diameter.

By the use of this values of experience fundamentally it should be taken into consideration, that in most cases the values have to be used as characteristic values in the sense of Eurocode 7. It also should be taken into consideration that the values can get inapplicable only in very disadvantageouse cases because these values are chosen very carefully. So a quite cost-intensive static load test can be economical especially for high loaded piles.

5.2 *Determination of the bearing capacity of single bored piles by values of experience given by DIN 4014*

The permitted use of the values of experience for the tip resistance and the skin friction given by DIN 4014 for bored piles depends on a few conditions. At first in a certain way "simple" ground situations should be given, that means that the profile of the subsoil should be similar in the whole area, for examples in certain parts of the construction site, and that the strength of cohesionless soils should be able to be determined by the cone penetration resistance q_c and the strength of cohesive soils should be able to be determined by the undrained shear strength c_u.

The second condition defined by DIN 4014 is, that the minimum embedded length of the pile should be 2.50 m in a soil of sufficient bearing capacity. DIN 1054 gives the recommendation that such a sufficient bearing capacity could be assumed, if a cohesive soil of rough semi-solid state or a cohesionless soil of sufficient bearing capacity can be investigated. In this sense semi-solid state means that the consistency index I_c should be equal or greater than 1.0:

$$I_c = \frac{w_L - w}{w_L - w_P} \geq 1.0 \tag{2}$$

where w_L is the liquid limit, w is the moisture content and w_P is the plastic limit. A cohesionless soil has a sufficient bearing capacity in the sense of the standards, if the density D as:

$$D = \frac{n_{max} - n}{n_{max} - n_{min}} \tag{3}$$

is $D \geq 0.4$ for uniformly soil ($U < 3$; with U = coefficient of uniformity) and $D \geq 0.55$ for not uniformly soil ($U \geq 3$). A further condition is that the thickness of the sufficient bearing soil layer underlying the pile foot has to be greater than three times the diameter of the pile foot, at least greater than 1.5 m.

The proceeding by DIN 4014 for the determination of the bearing capacity of single bored piles is as follows. The pile resistance R can be calculated by the both parts R_b as the base resistance and R_s as the total shaft resistance as follows:

$$R(s) = R_b(s) + R_s(s) \tag{4}$$

In the sense of DIN 4014 both parts, R_b and R_s, depend on the settlement s of the pile head, the diameter D or D_F of the pile (D_F for a base enlargement) and the strength of soil expressed by q_{ck} or c_{uk}. The strength of soil is classificated for cohesionless soils by the characteristic cone penetration resistance q_{ck} and for cohesive soils by the characteristic values of the undrained shear strength c_{uk}. The cone penetration resistance q_{ck} can be determined by a cone penetration testing (CPT) or a dynamic penetration testing. If it is impossible to carry out a cone penetration test for example in coarse-grained soils, the results of a heavy dynamic penetration test (DPH) can correlate by q_{ck} [MN/m²] $\approx N_{10}$ [blows/10 cm penetration]. The undrained shear strength c_{uk} can be determined for cohesive soils by cone penetration testing, by vane tests or by laboratory investigations. By using the vane test the correction factors by Bjerrum (1972) should be applied.

In DIN 4014 four tables are given comprising values of experience for the tip resistance $q_{bk}(s)$ [MN/m²] and the ultimate skin friction q_{s1k} [MN/m²]. One table summarizes the values for the tip resistance q_{bk} depending on the cone penetration resistance q_{ck} and one table for the ultimate skin frction q_{s1k} (q_{ck}) for cohesionless soils and also one table for q_{bk} depending on the undrained shear strength c_{uk} and last a table for the ultimate skin friction q_{s1k} (c_{uk}) for cohesive soils.

For cohesionless soils table 2 gives the values offered by DIN 4014 for the tip resistance $q_{bk}(s)$ and table 3 for the ultimate skin friction q_{s1k}. The values of $q_{bk}(s)$ varies between 0.7 and 4.0 MN/m². They depend on the strength of the soil, which is described by the cone penetration resistance q_{ck} and they also depend on the ratio settlement s to shaft diameter D. The values of the ultimate skin friction q_{s1k} varies between 0 and 0.12 MN/m².

Tables 4 and 5 comprise the values of experience for cohesive soils given by DIN 4014. A comparison of the values for cohesionless and cohesive soils shows that the values for the cohesionless soils have a tendency to be slightly greater.

Table 2. Characteristic tip resistance q_{bk} depending on the settlement index s/D and the average cone penetration resistance q_{ck} in cohesionless soils given by DIN 4014.

settlement index s/D or s/D_F	characteristic tip resistance q_{bk} [MN/m²] *) for an average cone penetration resistance q_{ck} [MN/m²]			
	10	15	20	25
0.02	0.70	1.05	1.40	1.75
0.03	0.90	1.35	1.80	2.25
$0.10 = s_g$	2.00	3.00	3.50	4.00

*) intermediate values are allowed to be interpolated linearily. Using a bored pile with base enlargement the values must be reduced to 75%.

Table 3. Characteristic ultimate skin friction q_{s1k} in cohesionless soils given by DIN 4014.

strength of the cohesionless soil determined by the cone penetration resistance q_{ck} [MN/m²]	characteristic ultimate skin friction q_{s1k} *) [MN/m²]
0	0
5	0.04
10	0.08
≥ 15	0.12

*) intermediate values are allowed to be interpolated linearily.

All these values defined by DIN 4014 are the result of a wealth of static load tests for cased and uncased bored piles. The investigations of Stocker (1980) and recently published results of further research work susbtantiate, that the values for the characteristic ultimate skin friction given by DIN 4014 have also an adequate safety for bored piles built with suspension support.

Table 4. Characteristic tip resistance q_{bk} depending on the settlement index s/D and the undrained shear strength c_{uk} in cohesive soils given by DIN 4014.

settlement index s/D or s/D_F	characteristic tip resistance q_{bk} [MN/m²] *) for an undrained shear strength c_{uk} [MN/m²]	
	0.1	0.2
0.02	0.35	0.90
0.03	0.45	1.10
$0.10 = s_g$	0.80	1.50

*) intermediate values are allowed to be interpolated linearily. Using a bored pile with base enlargement the values must be reduced to 75%.

Table 5. Characteristic ultimate limit skin friction q_{s1k} in cohesive soils given by DIN 4014.

strength of the cohesive soil determined by the undrained shear strength c_{uk} [MN/m²]	characteristic ultimate skin friction q_{s1k} *) [MN/m²]
0.025	0.025
0.1	0.04
≥ 0.2	0.06
*) intermediate values are allowed to be interpolated linearily.	

With this values of experience the base resistance $R_{bk}(s)$ and the limit shaft resistance R_{s1k} can be calculated:

$$R_{bk}(s) = q_{sk}(s) \cdot \pi \cdot D_F^2 / 4 \qquad \text{and} \qquad R_{s1k} = \pi \cdot D \cdot \Sigma\,(q_{si1k} \cdot l_i) \tag{5}$$

The base resistance $R_{bk}(s)$ grows up with an increasing settlement of the pile head to a maximum value for a limit settlement of:

$$s_g = 0.10 \cdot D \tag{6}$$

at the pile head. The skin friction q_{sik} grows up linear with increasing settlement index s/D until reaching the value of the ultimate skin friction q_{si1k} given by the tables for a settlement of the pile head of:

$$s_{sg} = 0.5 \cdot R_{s1k}\,[\text{MN}] + 0.5\ \text{cm} \le 3.0\ \text{cm} \tag{7}$$

So the load-settlement curve for a single bored pile can be calculated as:

$$R_k(s) = R_{bk}(s) + R_{sk}(s) = q_{bk}(s) \cdot \pi \cdot D_F^2 / 4 + \pi \cdot D \cdot \Sigma\,(\,q_{sik}(s) \cdot l_i) \tag{8}$$

demonstrated by figure 2. This calculation method neglects that in a multi-layer ground the ultimate skin friction q_{si1k} is mobilized for different values of the settlement of the pile head.

The characteristic value R_{1k} of the ultimate bearing resistance of a single pile determined by this method is assumed to be mobilized for the limit settlement of $s_g = 0.10 \cdot D$ as:

$$R_{1k}(s_g) = R_{b1k} + R_{s1k} = q_{bk}(s_g) \cdot \pi \cdot D_F^2 / 4 + \pi \cdot D \cdot \Sigma\,(\,q_{si1k} \cdot l_i) \tag{9}$$

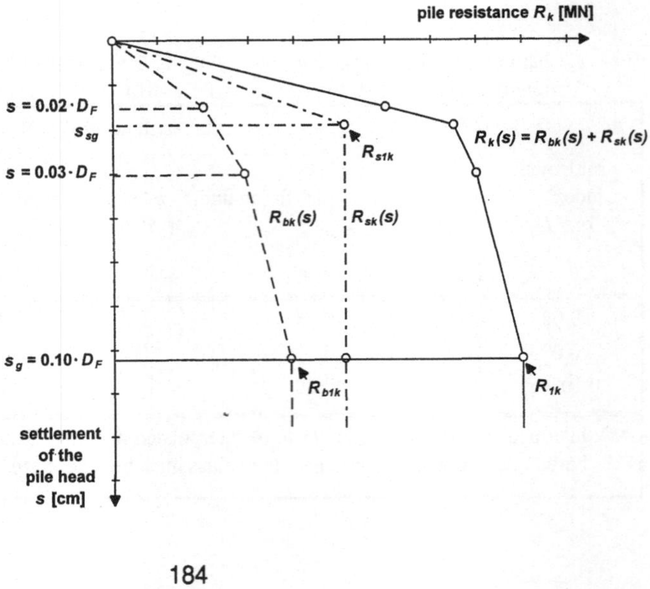

Figure 2. Example of a load-settlement curve for a single bored pile calculated with the values of experience given by DIN 4014.

This procedure to calculate a theoretical load-settlement curve takes the results of static load tests into account. The load tests usually show that the skin friction is mobilized with its ultimate value by settlements of only 1 or 2 cm while the tip resistance increases for settlements of 10 or more cm. Examinations demonstrate the influence of the pile diameter for mobilizing the tip resistance. Concerning the skin friction it could be proved by measurements that the settlement s_{sg} for mobilizing the ultimate skin friction is not only a function of the pile diameter but also a function of the ultimate shaft resistance R_{s1k}. So with an increasing ultimate shaft resistance the settlement s_{sg} for mobilizing the ultimate skin friction becomes greater.

With the theoretical load-settlement curve for a bored pile calculated in this way, for the allowable working pile load R_{work} two limit states have to be proved. First the settlement of the pile head have to be lower than the characteristic value of a permissible value s_{2k} for the settlement defined for example by the designing team of structural and geotechnical engineer:

$$R_{work} \leq R_k(s_{2k}) \tag{10}$$

This proof corresponds to the serviceability limit state design in the sense of Eurocode 7.

Second has to be proved that the working pile load R_{work} is lower than the ultimate pile resistance R_{1k} devided by the global safety factor η with R_{1k} as the result of the calculation with the values of experience given by the standard DIN 4014:

$$R_{work} \leq R_{1k} / \eta \tag{11}$$

This proof corresponds to the ultimate limit state design of EC7. The numerical values for η are given for different load cases by DIN 1054 (Table 1).

Figure 3 shows a simple example for the construction of the load-settlement curve of a single bored pile by DIN 4014. The exemplarily chosen bored pile has a shaft diameter of half a meter and a pile length of 7 meters. The soil is seperated in an upper layer of middle dense sand with sea silt inliers and with a thickness of 3 m and a lower layer of sand densely packed. The cohesionless soil has been investigated by a cone penetration test. The average cone penetration resistance q_{ck} of the middle dense sand is 5 MN/m² and of the dense sand 15 MN/m². The pile resistance R_{1k} is determined as the sum of the ultimate shaft resistance R_{s1k} and the maximum value of the base resistance R_{b1k} as 1.531 MN. The load-settlement curve of the bored pile is constructed with its values $R_{sk}(s)$, $R_{bk}(s)$ and the sum $R_k(s)$. The shaft resistance $R_{sk}(s)$ grows up to the limit value R_{s1k} and then keeps constant. The base resistance $R_{bk}(s)$ grows up with the settlement of the pile head to the maximum value R_{b1k} for a limit settlement $s_g = 0.10 \cdot D_F$. For the load case 1 the saftey factor η is 2, so the working load of the pile is $R_{work} = 768$ kN.

For piles bearing on rock DIN 4014 also offers values of experience to determine the bearing capacity of single bored piles. Depending on the uniaxial compressive strength q_{uk} of the rock table 6 gives the values for the ultimate tip resistance q_{b1k} and the ultimate skin friction q_{s1k}. If there is no static load test or other experiences from similar soil situation available, the ultimate bearing capacity of a single bored pile in rock can be calculated as:

$$R_{1k} = q_{b1k} \cdot \pi \cdot D_F^2 / 4 + \pi \cdot D \cdot \Sigma (q_{si1k} \cdot l_i) \tag{12}$$

Table 6. Characteristic values for the ultimate tip resistance q_{b1k} und ultimate skin friction q_{s1k} depending on the uniaxial compressive strength q_{uk} of the rocky soils for piles bearing on rock.

characteristic uniaxial compressive strength q_{uk} [MN/m²]	characteristic ultimate tip resistance q_{b1k} [MN/m²]	characteristic ultimate skin friction q_{s1k} [MN/m²]
0,5	1,5	0,08
5,0	5,0	0,5
20	10,0	0,5
intermediate values are allowed to be interpolated lineary.		

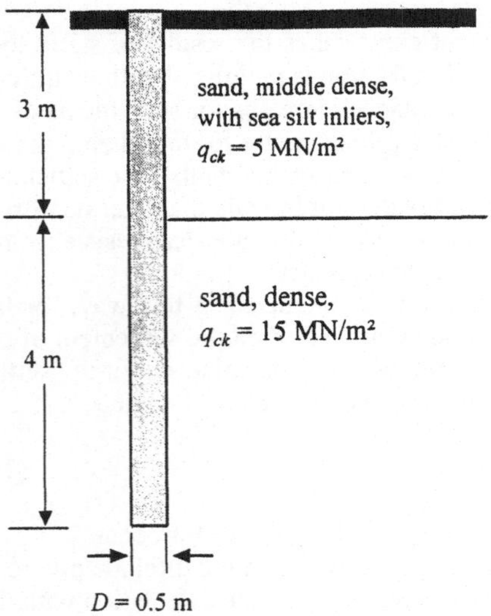

a) shaft resistance R_{s1k}

layer 1 *layer 2*

$R_{s1k} = \pi \cdot 0.5 \cdot (40.0 \cdot 3.0 + 120.0 \cdot 4.0) = 942$ kN

for $s_{sg} = 0.5 \cdot R_{s1k} + 0.5 = 1.0$ cm

b) base resistance R_{bk}

$s = 0.02 \cdot D = 1.0$ cm: $R_{bk} = \pi \cdot 0{,}5^2/4 \cdot 1.05 = 206$ kN

$s = 0.03 \cdot D = 1.5$ cm: $R_{bk} = \pi \cdot 0{,}5^2/4 \cdot 1.35 = 265$ kN

$s_{sg} = 0.10 \cdot D = 5.0$ cm: $R_{b1k} = \pi \cdot 0{,}5^2/4 \cdot 3.00 = 589$ kN

$R_{1k} = 942 + 589 = 1531$ kN

$R_{work} \leq R_{1k} / \eta = 1531$ kN / 2 = 768 kN (load case 1)

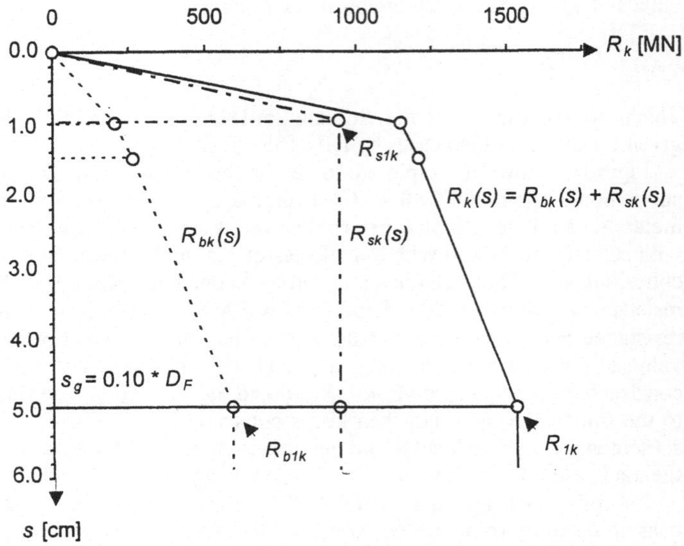

Figure 3. Example for the analytical calculation of a load-settlement curve of a bored pile by the values of experience given by DIN 4014

The recommendations for piles in rock as defined by DIN 4014 neglect the fact, that the bearing capacity of piles in rock is significantly influenced by the joint system of the rocky ground. So the values for the ultimate tip resistance and ultimate skin friction are not only a function of the uniaxial comressive strength q_{uk} of the rock but are also a function of the direction and properties of the joint-system.

DIN 4014 doesn´t offer explicit recommendations concerning the bearing behaviour of axially loaded pile groups. Considering the manifold interactions inside a pile group in code DIN 4014 it is mentioned that a fix value for a distance between adjacent piles, for which group effects could be neglected, doesn´t exist. In engineering practice it is often assumed that for a pile spacing of $3 \cdot D_F$ or more an interacion of neighbouring piles can be neglected also the results of recently done research work prove, that for a pile spacing of $3 \cdot D_F$ the bearing behaviour of a pile inside a pile group clearly differs from the bearing behaviour of a single pile (Katzenbach et al. 1996).

5.3 *Determination of the bearing capacity of single driven piles by values of experience given by DIN 4026*

The static load tests of driven piles usually show a clear failure so that the permissible design value of the bearing capacity is determined by an adequate safety against the bearing resistance. Concerning this for commonly used single driven piles DIN 4026 offers values of experience for the total working load R_{work} as permissible bearing capacity without considering the corresponding settlements. The given values for the working load R_{work} depend on the pile diameter D_F and on the embedded length of the pile in a soil of sufficient bearing capacity. The use of the values of experience offered by DIN 4026 depends on a few conditions. So so called "simple" ground situations should be given on site as defined by DIN 1054 and DIN 4014. The minimum embedded length of the pile must be at least 3 m in a cohesive soil of rough semi-solid state or in a cohesionless soil of sufficient bearing capacity and the pile length must be longer than 5 m for compression piles. Furtherone under the pile base the thickness of the bearing soil should be greater than $4 \cdot D_F \geq 1.5$ m or $2 \cdot b$ for a pile group (with b = width of the pile group). The last condition is that the minimum distance between adjacent piles must be $e \geq 3 \cdot D \geq 1\ \text{m} + D$ with e as the pile spacing as shown in figure 4.
It has to be taken into consideration, that the given values R_{work} for the bearing capacity at working loads are design values. So the procedure is different to the regulations in DIN 4014, where characteristic values are given. The values for the working load R_{work} are given by DIN 4026 for single timber piles, for steel piles and for reinforced-concrete piles.

A separation into the base resistance R_{bk} and the total shaft resistance R_{sk} is not been taken into consideration by DIN 4026. There are also no further relevant factors for the determination of the bearing capacity included by DIN 4026. Also the type or the strength of the soil is not exactly been taken into account or defined. So the rules given by DIN 4026 are quite simple

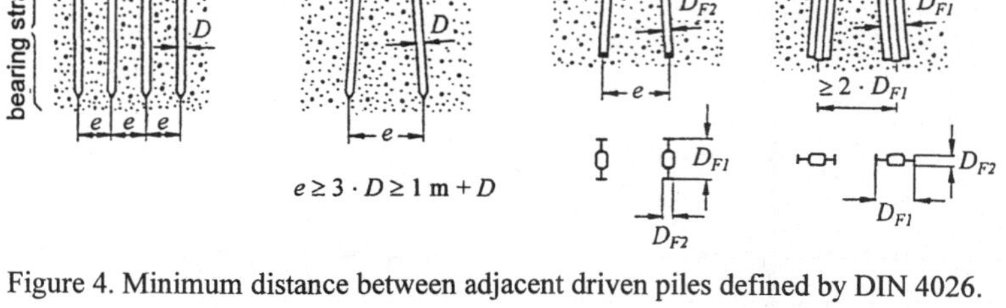

Figure 4. Minimum distance between adjacent driven piles defined by DIN 4026.

Table 7. Permissible working load R_{work} for single driven timber piles given by DIN 4026.

embedded length in a soil of sufficient bearing capacity [m]	bearing capacity R_{work} at working loads [kN] D_F [cm]				
	15	20	25	30	35
3	100	150	200	300	400
4	150	200	300	400	500
5	—	300	400	500	600

fashioned offering no informations about the safety factor included in the calculation or about the load-settlement behaviour of the piles.

Table 7 contents the values of the working load R_{work} for timber piles. The given values depend on the the pile diameter D_F and on the embedded length of the pile in a soil of sufficient bearing capacity.

Table 8 summarizes the values of the working load R_{work} for reinforced-concrete piles and at least in table 9 the values of the working load R_{work} for steel piles with different types of cross sections namley for I-beam piles, for steel-pipe piles and for steel-box piles are documented.

For cohesionless soils with a high strength, characterized by $D \geq 0.5$ for $U < 3$ or $D \geq 0.65$ for $U \geq 3$ or $q_{ck} \geq 10$ MN/m², or for cohesive soils of solid state ($I_C > 1{,}0$ or $c_{uk} \geq 200$ kN/m²) the values given in table 7, 8 and 9 can be increased by 25 % (DIN 4026).

Table 8. Permissible working load R_{work} for single driven reinforced-concrete piles with square-crossed section given by DIN 4026.

embedded length in a soil of sufficient bearing capacity [m]	bearing capacity R_{work} at working loads [kN]				
	length of cross section side a_s [1)] [cm]				
	20	25	30	35	40
3	200	250	350	450	550
4	250	350	450	600	700
5	—	400	550	700	850
6	—	—	650	800	1000

[1)] is also valid for approximate square-crossed sections with a_s as average length of side.

Table 9. Permissible working load R_{work} for single driven steel piles given by DIN 4026.

embedded length in a soil of sufficient bearing capacity [m]	bearing capacity R_{work} at working loads [kN]				
	I-beam pile[1)]		steel-pipe pile[2)] steel-box pile[2)]		
	breadth or height [cm]		D or a_s [3)] [cm]		
	30	35	35 or 30	40 or 35	45 or 40
3	—	—	350	450	550
4	—	—	450	600	700
5	450	550	550	700	850
6	550	650	650	800	1000
7	600	750	700	900	1100
8	700	850	800	1000	1200

[1)] Wide I-beam with height : breadth ≈ 1 : 1.
[2)] The table values are valid for piles with a closed point.
[3)] D = external diameter of a steel-pipe pile or the medium diameter of a radialsymmetric pile.
a_s = medium length of a side for approximate cross section or rectangular steel-box piles with same area.

The recommended values for the bearing capacity are carefully chosen values to eliminate damages, although for steel piles static load tests recently carried out demonstrate that the global safety factor included by the use of these values is only about η = 1,5 and that the validity of these values depends on a cone penetration resistance of $q_{ck} \geq 10$ MN/m² (Arz et al. 1991).

5.4 *Determination of the bearing capacity of injection piles by values of experience given by DIN 4128*

For injection piles with a diameter less than 30 cm DIN 4128 demands in principle static load tests. Only in special exceptional cases of lacking such test results values for the ultimate skin friction q_{s1k} given by DIN 4128 can be used as a guideline (Table 10). In this case the tip resistance is to be neglected.

6 PILE DESIGN BY EUROCODE 7 AND THE GERMAN NATIONAL APPLICATION DOCUMENT

The recently born Eurocode 7 (EC7) is mainly giving design principles, specifying the ultimate limit state design and the partial factoring, which in parts even have to be defined in the individual National Application Documents (NAD). EC7 does not contain design methods as such, nor is specifying relevant values or determination methods for the "characteristic values" for axially loaded piles. Even for static load tests, it is not defined in EC7 how the characteristic ultimate bearing resistance R_{1k} of a single pile should be derived.

In Germany the EC 7 completed by the German NAD was published as a prenorm in April 1996. This German NAD refers to a number of recently published German prenorms, which were created to realise the turn over of the German geotechnical standardisation actually based on a global safety concept to the probabilistic safety concept. Under the title "Soil · verification of the safety of earthworks and foundations" the precode DIN V 1054-100, which was published in April 1996 too, prescribes the basic rules for geotechnical analysis in accordance with the partial safety factor concept (Gudehus & Weißenbach 1996). Chapter 10 of DIN V 1054-100 deals with pile foundations and anchoring. For the ultimate limit state design (case 1) the design have to demonstrate that the following classes of limit states are sufficiently improbable:

- ultimate limit state of bearing resistance failure of the material of the piles (case 1B),
- ultimate limit state of bearing resistance failure of the piled foundation caused by failure of the soil surrounding the piles (case 1B),
- ultimate limit state of overall stability failure of the supported building caused by displacement of the piled foundation (cases 1B and 1C).

Furtherone the recommendations given by DIN V 1054-100 require that for pile foundations the serviceability limit state (case 2) has generally to be proved.

By DIN V 1054-100 the resistance of axially loaded piles has to be desribed by a load-settlement curve which is the result of one ore more than one carried out static load tests or based on equivalent experiences from static load tests under comparable circumstances. In contrast to the recommendations fixed in EC7 by DIN V 1054-100 the load-displacement graph measured by a static load test is considered to be already the characteristic load-displacement

Table 10. Characteristic ultimate skin friction q_{s1k} for injection piles by DIN 4128.

type of soil	characteristic ultimate skin friction q_{s1k} for compression piles [MN/m²]	characteristic ultimate skin friction q_{s1k} for tension piles [MN/m²]
medium and coarse gravel	0.20	0.10
sand and gravelly sand	0.15	0.08
cohesive soil	0.10	0.05

graph $R_k(s)$ without any additional factors. The defintion of the ultimate bearing resistance R_{1k} from the measured load-displacement line is equivalent to the regulations of DIN 1054 as reported above.

In those cases in which static load tests couldn´t have carried out the characteristic value of the axially bearing capacity of bored piles can be calculated by values of experience given by DIN V 1054-100. The offered values for the characteristic tip resistance q_{bk} and the characteristic value of the ultimate skin friction q_{sk} and the conditions for their use are absolutly identical with the reglementations of DIN 4014 described above, so that the determination of the characteristic value of the pile resistance, $R_k(s)$, is devided into components of base resistance R_{bk} and shaft resistance R_{sk} such that:

$$R_k(s) = R_{bk}(s) + R_{sk}(s) = q_{bk} \cdot A_b + \Sigma\, q_{sik} \cdot A_{si} \tag{13}$$

As in DIN 4014 it is not taken into account that in a multi-layered soil medium the ultimate limit skin friction q_{si1k} is mobilized along the embedded length of the pile for different values of the settlement of the pile head depending on the strength of the single layers.

A similar procedure to determine the characteristic value of the axially bearing capacity is described by DIN V 1054-100 for displacement piles. The values of experience known from DIN 4026 are identically part of the appendix of the new prenorm and can be used as characteristic values for the calculations under ultimate (R_{1k}) and serviceability limit state conditions (R_{2k}). To determine the characteristic value of the axially bearing capacity (R_{1k}) of displacement piles for the ultimate limit state (case 1B by DIN V 1054-100) the values q_{b1k} and q_{s1k} given in table 11 can also be used. The recommended characteristic values are valid for piles with a diameter or a length of a side ≤ 0.5 m.

Table 11. Characteristic values for the ultimate tip resistance q_{blk} and the ultimate skin friction q_{slk} for displacement piles given by DIN V 1054-100.

type of soil	area under surface of bearing horizon [m] [4]	characteristic ultimate skin friction q_{slk} [kN/m²]				characteristic ultimate tip resistance q_{blk} [MN/m²]			
		timber pile	rein-forced concrete pile	steel-pipe pile steel-box pile open	I-beam steel pile	timber pile	rein-forced concrete pile	steel-pipe pile[3]) steel-box pile open [1])	I-beam steel pile
cohesionless soil	< 5	20-45	20-45	20-35	20-30	2-3.5	2-5	1.5-4	1.5-3
	5-10	40-65	40-65	35-55	30-50		3.5-6.5	3-6	2.5-5
	> 10		60	50-75	40-75	3-7.5	4-8	3.5-7.5	3-6
cohesive soil I_c = 0.5-0.75 [5])		5-20							
I_c = 0.75-1 [5])		20-45				0-2			
boulder clay semi-solid to solid	< 5		50-80	40-70	30-50		2-6	1,5-5	1.5-4
	5-10			60-90	40-70		5-9	4-9	3-7.5
	> 10		80-100	80-100	50-80		8-10	8-10	6-9

[1]) wide steel-box or diameter pipe ≤ 500 mm.
[2]) webs have to be included for wide profil ≥ 350 mm.
[3]) for steel-box piles with closed foot see reinforced concrete piles.
[4]) length of pile for q_{slk}, embedded depth in bearing horizon for q_{blk}
[5]) I_c by DIN 18122-1. As the boulder clay's consistency index I_c is not determinable by DIN 18122-1 and DIN 4022-11 it is to estimate on the base of local experiences.

As regulated in DIN 4128 especially for injection piles with a relative small cross section (D < 0.3 m) static load tests are recommmended by DIN V 1054-100. For compression injection piles DIN V 1054-100 offers values of experience for the characteristic ultimate skin friction q_{s1k}, which are identical with the values q_{s1k} given by DIN 4128 for compression piles (table 10). So the characteristic value of the bearing capacity of a single compression injection pile can be calculated as:

$$R_{1k} = R_{sk} = \Sigma\, q_{si1k} \cdot A_{si} \tag{14}$$

By DIN V 1054-100 the results of dynamic load tests or driving formulae should only be used as a basis for the determination of the characteristic bearing capacity, if their validity has been demonstrated by static load tests in comparable situations.

With the determined axial bearing resistance R_{1k} it has to be demonstrate that the single pile will support the design load S_{1d} with an adequate safety against bearing resistance failure R_{1d}, so that the following inequality shall be satisfied (case 1B):

$$S_{1d} \leq R_{1d} \tag{15}$$

The design value of the bearing resistance of a single pile is:

$$R_{1d} = \eta \cdot R_{1k} / \gamma_P \tag{16}$$

The characteristic value R_{1k} derives from:

$$R_{1k}(s_1) = R_{bk}(s_1) + R_{sk}(s_1) \qquad \text{with } s_1 \leq 0.1 \cdot D_F \tag{17}$$

as shown above, while the factor η is defined as:

$$\eta = \eta_N \cdot \eta_z \tag{18}$$

where the factor η_N, considering the influence of the number N of carried out static load tests, varies from 1.00 (N = 1) to 1.35 ($N \geq 3$) with the exception of injection piles for which is in any case η_N = 1.0. The factor η_z decribes the reduction of the bearing resistance caused by cyclic loads. While under static loads the value of η_z is 1.0, it is decreased gradual to 0.4 for alternating stresses depending on the amount of load cycles. γ_P is the partial factor for piles and has a numerical value as shown in table 12.

Furtherone the recommendations given by DIN V 1054-100 require that for pile foundations the serviceability limit state (case 2) has to be proved by satisfying the following inequality:

$$S_{2d} \leq R_{2d} \tag{19}$$

The design value of the resistance R_{2d} derives directly from the characteristic value of the resistance R_{2k}. With an individual specified value of a characteristic settlement s_{2k} the characteristic value of the resistance $R_{2k}(s_{2k})$ of an axially loaded single pile can be determined from the characteristic load displacement line.

Table 12. Partial factor γ_P for single piles (compression and tension) by DIN V 1054-100.

case	partial factor γ_P for single piles for load case		
	1	2	3
1B	1.40	1.20	1.10
1C	1.60	1.40	1.20

In summary, one can say, that the design approaches for axially loaded piles recommended by DIN V 1054-100 as part of the German National Application Document of EC7 follows quite close the regulations known by DIN 1054 and known by the well-established pile codes DIN 4014, DIN 4026 and DIN 4128. So especially the values of experience for the characteristic values of the tip resistance q_{bk} and the skin friction q_{sk} for bored piles given by DIN 4014 will exist as a guideline for pile design in Germany furthermore.

7 TENSION PILES

To realize the excavation of deep impermeable pits in the centre of Berlin during the last years the use of tension piles to anchor underwater concrete floors have given some new experiences concerning this type of axailly loaded piles. An often used type of pile for this kind of highly loaded tension pile groups is the vibrating injection pile which can mobilize noticeable values for the ultimate skin friction as a result of grouting cement while driving the pile into the ground. Values of experience for the skin friction of this type of pile can be found in EAU (1990).

For other types of tension piles values to determine the skin friction are given by DIN 4014, DIN 4026 and DIN 4128. In some cases significant differents between the calculated and the really measured bearing resistance of tension piles have been watched. Therefore for tension piles static pile load tests should always be carried out. During the last two years several static load tests on tension pile groups of mostly five piles have taken place in Berlin. The measured results have given new knowledges about the bearing behaviour of tension pile groups.

8 EXAMPLE FOR THE DETERMINATION OF THE BEARING CAPACITY OF A SINGLE BORED PILE IN BERLIN BY A STATIC LOAD TEST

Concluding the application of the described German recommendations for the design of axially loaded piles will be illustrated by an actual example of a bored pile constructed in Berlin. In frame of the tremendous construction work actually done by Daimler Benz, Sony and other

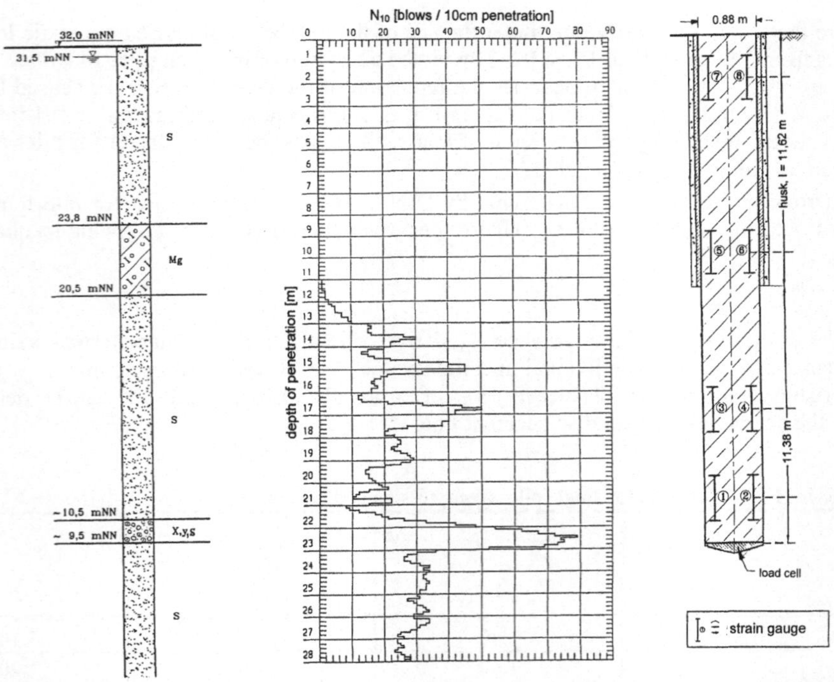

Figure 5. Ground situation and bored test pile-equiped with strain gauges and load cell · Potsdamer Platz · Berlin.

financiers at the Potsdamer Platz in the centre of Berlin the task arose to determine reliable values for the bearing capacity of bored piles in Berlin Sand. For the foundation of the Office tower of the so called Sony Center a Combined Piled Raft Foundation was designed. To optimize the calculation values for the skin friction and the tip resistance in April 1996 a static pile load test was carried out for a bored pile on site.

The test pile had a length of 23.0 m and a diameter of 0.88 m. In the upper part of the pile down to 11.6 m under the pile head a steel husk was located to eliminate the skin friction in this area (Figure 5). The ground situation was characterized by an upper sand layer of loose to medium dense packing underlying by a marl layer. Under the marl layer a second sand layer of medium dense to dense packing followed as documented by a dynamic penetration test with a middle number of blows of $N_{10} = 25$ along the bearing pile shaft . The dynamic penetration test was carried out from 11 m under the pile head downstairs. In altitude of the pile base a small shingle layer was investigated.

The test pile was equiped by the Institute of Geotechnics of the Darmstadt University of Technology with 8 strain gauges in DMS-technique which are installed in 4 different depths, each 2 m long and arranged in pairs vìs-a-vìs at the inner side of the reinforcement cage to measure the load distribution along the pile shaft. The base resistance was measured with a load cell while the load measurement at the pile head was effected by monitoring of the hydraulic pressure in the jacking system and additional by four load cells (Figure 6).
The measured load-settlement curve $R_k(s)$ is documented in figure 7 devided up in the both parts of the shaft resistance $R_{sk}(s)$ and of the base resistance $R_{bk}(s)$. The load transfer is mainly realized by mobilizing the skin friction. The pile-toe resistance bore only about 15 % of the pile load. By a limit settlement of the pile head defined by DIN 4014 of $s_g = 0.1 \cdot D$ the characteristic ultimate bearing resistance of the test pile could be determined to $R_{1k} = 8.5$ MN devided into components of characteristic limit shaft resistance $R_{s1k} = 7.1$ MN and characteristic ultimate base resistance $R_{b1k} = 1.4$ MN. In figure 7 the measured load-settlement curve is compared with the theoretical load-settlement curve calculated with the values of experience given by DIN 4014. Both curves show relative good correspondence although the theoretical calculated value of the characteristic ultimate bearing resistance of the test pile is with $R_{1k} = 7.5$ MN lower than the measured value.

The load transfer of the pile load is determined by the lower sand layer of dense packing. Figure 8 shows the distrbution of pile force and shaft resistance with the depth. By a limit settlement of the pile head of $s_g = 0.1 \cdot D$ the middle limit skin friction is $q_{s1k} = 220$ kN/m² with a significant concentration at the pile foot.

The measured load settlement line was the basis for the further numerical studies for the bearing behaviour of the Combined Piled Raft Foundation designed for the foundation of the high rise building of Sony Center (Richter et al. 1996).

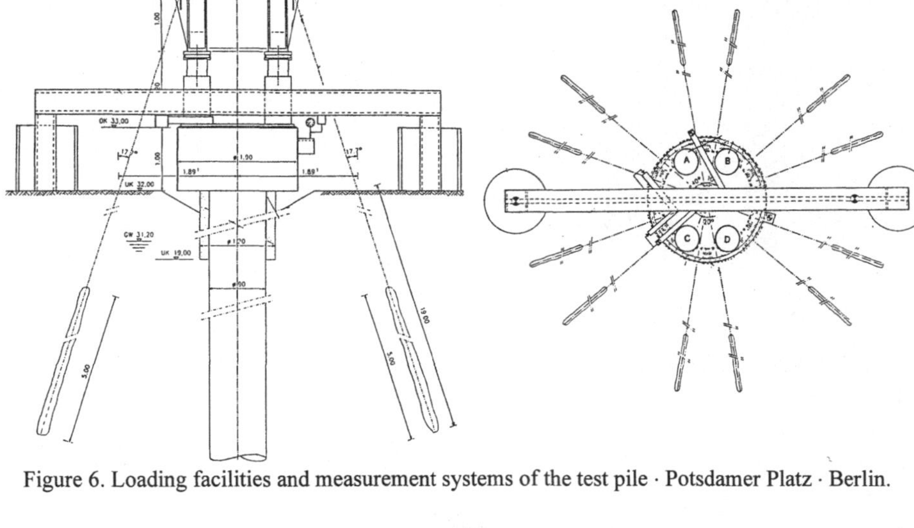

Figure 6. Loading facilities and measurement systems of the test pile · Potsdamer Platz · Berlin.

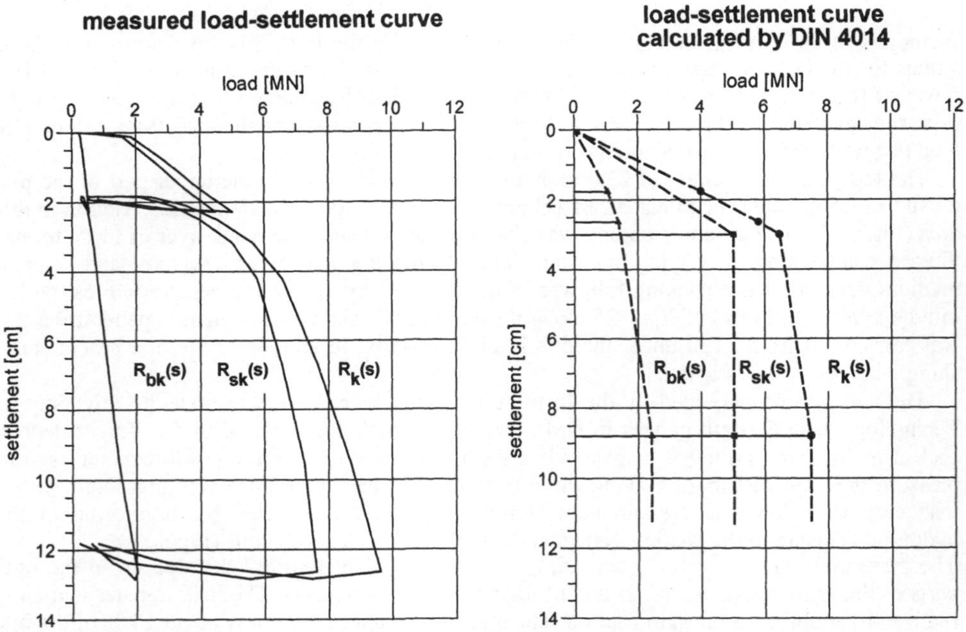

Figure 7. Load-settlement curve of the bored test pile · Potsdamer Platz · Berlin.

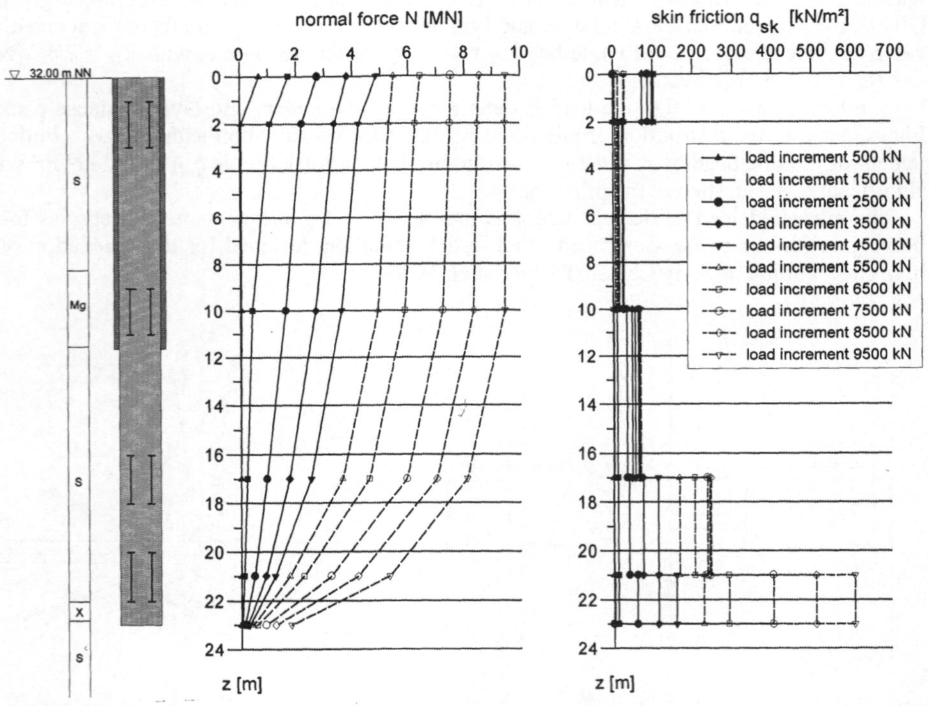

Figure 8. Distribution of pile force and shaft resistance of the bored test pile with the depth · Potsdamer Platz · Berlin.

9 BEARING BEHAVIOUR OF PILE GROUPS AND COMBINED PILED RAFT FOUNDATIONS · RESULTS OF RECENT RESEARCH WORK

The bearing behaviour of axially loaded pile groups demands special considerations concerning the pile-soil-interaction and the pile-pile-interaction between adjacent piles. DIN 1054 gives the guideline that pile foundations have to be designed in the way that the full load of the upper building is transfered into the ground only by the piles.

The German Standards don´t give detailed specifications for the calculation of pile groups. So the only recommendation given by DIN 1054 requires that the settlement of axially loaded pile groups should be determined by two components: the settlement of a comparable single standing pile and an additional settlement caused by group-effects. To calculate the second component the pile group should be considered as a deep shallow foundation with a theoretical area surrounded by the edge piles of the pile group and located at the level of the pile foots.

A relative new foundation concept is the Combined Piled Raft Foundation (CPRF) which acts as a composite construction consisting of the three bearing elements piles, raft and subsoil (Figure 9) (Cooke et al. 1981). According to its stiffness the raft distributes the load of the structure R_{tot} over contact pressure, represented by R_{raft}, as well as over the piles, generally represented by $\Sigma R_{pile,i}$ in the ground. In comparison with a conventional foundation design of a pile group as defined by DIN 1054 the CPRF indicates a total new dimension for the soil-structure-interaction because of the new design philosophy to use the piles up to a load level which is besides the permissible design value for the bearing capacity of a comparable single standing pile. This leads to an extremely economic foundation with rather low settlements especially if the stiffness of the soil is increasing with depth (Katzenbach 1993). The usefulness and serviceability of CPRF can be described by the Combined Piled-Raft Coefficient α_{CPRF} which is defined as the ratio between the amount of the pile loads $\Sigma R_{pile,i}$ and the total load of the building R_{tot} (Figure 10). For the design and the computation of the CPRF actually no standards and definit design strategies are available (Katzenbach et al. 1996). So actually additional research work is done based on measurements and numerical computer-simulations.

Numerical simulations give the opportunity to examine the bearing behaviour of pile groups or of CPRF in parametric studies varying for example the number of piles, the diameter and length of piles and the distance between adjacent piles for defined boundary conditions. The spatial interaction analysis of pile foundations in multi-layered soil medium is a very complex engineering problem in three dimensions which especially requires a realistic constitutive law for the soil.

Within the frame of this research work an elastoplastic constitutive model is used for simulating the nonlinear elastoplastic material behaviour of soil in numerical analysis. The constitutive model consists of two main yield surface segments: a pressure dependent Drucker-Prager

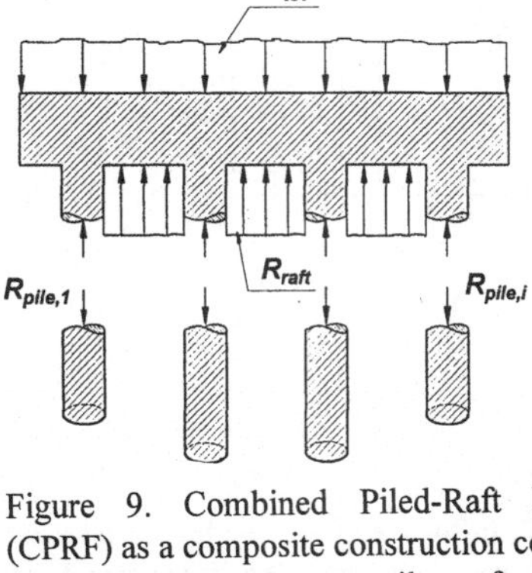

Figure 9. Combined Piled-Raft Foundation (CPRF) as a composite construction consisting of the three bearing elements piles, raft and subsoil.

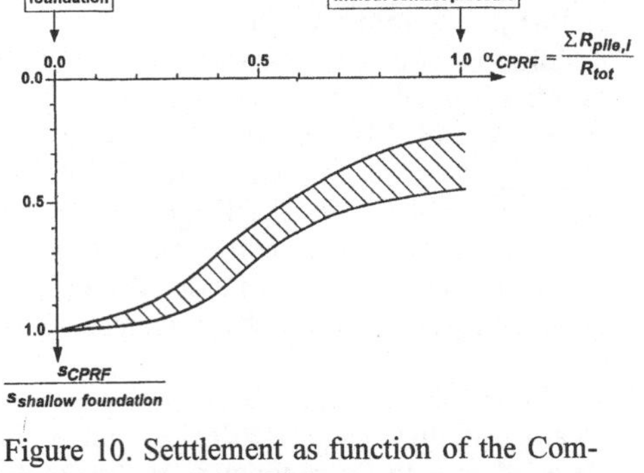

Figure 10. Setttlement as function of the Combined Piled-Raft Coefficient α_{CPRF}.

shear failure surface and a compression cap yield surface. The Drucker-Prager failure surface itself is perfectly plastic while the hardening/softening behaviour of the cap yield surface is a function of the volumetric plastic strain (Figure 11). The yield surface may change in size, position or shape as the soil is loaded successively to higher stress-levels. On the Drucker-Prager shear failure surface the material dilates while on the cap surface it compacts. The plastic flow on the Drucker-Prager shear failure surface produces inelastic volume increase, which causes the cap to soften. So this constitutive model gives the possibility for a realistic simulation of the stress-strain behaviour of soils which is nonlinear inelastic and dependent on stress path and previous stress history.

The aptitude of this numerical and constitutive model investigated at the Institute of Geotechnics at the Darmstadt University of Technology for reliable predicting the bearing behaviour of single piles and pile groups was verified by numerically simulating of experimentally carried out static pile load tests.

Figure 12 shows the results of a numerical analysis of two single bored piles in Frankfurt Clay (D = 1.5 m, l = 30.0 m). One pile (with the lower bearing capacity) is located concerning its pile head on the actual ground level while the other pile is located in an excavation pit with a depth of 20 m under the ground level. In consequence of the higher residudal stresses of the soil which increases with depth the second pile can mobilize a significantly higher value for the ultimate shaft resistance R_{s1k}. So the ultimate limit skin friction q_{s1k} of a pile is not only a function of the strength of soil as indicated by the standards but also a function of the residual stresses in the soil continuum.

The numerical model can also be used to analyse the bearing behaviour of pile groups and of Combined Piled-Raft Foundations (CPRF). Exemplarly for the same subsoil conditions two CPRFs which differ by the number of piles shall be examined (Figure 13.a). Model M1 has 64 piles with a pile spacing of $e = 3 \cdot D$ while model M2 has 16 piles with $e = 6 \cdot D$. The length of all piles is l = 30 m. The piles are combined with a square raft with 50 m as length of side. The raft is surrounded by a pit wall with an embedded length of 10 m.

For model M1 figure 13.b shows the numerically calculated load-settlement curve $R_{tot}(s)$ devided into the components of the total pile resistance $\Sigma\, R_{pile,i}(s)$, the raft resistance $R_{raft}(s)$

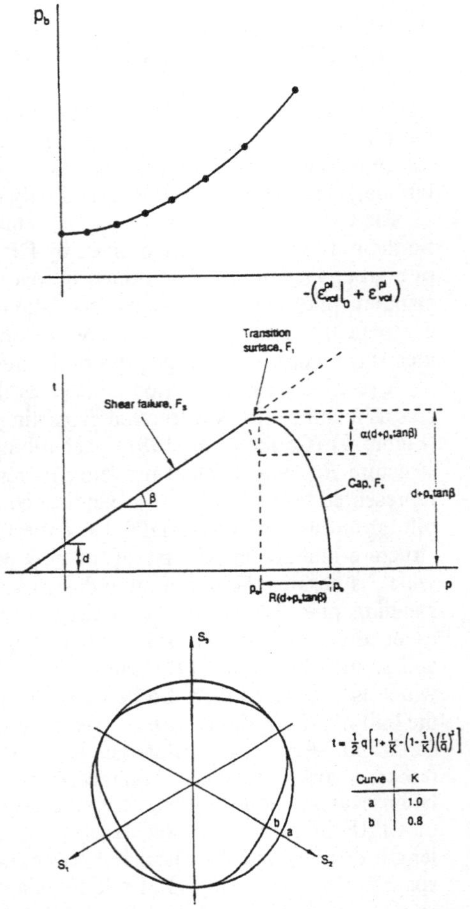

Figure 11. Elastoplastic constitutive law.

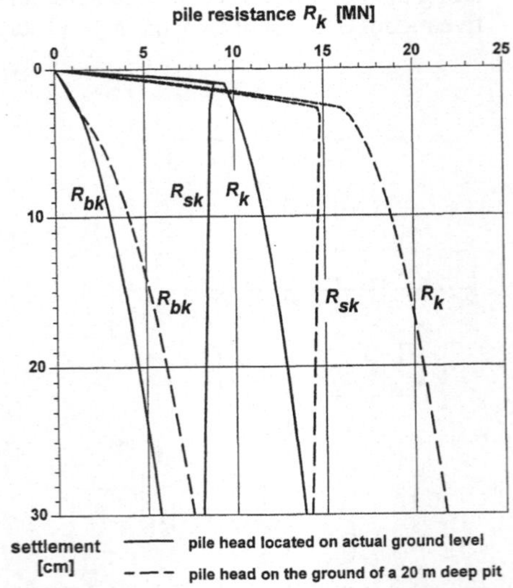

Figure 12. Influence of the residual stresses on the load-settlement curve of a single bored pile.

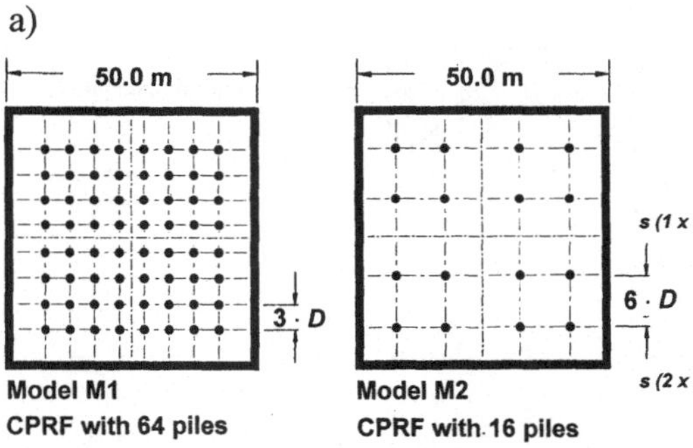

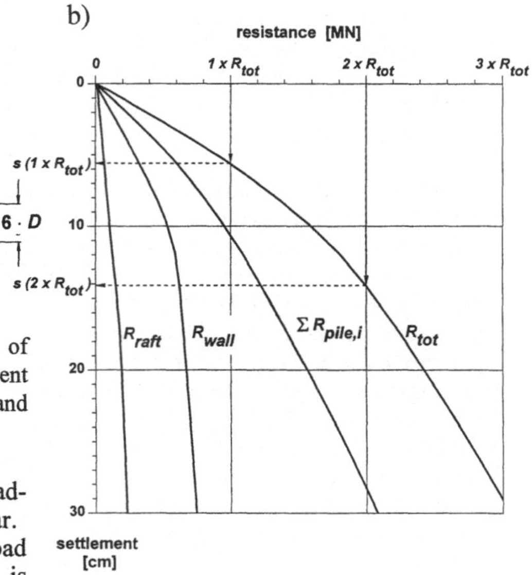

Figure 13. Numerical study for CPRF. a. system of model M1 and M2. b. computed load-settlement curve for model M1. c. load share between piles and raft for model M1 and M2.

and the resistance of the wall $R_{wall}(s)$. The load-settlement behaviour of the CPRF is not linear.

The percentage distribution of the total load of the building into the raft and the piles is illustrated for model M1 and M2 in figure 13.c. The load share between raft and piles is a non-linear function of the settlement of the combined foundation and is influenced by the number of piles of the Combined Piled-Raft Foundation.

For a CPRF the load transfer ist very different within the pile group. As exemplarly shown for model M1 at a settlement of $s(1xR_{tot})$ = 6 cm the centre piles carry 8 MN and the corner piles 13 MN (Figure 14). The different loading of the piles is a result of the different magnitude of the mobilized shaft resistance $R_{sk,i}$ of the piles as the tip resistance $R_{bk,i}$ varies only slightly (Figure 14). So as it is known from instrumented pile groups (Poulos 1993) also the pile loads of CPRF depend on the pile position inside the pile group. The pile loads decrease from outside to inside the pile group.

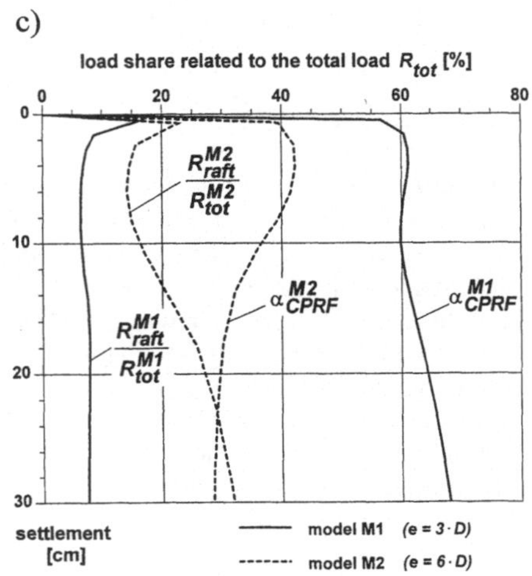

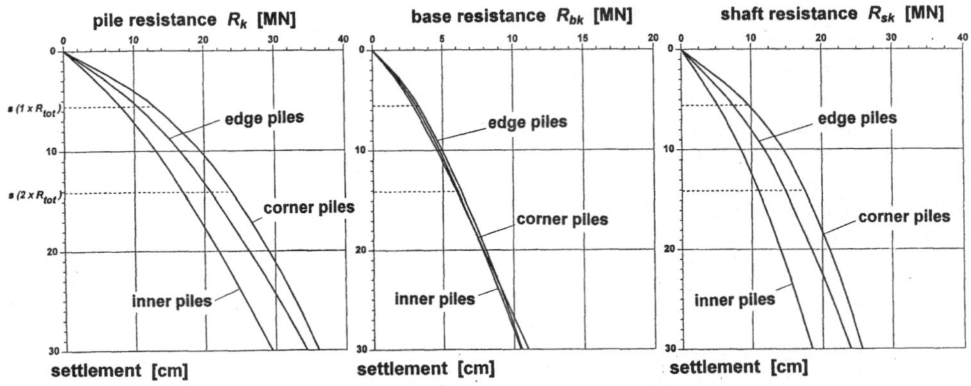

Figure 14. Computed load-settlement behaviour of the foundation piles of model M1.

The influence of the soil-structure interaction and especially of pile-pile-interaction of pile groups is demonstrated by figure 15. The three load-displacement curves describe the numerically determined load-settlement behaviour of a single standing pile, of a inner pile of model M1 ($e = 3 \cdot D$) and of a inner pile of model M2 ($e = 6 \cdot D$). Although all piles have the same length ($l = 30$ m) and diameter ($D = 1.5$ m) and are embedded in the same soil (Frankfurt Clay) the load-settlement behaviour and the load bearing capacity is significantly different. By a more and more decreasing pile spacing e inside a pile group the bearing resistance failure of the pile caused by failure of the surrounding soil as known for a single standing pile gets lost. Depending on the pile spacing e the pile-pile-interaction causes a significant reduction of the stiffness of the piles while the bearing capacity increases. The comparison proves that a pile distance of $e = 3 \cdot D$ indicates a remarkable pile-interaction inside the pile group.

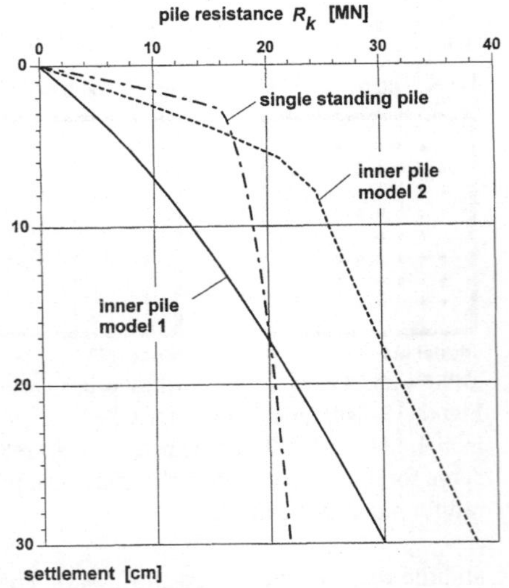

Figure 15. Comparison of the load-settlement behaviour of a single standing pile and an inner pile of model M1 and M2.

The numerical study indicates that the bearing behaviour of pile groups and CPRF is determined by factors which superpose the bearing behaviour of a single standing pile.

10 CONCLUSIONS

The German approaches for the design of axially loaded piles mainly base on static load testing and the use of values of experience as reported. While these design methods refer to single standing piles, measurements and numerical studies illustrate that the bearing behaviour of piles inside a pile group or inside a CPRF is influenced by interaction-effects between the different components of the foundation and the ground. An economic and safe design of pile groups and Combined Piled-Raft Foundations is only possible, if the soil-structure-interaction is considered with adequate and reliable calculation methods.

Besides an efficient control of the quality of constructed piles, the application of the observational method as outlined in EC7 is an important aspect for the successful design and construction of piles and pile groups.

DEFINITON OF USED SYMBOLS

A_b : cross section area of the pile foot
A_s : area of the pile shaft
A_{si} : part i of the area of the pile shaft in layer i
c_u : undrained shear strength
c_{uk} : characteristic value of the undrained shear strength
e : distance between adjacent piles
D : density
D : diameter of the pile
D_F : diameter of the pile foot
I_c : consistency index
l : pile length
n : porosity
n_{max} : porosity in loosest state
n_{min} : porosity in densest state
N : number of carried out static pile load tests

N_{10} : number of blows measured by a dynamic probing [blows/10 cm penetration]
q_c : cone penetration resistance of a cone penetration test (CPT)
q_{ck} : characteristic cone penetration resistance of a cone penetration test (CPT)
q_{bk} : characteristic value of the tip resistance of a pile
q_{b1k} : characteristic value of the ultimate tip resistance of a pile
q_{sk} : characteristic value of the skin friction of a pile
q_{sik} : characteristic value of the skin friction of a pile in layer *i*
q_{s1k} : characteristic value of the ultimate skin friction of a pile
q_{si1k} : characteristic value of the ultimate skin friction of a pile in layer *i*
q_{uk} : characteristic value of the uniaxial compressive strength
F_{1k} : ultimate limit state axial characteristic compression load
R : bearing resistance of a single pile
R_{1d} : design value of the ultimate bearing resistance of a single pile
R_{1k} : characteristic ultimate bearing resistance of a single pile
R_{2d} : design value of the bearing resistance of a single pile for the serviceability limit state
R_{2k} : characteristic bearing resistance for the serviceability limit state
R_b : base resistance of a single pile
R_{bk} : characteristic base resistance of a single pile
R_{b1k} : characteristic limit base resistance of a single pile
R_k : characteristic bearing resistance of a single pile
$R_{pile,i}$: charcacteristic value of the resistance of the pile *i* inside a pile group
R_{raft} : charcacteristic value of the resistance of a raft of a CPRF mobilized by contact pressure
R_s : shaft resistance of a single pile
R_{sk} : characteristic shaft resistance of a single pile
R_{s1k} : characteristic ultimate shaft resistance of a single pile
R_{tot} : characteristic value of the total load of a building or structure (for a CPRF)
R_{max} : test limit load as the maximum compression load applied in a static pile load test
R_{wall} : charcacteristic value of the resistance of a wall as part of a combined foundation
R_{work} : permissible characteristic working compression load of a single pile defined as $R_{work} = R_{1k} / \eta$
S_{1d} : design value for the compression load of a single pile for the ultimate limit state
S_{2d} : design value for the compression load of a single pile for the serviceability limit state
s : settlement of the pile head
s_1 : settlement of the pile head for the ultimate limit state
$s_{1,pl}$: limit plastic settlement of driven piles for the ultimate limit state
s_{2k} : characteristic value of the permissible settlement for the serviceability limit state
s_g : limit settlement of the pile head defined by DIN 4014
s_{sg} : limit settlement of the pile head defined by DIN 4014 for mobilizing the ultimate skin friction q_{s1k}
U : coefficient of uniformity
w : moisture content
w_L : liquid limit
w_P : plastic limit
γ_P : partial safety factor for piles defined by DIN V 1054-100
η : global safety factor defined by German standards
η_N : factor given by DIN V 1054-100 to consider the influence of the number *N* of carried out static load tests on the bearing capacity
η_z : factor given by DIN V 1054-100 to consider the influence of cyclic loads on the bearing capacity

REFERENCES

DIN-codes and recommendations:
DIN 1054. *Baugrund · Zulässige Belastung des Baugrunds; (Subsoil; permissible loading of subsoil).* German code, November 1976. Berlin: Beuth Verlag.

DIN V 1054-100. *Sicherheitsnachweise im Erd- und Grundbau · Teil 100: Berechnungen nach dem Konzept mit Teilsicherheitsbeiwerten*; *(Soil · Verification of the safety of earthworks and foundations · Part 100: Analysis in accordance with the partial safety factor)*. German code, April 1996. Berlin: Beuth Verlag.

DIN 1055. *Lastannahmen für Bauten; Lagerstoffe, Baustoffe und Bauteile; Eigenlasten und Reibungswinkel; (Design loads for buildings; store materials, building materials and structural members, dead loads and angle of friction)*. German code, July 1978. Berlin: Beuth Verlag.

DIN 4014. *Bohrpfähle - Herstellung, Bemessung und Tragverhalten; (Bored piles; construction procedure, design and bearing behaviour)*. German code, March 1990. Berlin: Beuth Verlag.

DIN 4020. *Geotechnische Untersuchungen für bautechnische Zwecke; (Geotechnical investigations for civil engineering purpose)*. German code, October 1990. Berlin: Beuth Verlag.

DIN 4026. *Rammpfähle · Herstellung, Bemessung und zulässige Belastung; (Driven piles; construction procedure, design and permissible loads)*. German code, August 1975. Berlin: Beuth Verlag.

DIN 4026 - Beiblatt. *Rammpfähle · Herstellung, Bemessung und zulässige Belastung - Erläuterungen; (Driven piles; construction procedure, design and permissible loads; comment)*. German code, August 1975. Berlin: Beuth Verlag.

DIN 4126. *Ortbeton-Schlitzwände · Konstruktion und Ausführung; (In situ concrete diaphragm walls; design and construction)*. German code, August 1986. Berlin: Beuth Verlag.

DIN 4128. *Verpreßpfähle (Ortbeton- und Verbundpfähle) mit kleinem Durchmesser · Herstellung, Bemessung und zulässige Belastung; (Injection piles (in-situ concrete piles and composite piles) with small diameter; construction procedure, design and permissible load)*. German code, April 1983. Berlin: Beuth Verlag.

DIN 18301. *VOB Verdingungsordnung für Bauleistungen · Teil C: Allgemeine Technische Vorschriften für Bauleistungen · Bohrarbeiten; (Contract procedure for building works; Part C: General technical specifications for buildings works; drilling works)*. German code, December 1992. Berlin: Beuth Verlag.

DIN 18304. *VOB Verdingungsordnung für Bauleistungen · Teil C: Allgemeine Technische Vorschriften für Bauleistungen · Rammarbeiten; (Contract procedure for building works; Part C: General technical specifications for buildings works; driving works)*. German code, December 1992. Berlin: Beuth Verlag.

DIN V ENV 1997-1. *Eurocode 7 · Entwurf, Berechnung und Bemessung in der Geotechnik · Teil 1: Allgemeine Regeln · Deutsche Fassung ENV 1997-1 : 1994*; *(Eurocode 7: Geotechnical design · Part 1: General rules; German version ENV 1997-1 : 1994)*. German prenorm, April 1996. Berlin: Beuth Verlag.

DIN EN 1536. *Ausführung von besonderen geotechnischen Arbeiten (Spezialtiefbau) · Bohrpfähle (execution of special geotechnical works: bored piles)*. German prenorm, April 1996. Berlin: Beuth Verlag.

EAU 1990 · *Empfehlungen des Arbeitsausschusses „Ufereinfassungen" · Häfen und Wasserstraßen*. Edited by the Hafenbautechnische Gesellschaft e.V. und der Deutschen Gessellschaft für Erd- und Grundbau e.V., 8th edition. Berlin: Ernst & Sohn.

Other references:

Arz, P., H.G. Schmidt, J.M. Seitz & S. Semprich 1991. *Grundbau (Foundation engineering)*. Special issue taken from Beton-Kalender 1991. Berlin: Ernst & Sohn.

Balthaus, H.G., H. Meseck & J.M. Seitz 1985. Dynamic Pile Tests - German Practice. *ICSMFE 1985*. San Francisco, 4/A/4: 1341-1346.

Bjerrum, L. 1972. Embankments on soft ground. *American Society of Civil Engineers. Specialty Conference on Performance of Earth and Earth-Supported Structures*. Purdue Univ., Lafayette, Ind. 1972. Proceedings, Vol. 2: 1-54.

Cooke, R.W., D.-W. Bryden-Smith, M.N. Gooch & D.F. Sillett 1981. *Some observations of the foundation loading and settlement of a multi-storey building on a piled raft foundation in London clay*. Proc. Instn. civ. Engrs., Part 1.

Deutsche Gesellschaft für Erd- und Grundbau · DGEG (1993). Statische axiale Probebelastungen von Pfählen · Empfehlungen des Arbeitskreises 5 der Deutschen Gesellschaft für Erd- und Grundbau e.V. *Geotechnik* 16 (3): 124-136.

German Society for Getotechniques (DGGT) 1997. *Recommendations for Static Axial Pile Load Testing.* Worked out by DGGT · Working Group 2.1, will be published in autum 1997.

Gudehus, G. & A. Weißenbach 1996. Limit state design of structural parts at and in the ground. *Ground Engineering* 29(9): 42-45.

Hartung, M. 1995. Einflüsse der Herstellung auf die Pfahltragfähigkeit in Sand. *Pfahl-Symposium '95*, Fachseminar 23./24. Februar 1995. Mitteilungen des Institutes für Grundbau und Bodenmechanik, TU Braunschweig, Heft No. 48: 137-153. Braunschweig.

Katzenbach, R. 1993. Zur technisch-wirtschaftlichen Bedeutung der Kombinierten Pfahl-Plattengründung, dargestellt am Beispiel schwerer Hochhäuser; (The technical and economical importance of the Combined Piled-Raft Foundation, represented by example of high rise buildings). *Bautechnik* 70(3): 161-169.

Katzenbach, R., U. Arslan & J. Gutwald 1994. A numerical study on pile foundation of the 300 m high Commerzbank Tower in Frankfurt am Main. *Numerical Methods in Geotechnical Engineering*: 271-277. Rotterdam: Balkema.

Katzenbach, R. & U. Arslan 1995. Foundations of tall buildings in Frankfurt Clay and Frankfurt Limestone - Examples for soil-structure-interaction. *Proc. Intern. Symposium "70 Years of Soil Mechanics".* Turkish National Committee of ISSMFE, Istanbul.

Katzenbach, R., U. Arslan, U. & Chr. Moormann 1996. Nachweiskonzept für die Kombinierte Pfahl-Plattengründung (KPP); (Design and safety concept for the Combined Piled-Raft Foundation (CPRF)). *Geotechnik* 19(4): 280-290.

Katzenbach, R.; H. Quick, U. Arslan, J. Gutwald & J. Holzhäuser 1997. Soil-structure-interaction of the 300 m high Commerzbank tower in Frankfurt am Main - measurements and numerical studies. *XIVth International Conference on Soil Mechanics & Foundation Engineering*, Hamburg. Will be published in September 1997.

Klingmüller, O. 1993. Dynamische Integritätsprüfungen und Qualitätssicherung bei Bohrpfählen; (Dynamic integrity test and quality controlling with bored piles). *Geotechnik* 16(2): 72-80.

Nendza, H. & D. Placzek 1988. Die Erhöhung der Pfahltragfähigkeit durch gezieltes Nachverpressen - Stand der Erfahrungen; (The improvement of the bearing capacity of piles by a specific re-injection - latest experience). *Vorträge der Baugrundtagung 1988 in Hamburg.* Deutsche Gesellschaft für Erd- und Grundbau, 323-339.

Poulos, H.G. 1993. Settlement prediction for bored pile groups. *Deep foundation on Bored and Auger Piles:* 103-117. Rotterdam: Balkema.

Petermann, Lackner & Schenck 1967. *Tragfähigkeit von Pfählen, Großversuche und ihre Auswertung.* 2. Versuchsreihe 1958/57. Berichte aus der Bauforschung, Heft 49, Berlin: Verlag Ernst & Sohn.

Richter, Th., S. Savidis; R. Katzenbach & H. Quick 1996. Wirtschaflich optimierte Hochhausgründungen im Berliner Sand. *Vorträge der Baugrundtagung 1996 in Berlin.* Deutsche Gesellschaft für Geotechnik, 129-146.

Schnell, W. 1996. *Verfahrenstechnik der Pfahlgründungen.* Stuttgart: B.G. Teubner.

Seitz, J.M. 1992. Low-Strain Intergritätsprüfungen bei Bohrpfählen. *Bautechnik* 69(10): 551-557.

Stocker, M. 1980. Vergleich der Tragfähigkeit unterschiedlich hergestellter Pfähle. *Vorträge der Baugrundtagung 1980 in Mainz.* Deutsche Gesellschaft für Erd- und Grundbau, 565-590.

Wend, R. & J.M. Seitz 1995. Tiefgründung mit teleskopierten Großbohrpfählen (∅ 1.80 m/ ∅ 1.50 m) bei der Commerzbank in Frankfurt am Main. *Pfahl-Symposium '95*, 23./24.2.1995, Mitteilungen des Institutes für Grundbau und Bodenmechanik, TU Braunschweig, No. 48: 213-230. Braunschweig.

Design of Axially Loaded Piles – European Practice, De Cock & Legrand (eds)
© 1997 Balkema, Rotterdam, ISBN 90 5410 873 8

Design of axially loaded piles – Irish practice

B.M.Lehane
Trinity College, Dublin

ABSTRACT: This paper provides the author's overview of piling practice in the Republic of Ireland. The review concentrates on pile design and construction procedures adopted by most piling contractors and practising geotechnical engineers and, as such, does not consider unusual contracts where non-standard construction or design procedures were used.

1. REGIONAL GEOLOGY

The use of piled foundations is, as one would expect, most common in urban areas with most activity occurring in the cities of Dublin, Cork, Limerick and Galway; see Figure 1. The stratigraphy in these areas commonly comprises less than 10m of relatively soft soil overlying dense gravel or rock. As a consequence, many piles are relatively short and derive a large proportion of their resistance to compression loading in end bearing. Piles are generally socketed into rock if they are subjected to significant tension loads.

The superficial/drift deposits range from soft alluvial clays and silts close to the rivers in the major cities to stiff/hard glacial till in much of the Dublin region or medium dense sandy gravels in Cork city. The superficial deposits which occasionally demand the use of piles outside of the major cities include peats, soft clays/silts and calcareous tufa; piles in these deposits are often used to provide support to bridge abutments.

Carboniferous limestones, which form the bedrock over most of central Ireland, are known to be highly karstified in some areas. A high degree of redundancy is built into the foundation design if piles are used in these areas.

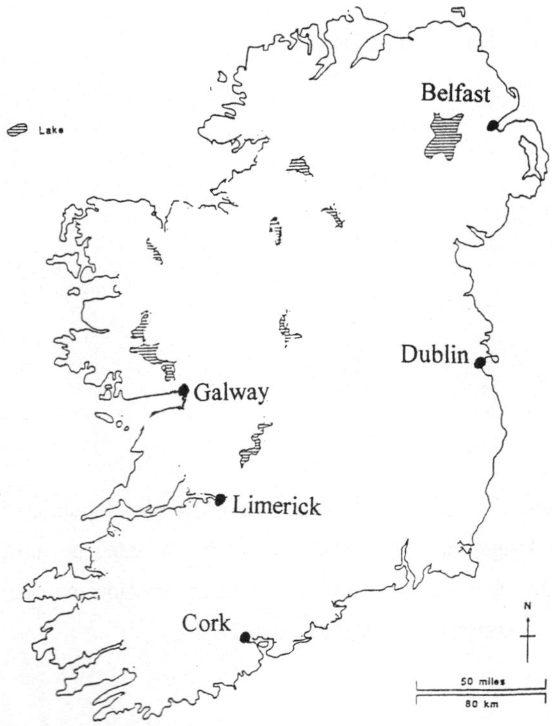

Figure 1 Major cities in Rep. of Ireland

2. SITE INVESTIGATION

Most site investigations for projects which potentially require piling investigate the stratigraphy and its properties using cable percussive and/or rotary cored boreholes. An important objective for each site investigation is to determine the depth to rockhead; cobbles and boulders present above rockhead often hamper an accurate estimate of this depth.

Standard classification tests (e.g. gradings and Atterberg limits) are performed on 50 ±20% of samples recovered from boreholes. The most frequently employed strength tests are shear box tests and unconsolidated undrained (UU) triaxial tests on driven 100mm diameter tube cohesive specimens.

By far the most commonly used index for the assessment of soil consistency is the Standard Penetration Test N value (blows/300mm); this N value is used directly in pile design calculations.

Cone Penetration testing (CPT) has generally been relatively unpopular because of the presence of coarse gravel and cobbles in the glacial till of Dublin and the gravels of Cork city. It is anticipated, however, that the recent purchase by Trinity College Dublin of a CPT rig will lead to renewed interest in its application to piling.

3. PILE TYPES

Piles in popular use in Ireland are generally of small diameter. This is because developments, up to recently, have been relatively small scale and hence contractors find it uneconomic to invest in heavy plant which may only occasionally find application.

Creed (1993) compiled a table, reproduced in Table 1, of pile types currently in use in Ireland. The terms micro, mini, small and large refer to piles with diameters less than 150mm, between 150mm and 300mm, between 300mm and 600mm and greater than 600mm respectively.

Details of the three most commonly used piles are summarised in the following sub-sections. Rotary and percussion bored replacement piles (without the use of bentonite) and top driven steel tubular and H piles are used but find fewer applications than these three.

3.1 Odex mini-pile

The Odex mini-pile is a 150mm to 350mm diameter cement-grout replacement pile in which the hole is formed by rotary percussive action using an eccentric reamer. As illustrated on Figure 2, this reamer drills a hole larger than the outer diameter of the casing, hence facilitating insertion of the casing. A reversed rotation of the drill causes the reamer to retract into the concentric position

Table 1. Pile types in Ireland (Creed 1993)

<table>
<tr><th colspan="4">REPLACEMENT PILES</th><th colspan="2">DISPLACEMENT PILES</th></tr>
<tr><th>Micro diameter</th><th>Mini diameter</th><th>Small diameter</th><th>Large diameter</th><th>Small diameter</th><th>Large diameter</th></tr>
<tr><td colspan="2"></td><td>Percussion bored</td><td></td><td>Driven segmental precast concrete</td><td rowspan="2"></td></tr>
<tr><td colspan="2"></td><td colspan="2">Rotary bored</td><td>Driven steel H piles</td></tr>
<tr><td colspan="2"></td><td colspan="2">Cont. flight auger concrete injected</td><td colspan="2">Driven steel tubes, open ended</td></tr>
<tr><td></td><td colspan="2">Cont. flight auger grout injected</td><td></td><td colspan="2" rowspan="4"></td></tr>
<tr><td colspan="2">Segmental CFA grout injected</td><td colspan="2"></td></tr>
<tr><td colspan="2">Drilled & grouted</td><td></td><td></td></tr>
<tr><td colspan="2">Drilled & toe grouted</td><td></td><td></td></tr>
</table>

and allows its removal. Reinforcement is then inserted and the grout is tremied as the casing is jacked out.

Odex mini-piles are often base grouted using a single sleeved tube à manchettes located at the base; this sleeve is injected with grout 24 hours after the pile has been constructed.

3.2 Precast concrete pile

Segmental precast high strength concrete piles (f_{cu} =50N/mm^2) are the most common form of driven pile used in Ireland. The piles are of square cross-section (widths varying between 250mm and 350mm) which, for pile lengths in excess of about 12m, are assembled in segments using a

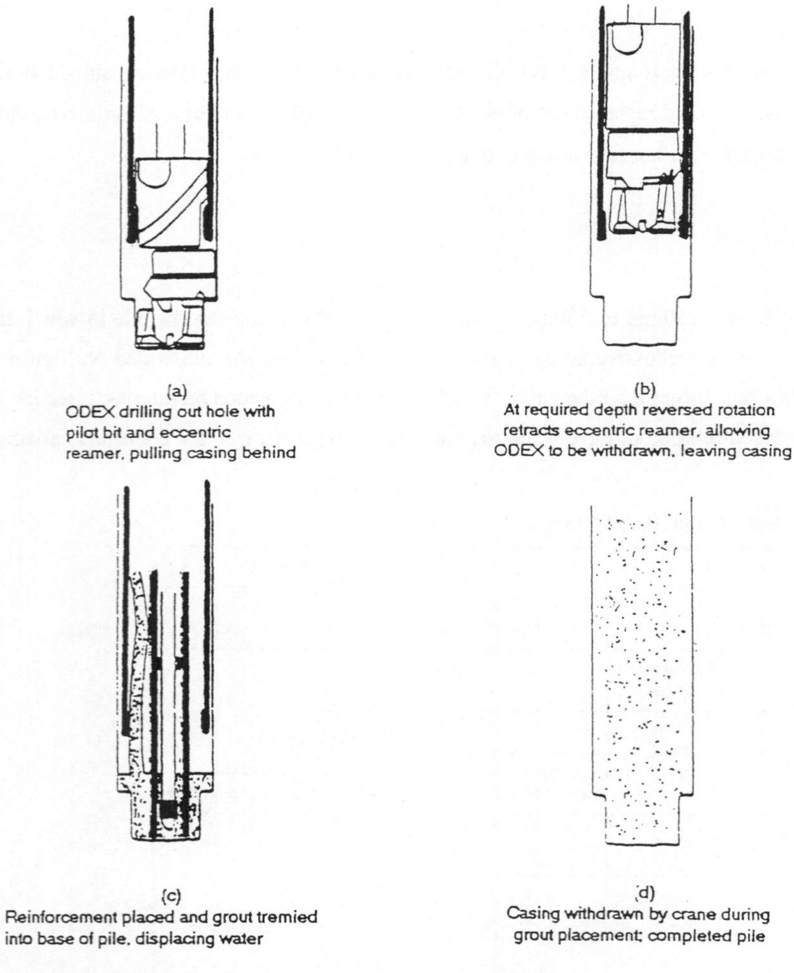

Figure 2. The Odex drilled pile (Higgins & Mason 1989)

Balken/Hercules bayonet connection or spigot & socket type connection. The piles are driven using a variety of hammers with impact energies of between 10 kNm and 50 kNm. Diesel hammers are not employed because of the increased risk of pile damage associated with their use with these piles in hard-driving conditions.

Cobbles and boulders in the soil overlying the bearing stratum pose the biggest obstacle to the successful installation of the piles. A hardened steel point is often fitted to the toe of the pile in these conditions. Pre-boring a smaller diameter hole is occasionally carried out using a heavy steel mandrel.

3.3 Continuous flight augered piles

The continuous flight augered (CFA) piles in use in Ireland vary in diameter (D) from 150mm to 750mm. As shown on Figure 3, the hollow-stemmed auger is rotated into the ground to the desired depth before grout (for D <350mm) or concrete (for D >450mm) is pumped under pressure via the stem to an outlet at the auger's base. The auger is withdrawn at a rate ensuring that the grout/concrete at the base of the auger is maintained at a sufficiently high pressure and finally the reinforcement is placed.

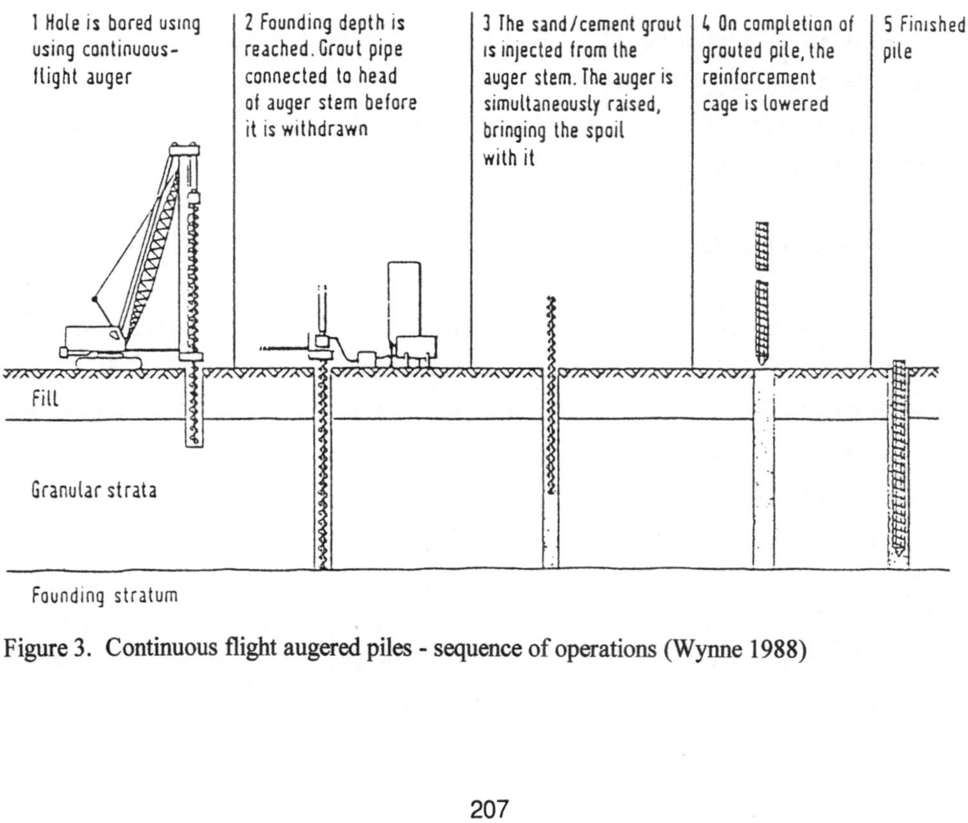

Figure 3. Continuous flight augered piles - sequence of operations (Wynne 1988)

To allow greater flexibility in situations where conventional CFA pile installation may run into difficulties, one Irish contractor (HMC Ltd.) has a rig adapted to allow installation of a temporary casing at the same time that the CFA is inserted. The insertion of the casing is assisted by using water flush pumped under pressure down the hollow auger. If an obstruction is encountered (this can happen quite frequently in Irish conditions), the auger is withdrawn and the hole is advanced by a down-the-hole hammer (with air flush) and by drilling the casing. The hole may subsequently be re-augered and the CFA pile constructed in the usual way.

4. DESIGN APPROACH

4.1 Design philosophy

Most designs are performed by the piling contractor who tenders a price to the Client/Engineer for the construction of piles with a prescribed working load. In this way, the piling contractor carries most of the liability for design and construction. The Contractor's tender must therefore represent a compromise between the need to be competitive and the need to have a design that will work and be profitable. The Client/Engineer must provide sufficient information on the stratigraphy and soil/rock properties to enable the Contractor to carry out the design and to limit the possibility of claims on the basis of unforeseen ground conditions.

As already stated, piles subjected to significant uplift forces are usually socketed into rock (usually by a distance of 2 to 3m). The design of these piles is similar to that of a structural element and therefore attention in the following sections will focus on aspects associated with piles loaded in compression.

4.2 Codes of practice & relevant publications

The British Standard code of practice for foundations (BS 8004,1986) is used to provide general guidance on pile design and construction and the British Institution of Civil Engineers 'Specification for piling' (ICE 1988) is often referred to in specifications for static load tests. The (new) 1996 ICE specification is currently being assessed and is likely make an impact on testing procedures in the coming years.

Current popular practical references particular to Irish conditions include Collins & Mitchell (1990) for piles in gravel, Looby et al (1995) for driven piles in boulder clay and Long (1995) for driven piles in mudstones. In the absence of local knowledge and experience, the textbook by Tomlinson (1994) is frequently referred to at the preliminary design stage.

5. NATIONAL DESIGN METHODS

5.1 General

Design is essentially reliant on the experience of the specialist piling contractor who, in turn, bases his proposals on the results from specific static load tests performed on the site and/or from load tests in similar conditions at other sites.

On large contracts (> 250 piles), after selection and agreement by all parties of the most appropriate pile type, preliminary static load tests which load piles to at least twice the design working load are carried out. These tests, which usually number about 1% of the total number of piles, are used to verify the design. A further 1% of contract piles are load tested to 1.5 times the working load.

On medium sized projects (50-250 piles), preliminary static load tests are not performed and reliance is placed on tests on contract piles (loaded to 1.5 times the working load).

On small contracts, it is often seen to be more cost-effective not to perform a static load test and to design the piles more conservatively. Dynamic load testing would usually be carried out in these instances.

5.2 Definitions

The following definitions are used in design terminology for axially loaded piles.

> Ultimate capacity: The pile head load required in a static (maintained) load test to induce a pile head settlement of 10% of the pile diameter.
>
> Factor of safety: The ultimate capacity divided by the pile design working load.
>
> Creep rate: The rate at which a pile settles after application of a load increment in a maintained load test.

5.3 Design on basis of static load tests

Pile tests are usually located in areas of the construction site where ground conditions are suspected to be poorer than average. Tests generally follow the maintained load test procedure (constant rate of penetration tests are rarely performed) and, in preliminary tests, the load is applied in typically about four cycles alternating from zero to nominal design working load (WL), zero to 1.5 x WL, zero to 2 x WL and finally to 2.5 x WL. Loads are maintained at each level until the creep rate reduces to below 0.1mm/hour (or 0.25mm/hour, in some contracts).

A satisfactory working load for the test pile is decided upon after examination of the pile load-deflection curve and the response of the pile on un-loading during the load cycling. Allowable pile head deflections at working load rarely exceed 1.5% of the pile diameter and the pile recovery on unloading must be at least equivalent to the anticipated elastic shortening of the pile.

Site specific correlations of ultimate shaft and base resistances with SPT N values are occasionally developed from test pile results at sites where soil conditions are variable, although predictable, and hence where there is scope for optimising designs. In general, however, there is limited opportunity for this practice in Ireland as most piles derive most of their resistance from the base. There is consequently no other economic option other than to found the piles in the bearing stratum.

5.4 Design on the basis of site investigation results

Preliminary designs are often verified on the basis of correlations in the literature with SPT N values and sometimes with UU undrained shear strengths (c_{u0}). It is appreciated by all that these design calculations provide nothing more than a guide to pile capacity.

The summary provided by Poulos (1989) is a good reflection of the *type* of SPT N correlations currently employed and is reproduced here in Tables 2 & 3 for completeness. Note that the symbols f_b and f_s refer to ultimate base pressure and ultimate average shaft shear stress respectively so that the pile capacity, P_{max}, is given as:

$$P_{max} = f_b A_b + f_s A_s$$

where A_b is the area of the pile base and A_s is the area of the pile shaft.

Typical values of f_s and f_b adopted when correlations are based on design c_{u0} values are listed in Table 4. Note that c_{u0} is its average value along the pile shaft when calculating f_s and the value at the pile base when calculating f_b. The correlations in Table 4 tend to be conservative by comparison with those implied by the correlations in Tables 2 and 3.

The application to Irish conditions of recently published new pile design approaches is currently being examined by researchers at Trinity College Dublin; these include those developed at Imperial College London for driven piles in sands/gravels and in clays (Lehane et al 1994,

Table 2 Ultimate base capacity fb (MPa) = K N Poulos (1989)

Pile type	Soil type	K	Remarks
Driven displacement	Sand	0·45	N = average SPT value in local failure zone
	Sand	0·40	
	Silt, sandy silt	0·35	
	Glacial coarse to fine silt deposits	0·25	
	Residual sandy silts	0·25	
	Residual clayey silts	0·20	
	Clay	0·20	
	Clay	0·12	
	All soils	0·30	For $L/d \geqslant 5$ If $L/d < 5$, $K = 0{\cdot}1 + 0{\cdot}04\ L/d$ (closed-end piles) or $K = 0{\cdot}06\ L/d$ (open-ended piles)
Cast in place	Cohesionless		$f_b = 3{\cdot}0\ \text{MN/m}^2$
		0·15	$f_b \not> 7{\cdot}5\ \text{MN/m}^2$
	Cohesive	—	$f_b = 0{\cdot}09\ (1 + 0{\cdot}16z)$ where z = tip depth (m)
Bored	Sand	0·1	
	Clay	0·15	
	Chalk	0·25 0·20	$N < 30$ $N > 40$

Table 3 Ultimate shaft capacity fs (kPa) = α + β N Poulos (1989)

Pile type	Soil type	α	β	Remarks
Driven displacement	Cohesionless	0	2·0	f_s = average value over shaft $\bar{N}$ = average SPT along shaft Halve f_s for small displacement pile
	Cohesionless & cohesive	10	3·3	Pile type not specified $50 \geqslant N \geqslant 3$ $f_s \not> 170\ \text{kN/m}^2$
	Cohesive	0	10	
Cast in place	Cohesionless	30	2·0	$f_s \not> 200\ \text{kN/m}^2$
		0	5·0	
	Cohesive	0	5·0	$f_s \not> 150\ \text{kN/m}^2$
		0	10·0	
Bored	Cohesionless	0	1·0	
		0	3·3	
	Cohesive	0	5·0	
	Cohesive	10	3·3	Piles cast under bentonite $50 \not> N \not> 3$ $f_s \not> 170\ \text{kN/m}^2$
	Chalk	−125	12·5	$30 > N > 15$ $f_s \not> 250\ \text{kN/m}^2$

Table 4. f_s and f_b correlations with c_{u0} used in Ireland

	Bored/CFA pile	Driven piles
f_s	$\approx 0.5\ c_{u0}$	$\approx 0.4\ c_{u0}$ for c_{u0} >100 kPa $\approx 1.0\ c_{u0}$ for c_{u0} <30 kPa
f_b	$\approx 9\ c_{u0}$	$\approx 9\ c_{u0}$ in alluvium $\approx 20\ c_{u0}$ in glacial till

Table 5. Pile details and set criteria used by Irish driven pile contractors

Contractor	Pile Dimension (mm).	Typical Working Loads (kN)	Pile Impact Energy (kN m)	Set Criteria (Blows/ 25mm)
Murphy International Ltd.	250 x 250 300 x 300 350 x 350	600 1000 1500	14 14 17.5	20-25
Lowry McKinney Ltd.	250 x 250 285 x 285	600 - 800 800- 1000	45	10
Taranto Ltd.	235 x 235 275 x 275	650 1000	18 25	15 23
F-K Piling Ltd	275 x 275	1000	30	12
Pierse Bachy Ltd.	275 x 275 350 x 350	750-1000 1200 - 1500	25	25-30

Lehane & Jardine 1994 and Jardine & Chow 1996). It is likely, however, that, the more traditional (f_s,f_b) approach will retain its current popularity because of its simplicity.

5.5 Design on the basis of driving formulae

A recent survey revealed that over 80% of Irish piling contractors use the 'Hiley' or 'Engineering News' dynamic formulae for the preliminary evaluation of driven pile capacity. The pile details disclosed by the survey for precast concrete piles in glacial tills are summarised in Table 5.

The working loads for the precast piles in glacial till correspond to an axial stress of $\approx$12 N/mm^2 and are based on the assumption that achieving the value of the 'set' given in the table ensures that the piles are embedded sufficiently in relatively rigid strata. The compatibility between the set criteria adopted by the various contractors for this design condition is illustrated on Figure 4 by the relatively unique relationship between the set value adopted by a given contractor and the energy imparted to the piles.

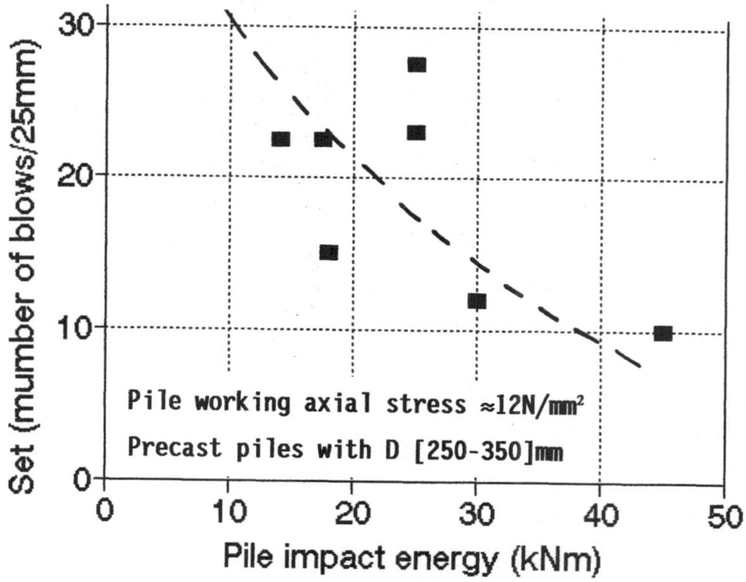

Figure 4. Pile set required by Irish contractors as a function of the impact momentum of their respective hammers.

5.6 Design on the basis of wave equation analysis

Dynamic testing to predict dynamic pile capacities and the load-displacement behaviour of the piles usually supplements information obtained from static load tests in large contracts and represents the only method of testing in small contracts. Its employment in all sizes of contracts is growing in popularity on account of its relatively low cost - with the result that a smaller percentage of static load tests are now being performed. Wave equation analyses are most often carried out for driven piles (CASE and CAPWAP analyses), although the use of the SIMBAT method (where a 10kN is dropped on the pile) is finding increased application for replacement piles.

A recent example of a comparison between capacities predicted by wave equation analyses (using CASE) on driven piles and those measured on the same piles in static load tests is provided in Table 6. Reasonable agreement is evident although the example illustrates the sensitivity of the dynamic solution to the assumed damping factor. The somewhat lower ultimate capacities predicted in the dynamic tests are thought to be due to the fact that the blows applied to the piles did not induce a permanent displacement sufficient to mobilise all available base resistance.

Comparisons such as those shown on Table 6 are providing improved confidence in predictions from wave equation analyses. There is still, however, some unease because of an over-reliance on the expertise of the analyst carrying out the interpretation.

Table 6. Comparison of measured ultimate capacities with those predicted using using wave equation analyses (Piles were driven piles in Dublin boulder clay).

Piling Contract	Pile Dimension (mm)	Working Load (kN)	Static Test Ultim. Load (kN)	Dynam. Test Ultim. Load Prediction (kN)	Case, Damping Factor, J_c
Croke Park	300 x 300	800	=2900	1730 2030	1 0.7
TCD Resid.	300 x 300	900	=3100	2532	1
Geog. Dock 2	285 x 285	1000	=3400	2911	.7

5.7 Factors of safety and rules for serviceability

Most piles in Ireland, as they are predominantly end-bearing, are designed with a global factor of safety of 3. This factor is occasionally reduced to 2.5 in materials such as Dublin Boulder Clay which are relatively well understood.

In practice, however, the limiting condition is that the pile settlement in service is acceptable to the structural designers.

6.0 PILE DESIGN CALCULATION EXAMPLES

The calculation of pile working loads is trivial, as is evident from Section 5.4. A brief summary of the 'static' calculation procedures for the most common pile types is summarised below.

(i) Precast concrete pile driven to a set greater than or corresponding with the trend line shown on Figure 4 is allowed to carry an axial stress of up to $f_{cu}/4$, where f_{cu} is the cube strength of the concrete. The allowable working load (P_w) is simply this stress times the cross-sectional area of the pile.

(ii) Bored/CFA piles are designed for working loads:

$$P_w = (f_b A_b + f_s A_s)/3$$

where f_b and f_s are related to the SPT N or c_u value (as in Tables 2 to 4). Given their crude nature, no attempt is made to assume that these correlations for *average* skin friction can be applied separately to the different strata penetrated by the pile shaft.

Checks are made, where possible, to ensure compatibility between the N and c_u approaches.

(iii) The allowable axial stress in piles founded on rock is usually controlled by the strength of the pile material. When these piles are socketed into rock, the maximum working tension load is given as:

$$P_w = (f_s)_{all}\, A_s$$

where $(f_s)_{all}$ is the allowable skin friction in the rock socket and A_s is the area of the shaft in the rock socket. Values of $(f_s)_{all}$ used for reasonable quality Irish Limestone (RQD >50%) vary from between 350 kPa to 700 kPa, depending on the designer. Higher values of $(f_s)_{all}$ are used if the socket is pressure grouted.

(iv) Negative skin friction forces are calculated, where appropriate, assuming downward local shear stresses are about $0.3s'_{v0}$, where s'_{v0} is the free-field vertical effective stress. Reference is made to examples in Tomlinson (1994) for more detailed approaches.

7.0 QUALITY CONTROL AND MONITORING

Aside from standard on-site supervision of the piling works and static/dynamic load tests of contract piles, the most popular additional means of quality control is integrity testing. Almost all bored and driven piles (with D >250mm) currently constructed in Ireland are integrity tested. The French Transient Dynamic Response (TDR) system is the most popular but the Dutch TNO system is also used. Integrity tests are used extensively because they are inexpensive, quick to execute and improve the client's confidence in the piling workmanship. The reliability of integrity testing is questioned less often than it was during its initial introduction to Ireland in the 1980's.

8.0 PARTICULAR NATIONAL EXPERIENCES & RESEARCH NEEDS

Most piles in Ireland are end-bearing and, as such, their performance is very much dependent on the quality of the procedure used in the construction of the pile. Integrity testing has had a high success rate in Ireland and has improved confidence levels. Problems with end-bearing piles are therefore generally very rare and, when they occur, are usually related to poor quality rock at or just beneath the pile base.

As more sites in Ireland continue to be developed, there has been an increased need to employ piles which are not predominately end-bearing and hence make full use of the available shaft capacity. Some of the questions raised by piling contractors following recent experiences with these piles include:

- Why are the sets of piles driven into glacial till often lower during re-drives ?
- Why was there only a relatively minor rate of increase in axial capacity with pile length in piles installed in gravels ?
- How do the base capacities of driven and bored piles differ ?
- How do dynamic and static capacities vary for various pile types in a range of soil conditions ?

Piling contractors are currently funding or have funded some instrumented pile test programmes to resolve these and other issues. An illustration of the type of data obtained from instrumented pile test programmes is shown on Figure 5 which plots the variation of base resistance with pile base displacement measured in tests on a 273mm diameter pile driven into Dublin Boulder Clay to a relatively modest set of 12 blows/25mm for an impact momentum of 12 kNm; see Figure 4. Test 1C, 2C and 3C were conducted 2.5 hours, 2 days and 17 days respectively after pile driving.

Figure 5 shows that, in contrast to conventionally made local assumptions, base capacities and stiffnesses increase markedly with time and, in the long term, f_b approaches a value of 25 MPa. Despite the relatively low set of the test pile, this capacity is significantly larger than an f_b value of 9 MPa estimated using the correlation provided in Table 4 with c_{u0} =450 kPa.
It is anticipated that high quality test results such as those shown on this figure will greatly assist the design of piles in Ireland.

9. CONCLUSION

Pile design and construction in Ireland has been relatively straight forward compared to other parts of Europe because of the high level of rockhead or very dense soil in the urban areas. However, as sites with less familiar geologies are developed and more demanding structures are constructed, designers are more frequently needing to examine 'outside' experiences such as those described in the proceedings of this seminar. It is encouraging to report that Irish piling contractors are showing great adaptability to the new challenges they are faced with.

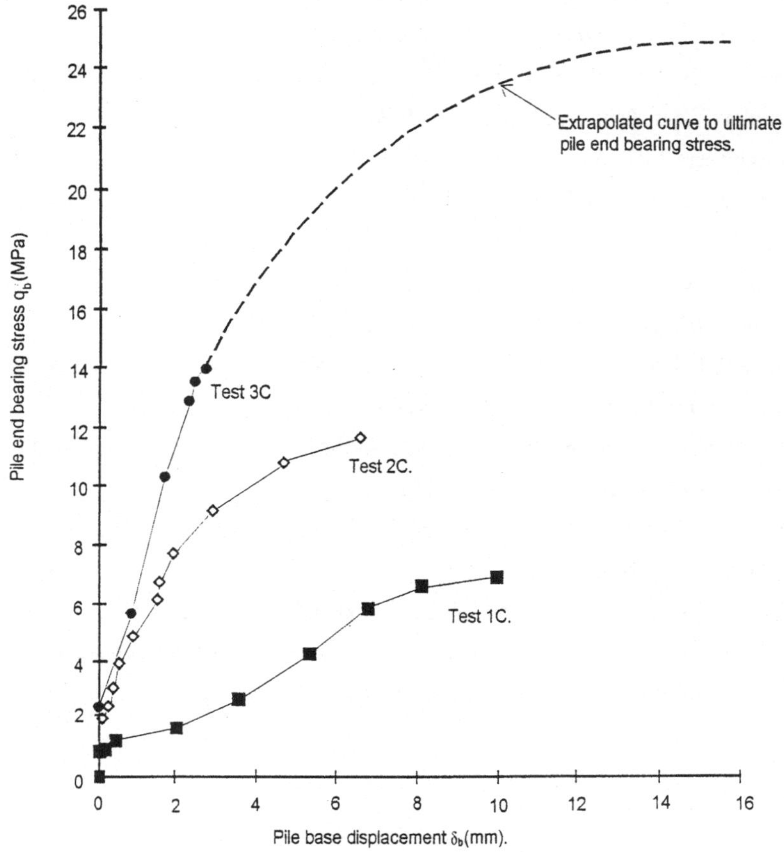

Figure 5. Pile end-bearing characteristics measured on a driven pile in Dublin Boulder Clay

ACKNOWLEDGEMENTS

The author gratefully acknowledges the assistance of Mr. Frank Callanan, Dr. Michael Creed, Dr. Eric Farrell, Mr. Michael Long, Mr. Michael Looby and Mr. Tim Paul during compilation of this paper.

REFERENCES

BS8004 1986. Foundations, British Standard Code of practice. *British Standards Institution.*

Collins F. and Mitchell J.M. 1990. Piling in gravels. *Trans. Instit. of Engineers of Ireland*, 114.

Creed M.J. 1993. Piled foundations - a review of the Irish scene. *Proc. Seminar on Foundations, The Irish Concrete Society, Galway.*

Farrell E.R., Lehane B.M. and Looby M. 1997. Instrumented pile tests in Dublin Boulder Clay. submitted to *Geotechnical Engineering.*

Higgins J. & Mason S. 1989. Geotechnical aspects of the re-development of the Custom House Docks area in Dublin. *Trans. Instit. of Engineers of Ireland,* 113.

Institution of Civil Engineers 1988. *Specification for piling,* Thomas Telford, London.

Jardine R.J. & Chow F.C. 1996. New design methods for offshore piles. *UK MTD publication 96/103.*

Lehane B.M. and Jardine R.J. 1994. Shaft capacity of driven piles in sand: a new design method. *Proc. VII Int. Conf. on the Behaviour of Offshore Structures, Boston, BOSS '94,* 1:23-36

Lehane B.M., Jardine R.J., Bond A.J. and Chow F. 1994. The development of shaft friction on piles in clay. *Proc. XIII Int. Conf. on Soil Mech. and Fdn. Engng., New Delhi,* 2:473-476

Long M. 1995. The behaviour of driven tubular steel piles in the laminated mudstones of south-west Ireland. *Geotechnical Engineering,* 113:242-252

Looby M., Farrell E.R. and Lehane B.M. 1995. Driven piles in boulder clay- design and practice. *Trans. Instit. of Engineers of Ireland,* 120.

Poulos H.G. 1989. Pile behaviour - theory and application. *Geotechnique,* 39:365:415

Tomlinson M.J. 1994. *Pile design and construction practice, 4th ed.* Viewpoint publications.

Wynne C.P. 1988. A review of bearing pile types. *UK PSA/CIRIA Report PG1.*

Design of Axially Loaded Piles – European Practice, De Cock & Legrand (eds)
© 1997 Balkema, Rotterdam, ISBN 90 5410 873 8

Design of axially loaded piles – Italian practice

Alessandro Mandolini
Department of Civil Engineering, Second University of Napoli

ABSTRACT: The paper is aimed to summarise the current methodology used by Italian designers when dealing with pile foundations. To give the possibility to better understand the reasons on the basis of which some design criteria are more widespread than others, some (necessarily brief) information are given about regional geology, common practice for soil investigations and Italian pile market. To this aim, explicit reference to publications, papers and textbooks was necessary. Finally, some practical indications coming from relevant under way research activities in Italy will be shown in a due detail, with the hope that the results might be helpful to other European countries.

1. REGIONAL GEOLOGY

Geological maps (1:100000 scale) covering the entire Italian country are available, but these are often old and of little use in detailed planning of engineering works. Detailed geological information can be gathered from literature or from thematic maps recently set up by Universities and/or State owned organisation.

From the general point of view, most of the Italian peninsula is included in the Alpine Orogenic Belt which underwent heavy folding and uplifting processes; due to lithostratigraphic and structural complexities connected with tectonics, Italian geological features are difficult to schematise. Anyway, a rough but comprehensive idea can be derived from Figure 1:

a) intrusive and metamorphic rocks outcrop in the Alps and Pre-Alps mountain chains along the east-west axis, in eastern Sardegna, in Calabria and in north-eastern Sicilia. In the latter two these rocks are deeply altered and weathered;

b) the so-called structurally complex formations (AGI 1979a) is found in the Apennines mountain chains, along the north-south axis. These formations are essentially made by flysch, turbidic deposits, indurated clay and clay-shales with chaotic structure;

c) moraine deposits, originated during quaternary glaciations, are rather common on the Alps and on the northern border of the Po River Plain;

d) volcanic soils (lava, cemented tuff and loose pyroclastic materials) are present in the middle part of the Peninsula, as well as in Sardinia and in Sicily Islands;

e) alluvial plains and lowlands sediments are mainly formed by soft normally consolidated fine grained sediments (AGI 1979b).

Alps, Pre-Alps and Appenninan mountain chains cover ≈ 55% of the total surface of the country. The remaining 45% is covered mainly by alluvial plains and lowlands like Po River Plain, Pisa Valley in Tuscany and Fiumicino near Rome (≈ 25%).

Earthquake, landslides and floods are natural phenomena which are rather common in Italy.

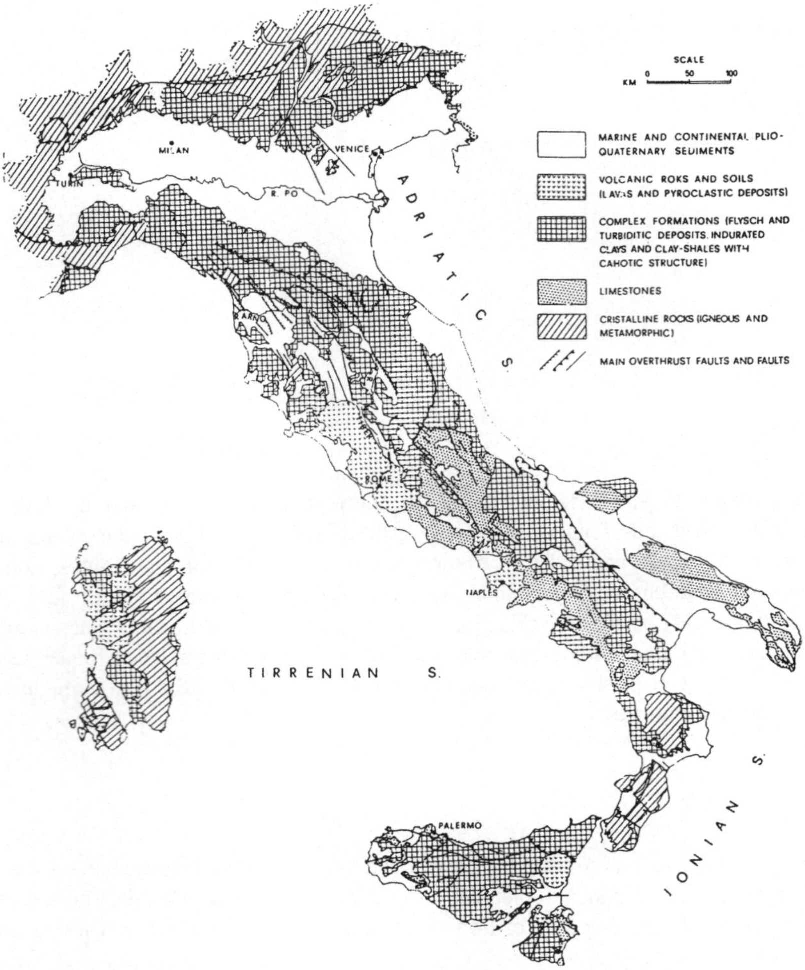

Figure 1. Schematic geological map of Italy (after Esu et al. 1984)

The geographic position of Italy make favourable a Mediterranean climate characterised by two rainfall maximum (in Autumn and Spring) and by a dry and hot long summer period. Due to the orientation of the long mountain ridges almost normal to the prevailing westerly winds, rainfalls turn often to storms causing severe flood damages.

Landslides occur in natural slopes of hilly regions throughout Italy, creating severe problems of maintenance of the transport network, and of stability of important towns and villages.

2. COMMON PRACTICE FOR SOIL INVESTIGATION

According to Lancellotta (1995) "*the ultimate aim of a subsurface investigation is to assess enough information to select the most appropriate foundation solution, to outline problems that could arise during construction and, on a more general scale, to highlight potential geological hazards in the examined area*".

The extension of the investigation programme depends on the importance of the project and on the stage of the design at which is referred.

With reference to a number of case histories recorded in Southern Italy (Napoli and suburbs), in

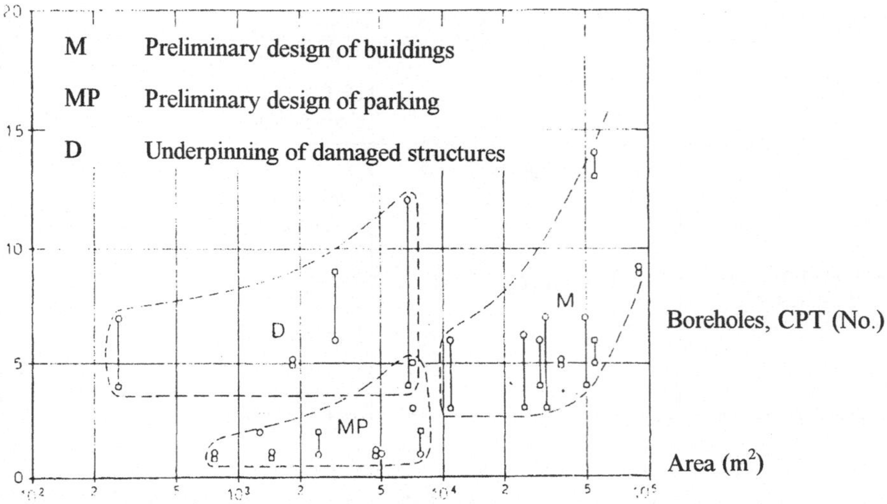

Figure 2. Correlation between size of the project and number of boreholes and/or CPT carried out in the Neapolitan area during the preliminary design of buildings, parking and underpinning of damaged structures (after Pellegrino 1996)

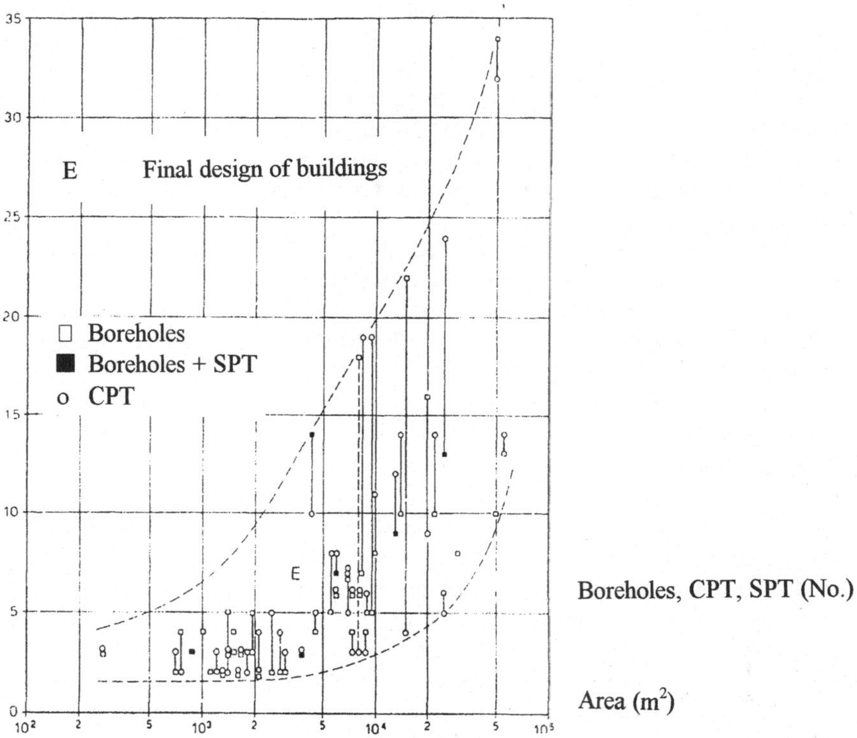

Figure 3. Correlation between size of the project and number of boreholes and/or CPT and SPT carried out in the Neapolitan area during the final design of buildings (after Pellegrino 1996)

Figure 2 the correlation existing between the size - assumed, generally speaking, proportional to the importance - of the project and the number of the boreholes and/or CPT carried out during the preliminary stage of the design is showed.

Table 1. List of the investigation methods commonly used in Italy (after Viggiani 1993a)

ESTABLISHED PURPOSE	COMMON INVESTIGATION METHODS	
soil profile	Direct Borehole Excavation (pit, trench, ...)	Indirect CPT Geophysical
physical and mechanical soil properties	Laboratory tests on undisturbed samples	In situ Static penetrometer (CPT, CPTU) Dynamic penetrometer (SPT, DPT) Vane Pressiometer Flat Dilatometer Plate Loading tests Permeability tests Geophysical (CH, DH, SASW)

With reference to only buildings, Figure 3 clearly shows that need of more detailed information for the final design make density of the investigations - number of explored vertical divided by the area - significantly larger.

The basic rules for Geotechnical Investigations are given in the section B of the Italian Geotechnical Code (MM.LL.PP - D.M. 21.01.1981, successively updated by D.M. 11.03.1988), state that "*any geotechnical design must be based on the results of a thorough geotechnical study of the site*".

Even if general rules are given, the Code leaves to the designer a wide liberty of how planning and performing such study. To get more specific guidance, the designer can refer to two handbooks edited by AGI (1977, 1994), the former dealing with the practical instructions for planning and executing site investigations and in situ tests, the latter with the suggested procedures to be followed if physical and mechanical soil properties have to be correctly derived from laboratory tests.

In Table 1 a schematic list of the investigation methods commonly used in Italy is given.

It must be stressed that, recently, in situ tests have received a great deal of attention in Italy, essentially due to an improvement of the capability of the existing techniques, to the introduction of new devices, to development of more appropriate theoretical approaches and, especially for common structures, to their promptness and economy.

3. ITALIAN PILE MARKET

Due to the high variable subsoil conditions, a vast spectrum of technology is applied when the decision of using piles as foundation system is taken; evidently, sometimes the choice of the pile type must be adapted to the requirements of the supported structure (i.e. high load concentration).

In the following some tables summarising the results of a detailed questionnaire that AGI submitted to the attention of all the public and private companies are shown (Baldovin 1989[1]).

The data were grouped in three different main categories: driven (table 2), bored (table 3) and mini (table 4) piles. For every category, more specific information (technique, type of structures for which they are used, maximum values of size and load, etc.) are given.

Since no data referred to continuous flight auger piles were collected at that time, information were taken from the paper by Trevisani (1992).

[1] In some cases, small changes were made on the basis of the updated collected information.

Table 2. Driven piles (modified from Baldovin 1989)

PILE TYPE	STRUCTURES				MAXIMUM VALUES			PATTERN	
	building	hydraulic	road railway	maritime	dia. [cm]	length [m]	axial load [MN]	vertical	raked
1 - TIMBER	X	X		X	30	10	0.2	↓	↘
2 - STEEL									
tubular	X	---	X	X	100	50	5.0	↓	↘
structural	X	---	X	X	50	20	0.6	↓	↘
offshore	---	---	---	X	300	150	>50.0	↓	↘
3- CONCRETE									
precast									
constant section	X	X	X	X	30/40	20	0.8	↓	↘
segmental	X	X	X	X	30/40	20/30	0.8	↓	---
conic section	X	X	X	X	60	15	0.5	↓	↘
prestressed	X	---	X	X	80	15/30	1.1	↓	↘
cast in place									
with temporary casing	X	X	X	---	50/80	20	0.9	↓	---
with permanent casing	X	---	X	X	40	40	1.1	↓	---

Table 3. Bored piles (modified from Baldovin 1989)

PILE TYPE	STRUCTURES			MAXIMUM VALUES				PATTERN	
	general	maritime	landslide	dia. [cm]	length [m]	axial load [MN]	enlarged base	vertical	raked
1 - SMALL TO MEDIUM DIAMETER (< 60 cm)									
rotary bored	X	X	X		30	1.0	X	↓	---
percussion bored	X	X	X		30	1.0	---	↓	---
2 - LARGE DIAMETER (> 80 cm)									
rotary bored	X	X	X	150/200	70/80	> 10.0	---	↓	---
percussion bored	X	X	X	150/200	70/80	> 10.0	---	↓	---

Table 4. Mini piles (modified from Baldovin 1989)

PILE TYPE	STRUCTURES			MAXIMUM VALUES			TYPE OF LOAD		
	general	maritime	landslide	dia. [cm]	length [m]	axial load [MN]	compressive	uplift	shear
1 - WITH STEEL TUBE	X	---	X	30	> 30	0.3	X	X	X
2 - WITH REINFORCEMENT CAGES	X	---	X	30	> 30	0.2	X	X	---

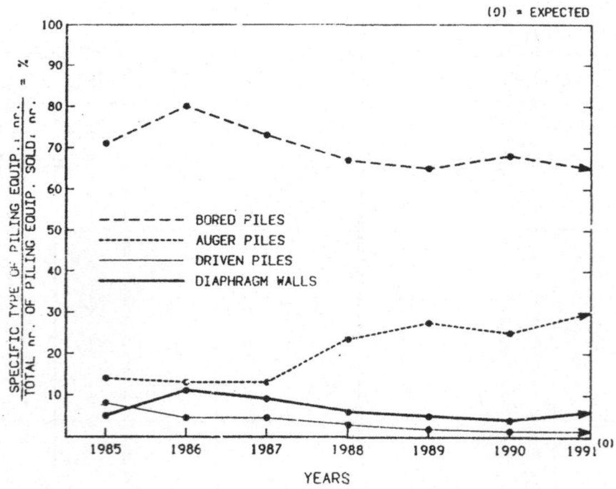

Figure 4. Trend of piling techniques in Italy (after Trevisani 1992)

In the alluvial plains and lowlands precast centrifuged and cast in situ concrete driven piles are generally used for low and medium bearing capacity foundation systems; large diameter bored piles are adopted when high loads are applied.

The Orogenetic areas of the Apennines and the Pre-Alps are characterised by landslides and instability of eluvial and colluvial layers. In these cases, high capacity piles are generally used also for their large stiffness to the lateral loadings.

Lands covered by volcanic sediments, being characterised by spatial heterogeneity and variability, claim for a high flexibility of the installation techniques, making bored piles preferable in this subsoil conditions.

In Figure 4 the yearly trend from 1985 to 1991 is showed, based on the sold equipment to perform different pile types.

From the figure it can be seen that, as a consequence of the previously mentioned reasons, bored piles are predominant in Italian market (from 65% to 80%); CFA piles are used more extensively than driven piles (14÷30% the former, 3÷11% the latter).

Furthermore, simply looking at the gradients, it could be deduced that CFA piles are gaining the market against driven piles, while bored piles follow an approximately steady trend. This is probably due to the fact that until now CFA piles have been produced with diameter ranging between 350 and 600 mm, hence without real competition with large diameter bored piles.

In the next future, CFA piles will probably gain the market also against the smaller size (800÷1200 mm) of the large diameter bored piles. These latter represent almost a 40% of the market, 80% of which performed with static bentonite mud.

About driven piles, almost 30% of their market is due to precast concrete piles; the remaining 70% is essentially due to cast in situ piles (50% with temporary steel casing, 20% with permanent steel casing).

According to Trevisani (1992), mini piles had an unexpected success in the last few years, rapidly doubling their market (up to 20% of the total) in the last few years. This kind of pile, originally conceived in Italy, is used for various engineering works (motorways in mountain area, subways, tunnels, historical buildings restoration, etc.), mainly because it can be installed in any soil and rock by using equipment with small overall size and limited power.

4. CODES AND RECOMMENDATIONS

The most important Code governing the subject is the D.M. 21.01.1981, successively updated

(D.M. 11.03.1988), issued by Ministry of Public Works. After a number of general prescriptions (section A) and rules for soil investigations (section B), it deals with different kind of geotechnical works: foundations (C), retaining structures (D), earth structures (E), underground structures (F), natural slopes and excavations (G). Furthermore, some indications about environmental geotechnical engineering, waste disposals, pumping, soil strengthening, drainage, filters and ground anchors are given.

Pile foundations are dealt in the section C.5 of the Code. It consists of six sub-sections, namely: design criteria, specific soil investigations, failure and working load of single pile, working load of pile group, load tests, pile cap.

The design of single pile must be based on one or more of the following methods:

- analytical;
- based on the results of in situ tests;
- based on the results of load tests;
- analysis of the pile driving.

No indications are provided about which method has to be used for a given pile type penetrating a given subsoil, leaving to the designer a complete freedom.

Load tests on single pile or pile group are mandatory whether when very important structures must be designed or when the available soil investigations don't allow for a confident prediction of the pile behaviour. Concerning with acceptance tests, 1% of the total number of the pile must be tested, with a minimum number of 2.

For small to medium pile diameter ($d < 80$ cm), the global safety factor SF for single pile (defined as the ratio between failure load Q_{lim} and working load Q_w[2]) must be larger than 2.5. If results of load tests up to failure and careful soil characterisation are available, SF can be reduced up to 2.

For large diameter piles ($d \geq 80$ cm), SF can be chosen by the designer on the basis of the established allowable settlement for the structure.

No rules are given for the safety factor of the pile group. It is however established that its evaluation must take into account the geometry of the pile group, pile type, subsoil conditions and pile cap. Furthermore, the settlement of the pile foundation must be compatible with the safety and serviceability of the structure.

In addition to the aforementioned recommendations on site and laboratory investigations, AGI published in 1984 Recommendations on Pile Foundations in order to give guidance to engineers and contractors for selecting, designing and constructing piles.

The Recommendation consists of three main sections: in the first, once distinguished the piles in two broad categories (with and without soil removal), factors influencing the choice of the pile type are highlighted and details about some predominant construction techniques are given; in the second one, well-known analytical and empirical methods aimed to the evaluation of the failure load and settlement of piles (either isolated or in a group) are suggested for everyday practice; in the third and final section, equipment, instruments and procedures for carrying out load tests on pile are described.

Concerning with the safety factor SF on single piles, different values are suggested as derived from the national experience, depending on the way by which the failure load is evaluated: by adding ultimate shaft and base resistance in their entirety ($2.5 \leq SF \leq 3.5$), by adding ultimate shaft resistance in full and base resistance at a given displacement ($1.75 \leq SF \leq 2.5$), by carrying out load tests at a large displacements ($2.0 \leq SF \leq 2.5$).

At the present time, the Ministry of Public Works is preparing a revised version of the D.M. 11.03.1988. Furthermore, the National Research Council has planned the drawing up of Instructions that could become part of the body of rules to be obeyed by the contractors when involved in Public Works.

---

[2] The working load on single pile must be always evaluated without taking into account the contribution of the pile cap (raft, beam, etc.).

5. DESCRIPTION OF THE NATIONAL DESIGN METHODS

As simply understandable, the degrees of freedom left to the designers by the Code in force make the description of the current design methods a quite unlikely goal.

Therefore, in the following the design procedures suggested by AGI will be assumed to represent the National Design Methods. In some cases, suggestions coming from well-known textbooks and relevant publications will be also included.

5.1. *Current design philosophy*

The design of any foundation must meet the following requirements:

- the total structural load has to be transferred to the soil with an adequate safety factor against a bearing capacity failure;
- the total and differential settlements should be contained in order to guarantee the safety and the serviceability of the structure.

In this context, there are some cases for which the choice of a pile foundation is practically mandatory (i.e. bridge piers resting on a river-bed under erosion); in other cases, the choice can derive from a comparative analysis of alternative solutions (i.e. pile foundation against shallow foundation on a strengthened soil).

In any case, once the decision of adopting a pile foundation has been taken, the current design practice tends to concentrate on the first requirement. Therefore, a preliminary design is essentially aimed to define the type of pile as a function of the subsoil properties, average load per pile and requirements of the supported structure (i.e. high load concentration). Generally speaking, in this step the criteria on the basis of which the pile material is underwent to stress state close to the admissible one is largely used (especially for concrete piles).

Once the number and the cross section of the piles is fixed, the length of the piles is chosen in order to satisfy the established requirements (safety factor, allowable settlement, etc.).

Finally, a more rigorous analysis (if necessary) is performed with the aim to verify that the selected foundation system is “well-behaving”.

5.2. *Failure load of single pile*

5.2.1. *Analytical method*

The failure load Q_{lim} is conventionally considered due to the contribution of the base capacity Q_b and the shaft capacity Q_s:

$$Q_{lim} + W = Q_b + Q_s = A_b \cdot q_b + \pi \cdot \int_0^L d \cdot f_s \cdot dz \tag{1}$$

where: W is the weight of the pile, A_b is the area of the pile base where the unit point resistance q_b is acting, L and d are respectively the length and the diameter of the pile, f_s is the unit shaft friction at a given depth z along the pile.

The unit point resistance q_b is given by:

$$q_b = N_q \cdot \sigma_{vL} + N_c \cdot c \tag{2}$$

where: N_q and N_c are bearing capacity factor, σ_{vL} is the vertical stress acting at a depth L, c is the

Table 5. Values of k and μ for some pile types (after Viggiani 1993a)

PILE TYPE		k loose soils	k dense soils	μ
DRIVEN PILES	structural steel section	0.7	1.0	tan20°=0.36
	closed-end steel tube	1.0	2.0	tan20°=0.36
	precast concrete	1.0	2.0	tan(0.75Φ)
	cast in situ concrete	1.0	3.0	tanΦ
BORED PILES		0.5	0.4	tanΦ
CFA PILES		0.7	0.9	tanΦ

Table 6. Values of α for driven and bored piles (after Viggiani 1993a)

PILE TYPE	c_u [kPa]	α
DRIVEN	≤ 25	1.00
	$25 < c_u < 70$	$1.00 - 0.011 \cdot (c_u - 25)$
	≥ 70	0.50
BORED	≤ 25	0.70
	$25 < c_u < 70$	$0.70 - 0.008 \cdot (c_u - 25)$
	≥ 70	0.35

cohesion of the soil. For small to medium diameter piles, it is common to use the values of bearing capacity factors as suggested by Berezantzev et al. (1961); for large diameter piles, reduced values for N_q are requested in order to limit the settlement of the piles (Berezantzev 1965). The unit shaft resistance f_s is given by:

$$f_s = a + \mu \cdot \sigma_h \tag{3}$$

where: a is the adhesion, μ is a soil-pile friction coefficient and σ_h is the horizontal stress at a given depth z. Since the horizontal stress can be assumed proportional to the vertical stress by introducing a coefficient k ($\sigma_h = k \cdot \sigma_v$), the following equation is finally obtained:

$$f_s = a + \mu \cdot k \cdot \sigma_v \tag{4}$$

For effective stress analysis it is suggested to put $c = 0$ in eq. 2 and $a = 0$ in eq. 4. For total stress analysis is suggested to put $N_c = 9$ in eq. 2, $\mu = 0$ and $a = \alpha \cdot c_u$ in eq. 4, where α is a coefficient and c_u is the undrained shear strength of the soil.

In the tables 5 and 6 are listed the values for k, μ and α suggested by AGI (1984), slightly modified and updated by Viggiani (1993a).

Only for piles embedded in normally consolidated cohesive soils, it has been accepted that the unit shaft resistance can be evaluated by putting in eq. 4:

$$\begin{aligned} \mu &= \tan \Phi \\ k &= 1 - \sin\Phi \end{aligned} \tag{5}$$

Table 7. Values of K for driven piles in granular soils (after AGI 1984)

SOIL	K
silt and sandy silt	0.20
sand and silty sand	0.35
gravelly sand	0.50
sandy gravel and gravel	0.60

For end-bearing piles with the base in a weak rock (i.e. Neapolitan tuff), the unit shaft resistance f_s is generally assumed proportional to the uniaxial compressive strength of the rock, as for istance suggested by Carter and Kulhawy (1992).

5.2.2. *Methods based on the results of in situ tests*

Driven piles

For driven piles embedded in granular soils, the unit point resistance q_b can be evaluated by:

$$q_b = q_c \tag{6}$$

$$q_b\ [\text{MPa}] = K \cdot N \tag{7}$$

where q_c and N are the average values between the depths (L-4·d) and (L+d), K is a coefficient which values are listed in table 7.

On the basis of the literature, Viggiani (1993a) gives almost the same values, but distinguished for precast or cast in situ driven piles.

The unit shaft resistance f_s at a given depth or its mean value along the whole shaft is assumed as follows:

$$f_s = \alpha_q \cdot q_c \tag{8}$$

$$f_{s,av}\ [\text{MPa}] = \alpha_N + \beta_N \cdot N_{av} \tag{9}$$

where N_{av} is the mean value along the whole length of the pile, α_q, α_N and β_N are coefficients which values are listed in table 8.

Viggiani (1993a) lists a number of values for α_N and β_N taken from the literature, allowing for a distinguished estimate of the unit shaft resistance for precast or cast in situ driven piles.

The failure load of driven piles in cohesive soils is generally evaluated by transforming the available q_c-profile in a c_u-profile as follows:

$$c_u = \frac{q_c - \sigma_{vo}}{N_c} \tag{10}$$

where σ_{vo} is the geostatic vertical total stress and N_c a factor ranging between 15 and 25 (if c_u-profile is available from laboratory tests, N_c can be selected in order to reproduce it). Therefore, analytical methods are generally used.

Table 8. Values of α_q, α_N and β_N for driven piles in granular soils (after AGI 1984)

	q_c [MPa]	α_q	α_N	β_N
very loose	< 2	0.020		
loose	$2 \le q_c \le 5$	0.015		
medium	$5 \le q_c \le 15$	0.012	0	0.002
dense	$15 \le q_c \le 25$	0.009		
very dense	$q_c \ge 25$	0.007		

Bored piles

Small to medium diameter piles in granular soils are generally designed by using the analytical methods. The same applies for large diameter bored piles in cohesive soils (Rampello, 1994).

For large diameter piles in granular soils, methods based on SPT results are usually adopted (i.e. Wright & Reese 1979; Reese & O'Neill 1988): values of f_s and q_b are selected as a function of N, being the value of q_b referred to a displacement of the pile equal to $0.05 \cdot d$.

By comparing the values of f_s as deduced from 14 load tests on instrumented large diameter bored piles embedded in Neapolitan pyroclastic soils with those derived from the procedures by Wright & Reese (1979) and DIN 4014 (in this latter case by assuming q_c [MPa] $= 0.5 \cdot N$), Caputo et al. (1993) confirmed on the overall their reliability, even if quite conservative for pile shafts above the water table.

Concerning with the point resistance q_b, Jamiolkowski and Lancellotta (1988) used the results of 15 load tests taken from the literature to establish a correlation between $q_{b,crit}$ (value of q_b for a relative displacement w_s/d equals to 0.05) and q_c as follows:

$$q_{b,crit}\left(w_s/d = 0.05\right) = 0.2 \cdot q_c \tag{11}$$

On the basis of the results from 34 deep plate loading tests performed in the calibration chamber, Ghionna et al. (1993) confirmed such findings, showing that, even if the ultimate point resistance q_b of a pile is virtually independent from the installation procedure (being always equals to q_c), the construction procedure markedly affect the value of mobilised q_b at a given relative displacement w_s/d of the pile.

Auger piles

No specific information for the design of auger piles are available.

A preliminary methodology for evaluating their failure load in Neapolitan pyroclastic soils is described by Viggiani (1993b). Starting from the availability of 37 load tests, 9 of which up to failure, and CPTs, it was found that the unit point resistance q_b can be taken equal to the mean value of q_c between the depths ($L-4 \cdot d$) and ($L+d$); the values of $q_c > 80$ MPa are cut to 80 MPa.

The unit shaft resistance can be evaluated from eq. 10, assuming that α is still a function of q_c as follows:

$$\alpha = \frac{a_1 + a_2 \cdot q_c}{a_3 + a_4 \cdot q_c} \tag{12}$$

For CFA piles in pyroclastic soils, the following values for the coefficients a_i were found (using

MPa for q_c): $a_1 = 6.6$; $a_2 = 0.32$; $a_3 = 300$; $a_4 = 60$. The author stressed that further and more detailed evidence is needed before any definite conclusion may be drawn.

5.2.3. *Design on basis of driving formulae*

The recommendations by AGI (1984) states that, being the driving formulae based on oversimplification of the driving process, their use could be aimed to check the similarity between piles belonging to the same foundation and not as a design tool. In this latter case, it is suggested to derive empirical coefficients in order to reproduce failure loads as deduced from static load tests (Viggiani 1993a).

5.2.4. *Design on basis of wave equation analysis*

It is largely accepted that the evaluation of the failure load of a single pile by wave equation analysis is more and more satisfactory, from the conceptual point of view, than that coming from the driving formulae. Nevertheless, being the wave equation analysis not so simple as driving formulae is, its use is generally limited to very important structures (i.e. offshore structures) or for research purposes (Carrubba and Maugeri 1995).

5.2.5. *Design on basis of static load tests*

Static load tests on single pile are generally performed in order to control that the piles were constructed in a correct way and that there is no variability among piles belonging to the same foundation (acceptance tests). In such cases, the piles are generally loaded up to 1.5 times the maximum working load Q_w.

In all the cases whereby it is mandatory or believed necessary (for instance, if uncertainties exist on the relevant parameters to be used for the design), load tests up to failure on piles expressly constructed are usually performed. In this case, being the failure load Q_{lim} not known before the loading of pile, it is generally assumed that the maximum load test Q_{max} must be not lesser than $3 \cdot Q_w$; smaller values can be accepted only if the displacement of the tested pile is large enough to allow for a reliable estimate of the failure load (10% of the pile diameter).

Both acceptance and failure tests are generally performed by using maintained load (ML) test procedure.

Usually, each increment is assumed equal to $0.25 \cdot Q_w$; in such way, at least 6 load increments are applied before reaching Q_{max} for an acceptance tests, 12 for a failure tests. Cessation of movement criteria is then adopted to establish when the following load increment has to be applied; suggested values for the rate of movement are listed in table 9.

During any step, readings at all the installed instruments are taken at the following times from the beginning of the step: 2, 5, 10, 20, every 20 minutes. Very often, the maximum working load

Table 9. Rate of movement for ML tests (after Mandolini 1995)

Pile diameter	rate of movement [mm/min]
small	0.005
medium	0.010
large	0.015

Q_w and/or the maximum load test Q_{max} are maintained for a larger duration, after which the pile is unloaded in order to measure residual displacement.

As alternative to ML procedure, load test with a constant duration of each increment (say 30÷60 min) is suggested due to the fact that, besides its simplicity as well as the obvious practical advantage, the load-settlement curve is not distinguishable from that obtained from ML test (Mandolini 1995).

The interpretation of the test results in terms of failure load Q_{lim} is very simple if the load-settlement curve exhibit a well-defined maximum, otherwise it is commonly defined as the load that fulfils one of the following conditions (AGI 1984):

- increasing the load from $0.9 \cdot Q_{lim}$ to Q_{lim}, the settlement of the pile increases from w to $2 \cdot w$;
- the settlement of the pile head w is equal to $0.10 \cdot d$.

If none of the above conditions is fulfilled, Q_{lim} is extrapolated by using the hyperbolic fitting of the experimental data (Chin 1970), paying attention to the deduced value if Q_{max} is significantly smaller than Q_{lim} (that is $w_{max} << 0.1 \cdot d$).

At the present time, no advantages (reduced values for SF) can be derived if the tested pile is instrumented along the shaft in order to have information on the load transfer mechanism between pile and soil. In any case, the results can be used for an appropriate selection of the relevant parameters.

5.3. *Settlement of single pile*

AGI (1984) recommends to evaluate the settlement of single pile under working load Q_w from load tests. Alternatively, if no test results are available, analytical and numerical methods based on transfer curves, boundary element method (B.E.M.) and finite element method (F.E.M.) are used.

Very rough estimate of the settlement of single pile under working load are obtained by using some empirical relationships given in literature; their use should be limited only to conditions similar to those from which they were derived.

Regional collection of experimental data (pile types vs. subsoil conditions) would make the estimate of the settlement more reliable, and therefore it must be strongly pursued. In this context, on the basis of the results of load tests on 14 large diameter bored piles, Caputo et al. (1993) found that, up to load lesser than $0.6 \cdot Q_{lim}$ (that is SF ≥ 1.5), the following empirical relation can be used:

$$w_s = \frac{d}{60} \cdot \frac{Q}{Q_{lim}} = \frac{d}{60 \cdot SF} \tag{13}$$

5.4. *Failure load of pile group*

The estimate of the failure load $Q_{lim,G}$ of a group made by n piles, each of them having a failure load Q_{lim}, is based on the following equation:

$$Q_{lim,G} = \eta \cdot n \cdot Q_{lim} \tag{14}$$

in which η is the efficiency factor.

For piles in granular soils and spacing $s \geq 4 \cdot d$, it was found that $\eta > 1$. AGI (1984) suggests to adopt $\eta = 1$, except for those cases for which piles were bored without care.

Concerning with piles in cohesive soils, Viggiani (1993a) suggests to assume $\eta = 0.6 \div 0.7$.

Table 10. Bearing capacity factor N_c (after AGI 1984)

	N_c		
L/B_1	$B_2/B_1 = 1$	$1 < B_2/B_1 < 20$	$B_2/B_1 > 20$
0.25	6.7	$5.6 \cdot (1 + 0.2 \cdot L/B_1)$	5.6
0.50	7.1	$5.9 \cdot (1 + 0.2 \cdot L/B_1)$	5.9
0.75	7.4	$6.2 \cdot (1 + 0.2 \cdot L/B_1)$	6.2
1.00	7.7	$6.4 \cdot (1 + 0.2 \cdot L/B_1)$	6.4
1.50	8.1	$6.8 \cdot (1 + 0.2 \cdot L/B_1)$	6.8
2.00	8.4	$7.0 \cdot (1 + 0.2 \cdot L/B_1)$	7.0
2.50	8.6	$7.2 \cdot (1 + 0.2 \cdot L/B_1)$	7.2
3.00	8.8	$7.4 \cdot (1 + 0.2 \cdot L/B_1)$	7.4
> 4.00	9.0	$7.5 \cdot (1 + 0.2 \cdot L/B_1)$	7.5

For pile groups in cohesive soils with $s \leq 4 \cdot d$, AGI (1984) recommends to determine $Q_{lim,G}$ assuming a block failure mode, as originally proposed by Terzaghi and Peck (1967):

$$Q_{lim,G} = B_1 \cdot B_2 \cdot N_c \cdot c_{uL} + 2 \cdot (B_1 + B_2) \cdot L \cdot f_{s,av} \tag{15}$$

where B_1 and B_2 are respectively breadth and width of the group ($B_1 < B_2$), L is the length of the piles, c_{uL} is the undrained shear stength at depth L, $f_{s,av}$ is the average value of the shear resistance along the pile length, N_c is a bearing capacity factor which values are listed in table 10.
If the spacing $s \leq 3 \cdot d$, more rigorous analysis are requested by D.M. 11.03.1988. Alternatively, the evaluation of the failure load of the pile group must be made by considering an equivalent shallow foundation at a depth L.

5.5. *Average and differential settlement of pile group*

In the last twenty years (Burland et al. 1977) a number of compelling arguments for moving towards a settlement based design methodology for pile foundations were discussed and interesting developments have occurred in this direction. Nevertheless, a capacity based design is still largely used, probably due to the widespread belief that predicting deformations is more difficult and less reliable than predicting capacity (Randolph 1994).

In the last years, an extensive research program consisting of theoretical analysis and experimental observation on full scale pile foundations has been carried out at the University of Napoli. As a consequence, in the following the obtained results will be shown in some details.

5.5.1. *Empirical approach*

Once the settlement w_s of the single pile is known, the recommendations by AGI (1984) suggest to evaluate the settlement w_G of pile groups as follows:

$$w_G = R_s \cdot w_s \tag{16}$$

where R_s is the group settlement ratio as originally introduced by Whitaker (1957).

Therefore, the following relations for R_s are suggested:

$$R_s = \left(\frac{1.2 \cdot B_1 + 2.7}{0.3 \cdot B_1 + 4} \right)^2 \tag{17}$$

$$R_s = \left(\frac{0.6 \cdot B_1}{0.3 \cdot B_1 + 0.3} \right)^2 \tag{18}$$

the former being applicable for driven piles, the latter being applicable for bored piles.

Recently, Mandolini (1994b, c) introduced an empirical method applicable to very different combination of pile and soil types. After having postulated the following relationship:

$$R_G = \frac{R_s}{n} = F\left(\sqrt{\frac{L}{n \cdot s}} \right) = F\left(R^{-1}\right) \tag{19}$$

where R_G is the group reduction factor (Butterfield and Douglas 1981; Fleming et al. 1992) and R is the modified aspect ratio introduced by Randolph and Clancy (1993), the author collected 104 well documented experimental works (13 full scale foundations, 12 large scale field tests, 79 small scale laboratory tests); they refer to bored, auger and driven piles embedded in granular, cohesive and layered subsoil. All the available data were fitted with a hyperbolic function as:

$$R_G = 0.34 \cdot R^{-1} \cong \frac{1}{3 \cdot R} \tag{20}$$

On the overall, the following information were derived (Figure 5):

- in the range of the common values of R, the small scale model tests are quite representative of the settlement behaviour of large scale tests and full scale foundations;

- ≈ 80% of data fall within ±50% range.

Trials to separate the collected data as a function of pile technology and/or soil type did not significantly improve the fitting.

From eq. 20 it can be seen that, for a given ratio L/s, R_s is a function of $n^{0.5}$, as theoretically found by Poulos and Mattes (1971).

It must be stressed that, according to eq. 20, the amplification (R_s) of the settlement of the single pile (w_s) due to group action depends only on the geometry of the group. Technological factors do affect, obviously, the settlement of the group (w_G) by means of the value of w_s.

Even if differential settlement plays a much more significant role than the average settlement in influencing the overall performance of a structure, a limited amount of data are available in literature.

By using the few available data (Figures 6 and 7), Mandolini (1994c) showed that is possible to establish a rough relationship between the maximum differential settlement ratio R_{ds}, defined as:

$$R_{ds} = \frac{\delta_{max}}{w_G} \tag{21}$$

and the ratio B_2/B_1, already introduced in §5.4, as follows:

$$R_{ds} = 0.60 + 0.05 \cdot \left(B_2 / B_1 - 1\right) \tag{22}$$

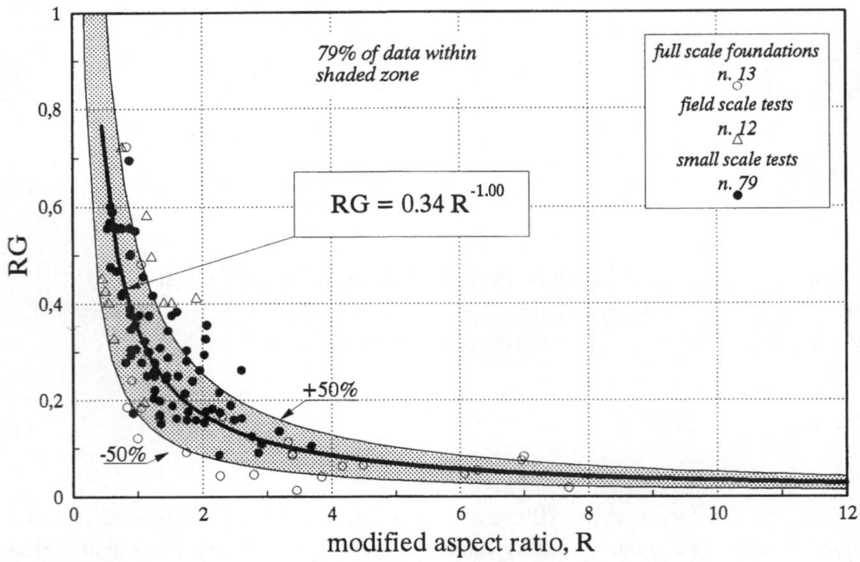

Figure 5. Relationship between group reduction factor and modified aspect ratio (after Mandolini 1994c)

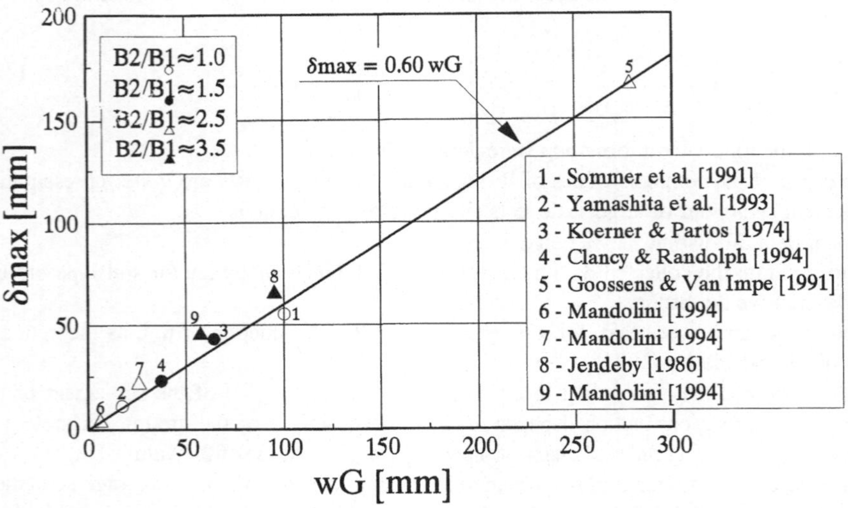

Figure 6. Measured average and differential settlements of full scale foundations (after Mandolini 1994c)

Being the data on which the eq. 22 is based referred to the measured settlement at the end of the construction of the structure, higher differential settlement are to be expected during the construction as a consequence of the reduced stiffening effect of the structure. In Figure 8 are reported the data collected from an extensive research program carried out in Napoli in the last years during the construction of some very important tall buildings founded on bored and auger pile groups ($n = 77 \div 323$; $L = 20 \div 42$ m; $d = 0.6 \div 2.2$ m) in Neapolitan pyroclastic soils.

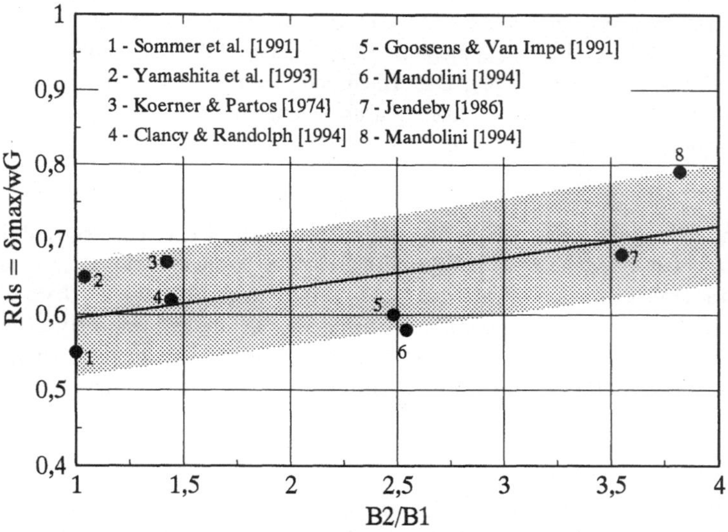

Figure 7. Observed differential settlement ratio for different values of B_2/B_1 (after Mandolini 1994c)

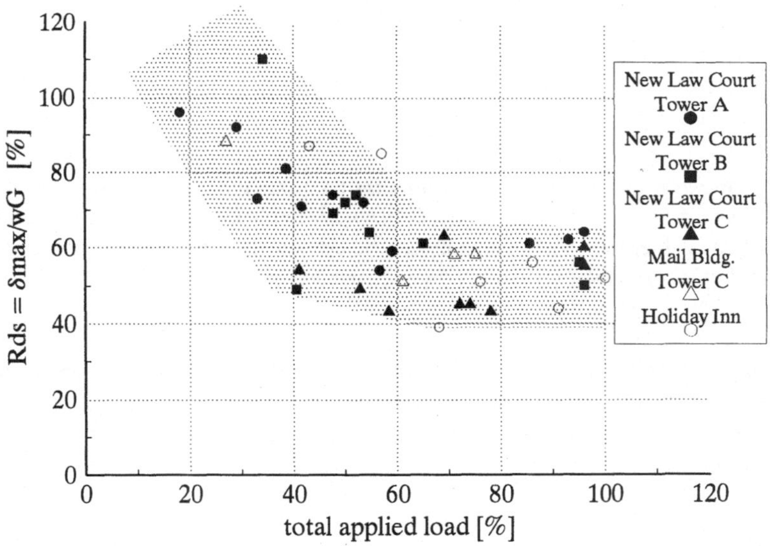

Figure 8. Observed differential settlement ratio during the construction (after Mandolini 1994c)

5.5.2. *Rational approach*

For complex foundations and/or for important structures, more rigorous analysis are required, very often based on simplified B.E.M. (Poulos and Davis 1980) or approximate analytical solutions (Randolph and Wroth 1979). Independent of the adopted tools for the analysis, according to Poulos (1993) the evaluation of the soil properties relevant to the prediction of the settlement of a piled foundations is the most difficult and uncertain step of the analysis.

This is probably true for all deformation problems in geotechnical engineering; in the case of piled foundations, however, the difficulties arising from the influence of technological factors connected to the installation of the piles make the evaluation still harder.

In order to reduce such uncertainties and to make the evaluation as objective as possible, Mandolini and Viggiani (1996) proposed a standard procedure in five steps, essentially based on computer codes (Banerjee and Driscoll 1978; Randolph 1987; Poulos, 1990; Mandolini 1994a) and/or available charts (Poulos and Davis 1980; Butterfield and Douglas 1981; Fleming et al. 1992), subsoil investigation and (when available) loading tests on single piles.
The procedure may be summarised as follows.

Step 1
All the available site and laboratory investigations are used to develop a model of the subsoil, in which the geometry is adapted to a scheme of horizontal layering; each layer is supposed to be linearly elastic and the ratio of the moduli of different layers is evaluated. The authors share the emphasis put by Randolph (1994) on the importance of considering the overall geometry of the foundation; engineers should be encouraged to prepare correct scaled drawings of the complete foundation and relevant soil stratigraphy, which will often reveal shortcomings in the proposed foundation scheme.

Step 2
The absolute value of the moduli is evaluated by fitting the results of load tests on single pile (if available) with the available computer codes and/or charts. The fitting is aimed to reproduce the measured initial axial pile flexibility I_{wo}.

Assuming a hyperbolic load-settlement relationship (Chin 1970), it can be shown that:

$$Q = \frac{w}{I_{wo} + \frac{1}{Q_{lim}} \cdot w} \tag{23}$$

It must be stressed that, when using the load test results for the evaluation of the stiffness of the spoil and the failure load of the pile, the technological effects are implicitly accounted for. Should pile load tests not be available, the same quantities can be estimated by the typical pile design approaches described before or, maybe preferably, by regional design methods.

Step 3
Once the subsoil model (geometry and properties) is fixed, the available computer codes and/or charts are used to derive a set of interaction factors α_{ij} for various spacing/diameter s_{ij}/d between piles i and j. Such values of α_{ij} and s_{ij}/d are then best-fitted in order to have a continuous function (linear, logarithmic, power, etc.).

Step 4
It is assumed that no interaction occurs for piles whose spacing s is larger than a limiting value s_{max} that is defined according to Randolph and Wroth (1978). If the piles belong to a group, s_{max} is increased by adding a term r_G as suggested by Randolph and Wroth (1979) for a square group, and extended by Mandolini (1994c) to any other shape of the group: For $s \geq s_{max} + r_G$, α_{ij} is assumed equal to zero.

Step 5
The non linearity of the behaviour is simulated as suggested by Caputo and Viggiani (1984). On the basis of the experimental evidence, they claim that the non linearity is essentially concentrated at the pile-soil interface, while the interaction may be represented by linear model with sufficient accuracy. Accordingly, in a method of analysis based on the interaction factors, they assume that all the factors α_{ij} $(i \neq j)$ are constant, irrespective of the load level, while the pile-pile interaction factors α_{ii} vary as follows:

$$\alpha_{ii} = \frac{1}{\left(1 - \frac{Q_i}{Q_{i,lim}}\right)} = \frac{1}{(1-\psi_i)} \tag{24}$$

where Q_i and $Q_{i,lim}$ are respectively the load acting on and the failure load of the i-th pile in the group (ψ_i = load level).

According to eq. 23, eq. 24 corresponds to a hyperbolic load-settlement relationship[3] for the single pile. With this assumption and considering the settlement of the single pile w_s under working load $Q_{w,i}$ as the sum of a linear elastic component $w_{s,le}$ and a non linear elastic component $w_{s,nl}$, it can be simply showed that:

$$w_s = \frac{I_{wo} \cdot Q_{w,i}}{\left(1 - \frac{Q_{w,i}}{Q_{i,lim}}\right)} = \frac{w_{s,le}}{(1-\psi_i)} \tag{25}$$

$$w_{s,nl} = w_s - w_{s,le} = w_{s,le} \cdot \frac{\psi_i}{(1-\psi_i)} \tag{26}$$

Concerning with groups of identical and equally loaded piles, the settlement of the pile group can be deduced in a very simple manner by superimposing the linear elastic displacement fields induced by nearby equally loaded piles in a group and adding the non linear component as follows:

$$w_G = R_s \cdot w_{s,le} + w_{s,nl} = w_{s,le} \cdot \left(R_s + \frac{\psi}{1-\psi}\right) \tag{27}$$

The same concepts are applicable when dealing with pile groups connected to a rigid cap.

It must be highlighted that the concept described in step 5 allows for a rational understanding of the role played by non linearity effects on the settlement of piled foundations at working load (SF $= 2 \div 3 \Rightarrow \psi = 0.33 \div 0.50$): for small pile group, for which the non linearity component $\psi/(1-\psi)$ is of the same order of magnitude of the group settlement ratio R_s, to neglect non linearity effects could imply a significant underestimate of w_G; on the other hand, for large pile group, R_s is so high that $\psi/(1-\psi)$ can be surely neglected.

The described approach was applied by Mandolini and Viggiani (1996) for the back-analysis of 19 well documented case histories taken from the literature. Some of the obtained results are showed in Figures 9 and 10.

The substantial correspondence between the values of the soil shear modulus obtained by shear wave velocity measurements and those backfigured from by the load test on piles in terms of initial axial pile flexibility I_{wo} (step 2) shows promising implications for both the interpretation of loading tests results in terms of soil properties and the prediction of settlement on the basis of geophysical measurements. It must be stressed that the data are referred to bored, driven and auger piles embedded in both granular and cohesive soils.
The non linear (NL) analysis, which essentially consists in adding the non linear component of the settlement of the single pile to the settlement of the group, obtained as in the linear elastic (LE)

[3] It is to point out that, either in eq. 23 or eq. 24, Q_{lim} is intended only as a geometrical parameter of the hyperbola fitting the load-settlement curve in the load range of interest. In some cases this value can significantly differ from the actual failure load.

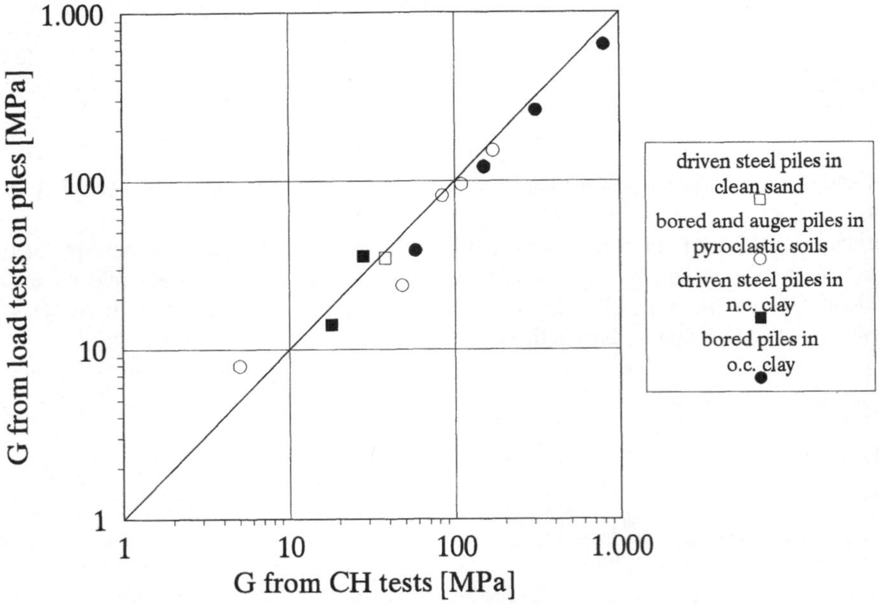

Figure 9. Comparison between the values of the soil shear modulus obtained by shear wave velocity measurements and those backfigured from by the load test on piles in terms of initial axial pile flexibility I_{wo} (after Mandolini and Viggiani 1995)

analysis, slightly improves the prediction of the average settlement in all the cases where the LE analysis was already successful. In those cases where the non linearity plays a significant role, NL analysis substantially improves the prediction.

In Figure 10 are also showed the results obtained by an equivalent linear elastic (ELE) solution, based on equivalent elastic soil properties backfigured from the observed load-settlement curve, considering the secant stiffness corresponding to the average working load of the piles in the group. This kind of analysis, which incorrectly amplifies both the elastic and plastic components of the settlement of the single pile, substantially overestimates the observed settlement.

5.6. *Rules for serviceability*

D.M. 11.03.1998 states that the designer must verify that the predicted settlement of any point of the foundation system is admissible for the structure; but no specific information are given about the way by which to establish some limiting values.

As a consequence, the designer generally refers to the existing criteria, generally based on limiting values for the angular distortion δ/L (δ = difference between the settlement of any two points spaced L) and curvature Δ/L (Δ = maximum movement from a straight line joining two reference points spaced L).

A significant contribution in this field was given by Ricceri and Soranzo (1985). The authors collected 69 Italian case histories published from the early '60, including various type of structures (steel, load-bearing brick walls, reinforced concrete), foundations (shallow and deep), soils (granular, cohesive, layered) and stiffness (very low as for steel oil tanks, intermediate as for office and industrial buildings, very high as for reinforced concrete cellular silos).

By using the available data in terms of maximum settlement and angular distortion (respectively w_{max} and δ/L in the Figure 11), they found a significant correlation between the two

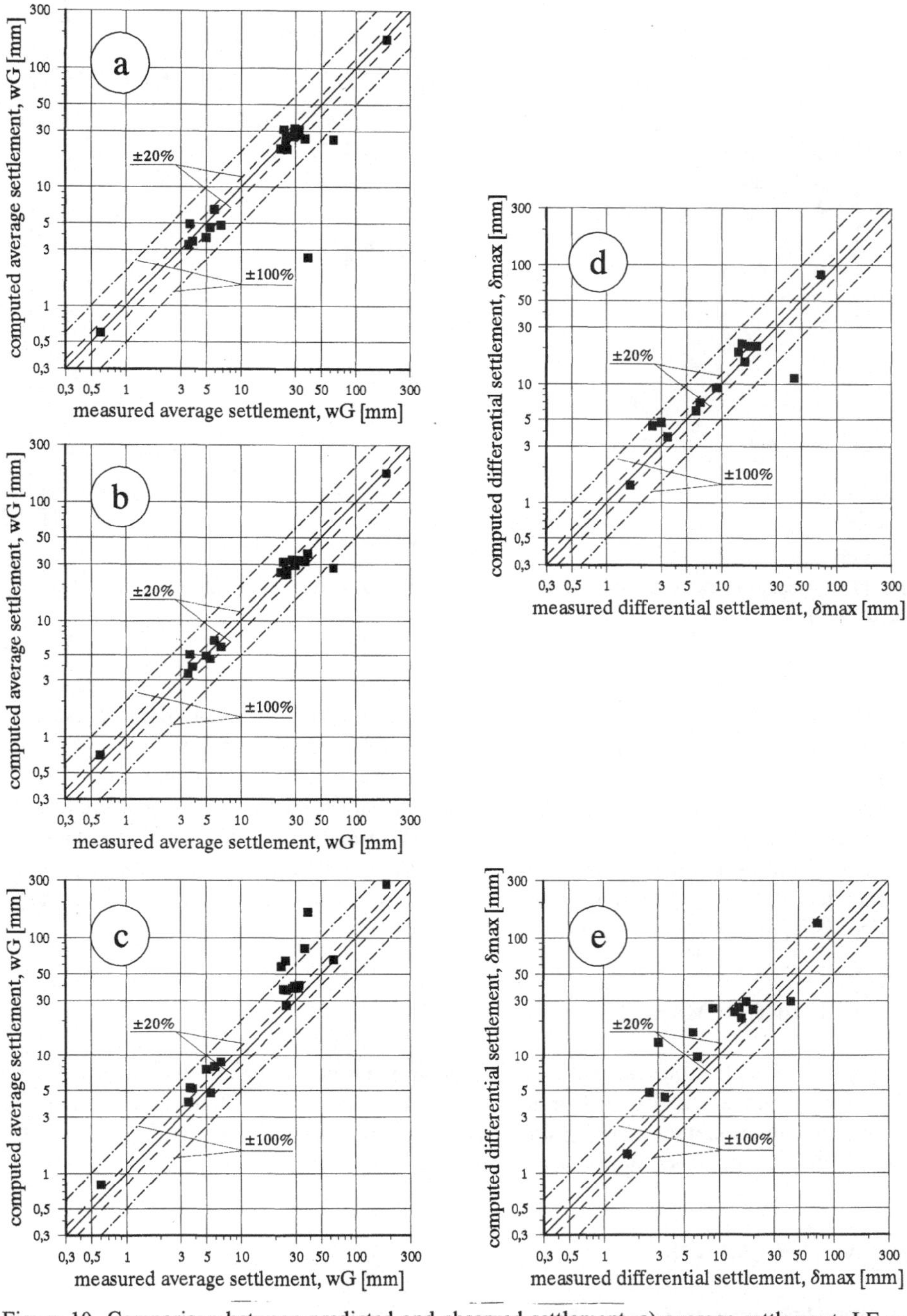

Figure 10. Comparison between predicted and observed settlement. a) average settlement, LE analysis; b) average settlement, NL analysis; c) average settlement, ELE analysis; d) maximum differential settlement, LE and NL analysis; e) maximum differential settlement, ELE analysis (after Mandolini and Viggiani 1996)

quantities, allowing for the possibility to check the admissibility of deformations directly in terms of the maximum settlement predicted, rather than attempting an estimate of the angular distortion which is definitely more complex. Similar results were obtained analysing the data from Skempton and MacDonald (1956). Very simple rules were therefore suggested:

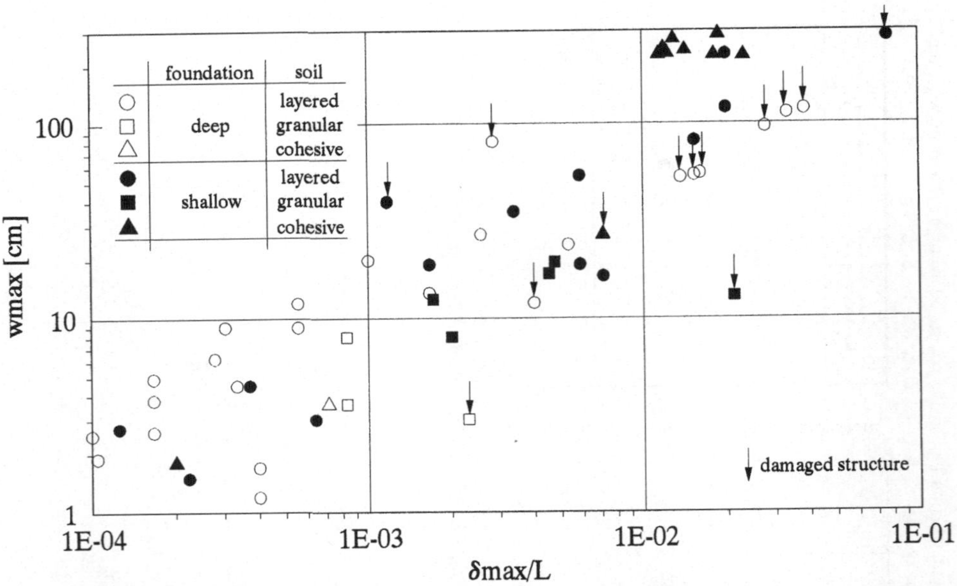

Figure 11. Correlation between w_{max} and δ/L for deep and shallow foundations (after Ricceri and Soranzo 1985)

- for $w_{max} \leq 8$ cm should not determine serious problems;
- for $w_{max} \geq 20$ cm are not tolerated by traditional structures and damage should be anticipated, the extent of which depends on the relative soil/structure stiffness;
- for 8 cm $< w_{max} <$ 20 cm, detailed soil-structure interaction analysis is required.

Obviously, as stressed by the authors, the quoted values must not be considered as rigid design rules but rather as indications for useful comparisons, according to the idea that "*each building or structure should be treated on its own merits, for its performance will depend on a large number of factors including construction materials, method and form of construction, type of cladding and brittleness of finishes*" (Burland 1977).

6. CONCLUDING REMARKS

In the paper it was attempted to represent the Italian point of view about the current design of axially loaded piles, either isolated or in a group. For several and obvious reasons (brevity, vastness of the topics, author's lack of knowledge, available time, etc.) the writer apologise to all that people thinking to be not represented their personal experiences in the paper.

ACKNOWLEDGEMENTS

Researchers belonging to some Italian Universities sent to the author a number of papers, publications and documents, widening his knowledge about the national "feeling" on pile problems. Among them, dr. P. Carrubba (University of Padova), Prof. M. Maugeri (University of Catania) and Prof. G. Scarpelli (University of Ancona) are worthy of special thanks.

Dr. C. Mascardi from Studio Geotecnico Italiano in Milano gave an essential contribution in writing the chapters related to piling technology (§3) and current design philosophy (§5.1).

The contribution of the author is mainly due to his belonging to the research group on piles of the Department of Geotechnical Engineering of the University of Napoli Federico II, masterly leaded by Prof. C. Viggiani.

REFERENCES

AGI 1977. Raccomandazioni per la programmazione e l'esecuzione di indagini in sito.

AGI 1979a. Some Italian experiences on the mechanical characterization of structurally complex formation. *Proc. 4th Int. Cong. Rock Mech., vol. 1, p. 827*

AGI 1979b. Experiences on the time-settlement behaviour of some Italian soft clays. *Proc. 7th E.C.S.M.F.E., vol. 1, p. 1*

AGI 1984. Raccomandazioni sui pali di fondazione.

AGI 1994. Raccomandazioni per l'esecuzione di prove di laboratorio.

Baldovin, G. 1989. Le strutture per fondazioni profonde in Italia: tipologie attuali, recenti innovazioni, prospettive. *Atti del XVII Convegno Nazionale di Geotecnica, vol. II, p. 141*

Banerjee, P.K., R.M. Driscoll 1978. Program for the analysis of pile groups of any geometry subjected to horizontal and vertical loads and moments, PGROUP (2.1). *HECB/B/7 Department of Transport, HECB, London*

Burland, J.B. 1977. Discussion. *Proc. IX I.C.S.M.F.E., Tokyo, vol. 3, p.381*

Burland, J.B., B.B. Broms, V.F.B. De Mello 1977. Behaviour of foundations and structures. *Proc. IX I.C.S.M.F.E., Tokyo, vol. 2, p. 495*

Butterfield, R., R.A. Douglas 1981. Flexibility coefficients for the design of piles and pile groups. *CIRIA Technical note 108, London, CIRIA*

Caputo, V., C. Viggiani 1984. Pile foundation analysis: a simple approach to non linearity effects. *Rivista Italiana di Geotecnica, vol. 18, No. 2, p. 32*

Caputo, V., A. Mandolini, C. Viggiani 1993. Large diameter bored piles in pyroclastic soils. *Proc. II Seminar on "Deep Foundation on Bored and Auger Piles", Balkema, p. 227*

Carrubba, P., M. Maugeri 1995. Modellazione di prove dinamiche su pali nel dominio del tempo. *Proc. "Metodi per il controllo non distruttivo dei pali di fondazione", Quaderno CIAS No. 3*

Carter, J.P., F.H. Kulhawy 1992. Socketed foundations in rock masses. *Eng. in Rock Masses, Ed. F.G. Bell, p. 509*

Chin, F.K. 1970. Estimation of the ultimate load of piles from tests not carried to failure. *Proc. II South-East Asian Conf. Soil Eng., Singapore, p. 81*

Esu, F., G. Manfredini, M. Maugeri 1984. Site investigation practice in Italy. *Rivista Italiana di Geotecnica, vol. 18, No. 1, p. 7*

Fleming, W.G.K., A.J. Weltman, M.F. Randolph, W.K. Elson 1992. Piling Engineering. *Blackie & Son, Glasgow, and John Wiley and Sons, New York*

Ghionna, V.M., M. Jamiolkowski, R. Lancellotta, S. Pedroni 1993. Base capacity of bored piles in sands from in situ tests. *Proc. II Seminar on "Deep Foundation on Bored and Auger Piles", Balkema, p.67*

Jamiolkowski, M., R. Lancellotta 1988. Relevance of in situ test results for evaluation of allowable base resistance of bored piles in sands. *Proc. I Seminar on "Deep Foundation on Bored and Auger Piles", Balkema, p. 107*

Lancellotta, R. 1995. Geotechnical Engineering. *A.A. Balkema Editor, 436 pages*

Mandolini, A. 1994a. Cedimenti di fondazioni su pali. *Ph.D. Thesis in Geotechnical Engineering, University of Napoli Federico II*

Mandolini, A. 1994b. Settlement of piled foundations. *Proc. VIII European Young Geotechnical Engineers Conference, p. 149*

Mandolini, A. 1994c. Modelling settlement behaviour of piled foundation. *Proc. Workshop on "Pile foundations - experimenal investigations, analysis and design", Napoli, p. 361*

Mandolini, A. 1995. Prove di carico su pali di fondazione. *Collana "Argomenti di Ingegneria Geotecnica", No. 1, Hevelius Edizioni, 61 pages*

Mandolini, A., C. Viggiani 1995. Il modulo G_o nella previsione dei cedimenti di fondazioni su pali. *Atti Gruppo Nazionale Coordinamento Studi di Iingegneria Geotecnica - S.G.E. Padova*

Mandolini, A., C. Viggiani 1996. Settlement of piled foundations. *Gèotechnique, paper No. G2414, accepted for pubblication*

MM. LL. PP - D.M. 21.01.1981. Norme tecniche riguardanti le indagini sui terreni e sulle rocce, la stabilità dei pendii naturali e delle scarpate, i criteri generali e le prescrizioni per la progettazione, l'esecuzione ed il collaudo delle opere di sostegno delle terre e delle opere di fondazione.

MM. LL. PP - D.M. 11.03.1988. Norme tecniche riguardanti le indagini sui terreni e sulle rocce, la stabilità dei pendii naturali e delle scarpate, i criteri generali e le prescrizioni per la progettazione, l'esecuzione ed il collaudo delle opere di sostegno delle terre e delle opere di fondazione. *Gazzetta Ufficiale No. 127, 1 giugno 1988*

Pellegrino, A. 1996. Indagini e caratterizzazione geotecnica del sottosuolo. *Manuale dell'Ingegnere Civile, Cremonese*

Poulos, H.G., N.S. Mattes 1971. Settlement and load distribution analysis of pile groups. *Australian Geomechanic Journal, G1, No. 1, p. 18*

Poulos, H.G., E.H. Davis 1980. Pile foundation analysis and design. *John Wiley & Sons, New York*

Poulos, H.G., 1989. Pile behaviour - Theory and Application. *The Rankine Lecture, Gèotechnique, vol. 39, No. 3, p. 365*

Poulos, H.G., 1993. Settlement prediction for bored pile groups. *Proc. II Seminar on "Deep Foundation on Bored and Auger Piles", Balkema, p. 103*

Poulos, H.G., 1990. DEFPIG - Deformation analysis of pile groups - User's Guide. *Centre for Geotechnical Research, University of Sidney, Australia*

Rampello, S. 1994. Observed behaviour of large diameter bored piles in medium to stiff clays. *Proc. Workshop on "Pile foundations - experimenal investigations, analysis and design", Napoli, p. 255*

Randolph, M.F. 1987. PIGLET - A computer program for the analysis and design of pile groups. *Research Report No. 87036, University of Western Australia, Perth, Australia*

Randolph, M.F. 1994. Design methods fopr pile groups and piled rafts. *Proc. XIII I.C.S.M.F.E., New Delhi, vol. 5, p. 61*

Randolph, M.F., C.P. Wroth 1978. Analysis of deformation of vertically loaded piles. *Proc. ASCE, JSMFE, p. 103*

Randolph, M.F., C.P. Wroth 1979. An analysis of the vertical deformation of pile groups. *Gèotechnique, vol. 29, No. 4, p. 423*

Randolph, M.F., P. Clancy 1993. Efficient design of piled rafts. *Proc. II Seminar on "Deep Foundation on Bored and Auger Piles", Balkema, p. 119*

Reese, L.C., M.W. O'Neill 1988. Drilled shafts: construction procedures and design methods. *U.S. Department Transportation, federal Highway Administration,FHW-HT-88042*

Ricceri, G, M. Soranzo 1985. An analysis of allowable settlements of structures. *Rivista Italiana di Geotecnica, vol. 19, p. 177*

Skempton, A.W., D.H. MacDonald 1956. The allowable settlements of buildings. *Proc. I.C.E., vol. 5, p. 727*

Terzaghi, K., R.B. Peck 1967. Soil Mechanics in Engineering Practice. *John Wiley & Sons, New York*

Trevisani, G. 1992. Keynote paper: Italy and Greece. *Proc. Conference on "Piling: European practice and worldwide trends", Institution of Civil Engineers, p. 17*

Viggiani, C. 1993a. Fondazioni. *Ed. CUEN, 472 pages.*

Viggiani, C. 1993b. Further experiences with auger piles in Naples area. *Proc. II Seminar on "Deep Foundation on Bored and Auger Piles", Balkema, p. 445*

Withaker, T. 1957. Experiments with model piles in sand. *Gèotechnique, vol. 7, p. 147*

Wright, S.J., L.C. Reese 1979. Design of large diameter bored piles. *Ground Engineering, vol. 12, n. 8*

Design of Axially Loaded Piles – European Practice, De Cock & Legrand (eds)
© 1997 Balkema, Rotterdam, ISBN 90 5410 873 8

Dutch national codes for pile design

H.J. Everts & H.J. Luger
Delft Geotechnics

ABSTRACT: Since october 1992 the new Dutch law *Bouwbesluit* is operative. Part of this *Bouwbesluit* is a set of NEN codes, which prescribes in detail the requirements and design methods for buildings and their foundations. The safety philosophy is based on the partial safety factor approach. These codes ascertain that each structure fulfils certain minimum requirements with respect to stability (ultimate limit state) and deformations (serviceability limit state). The calculation method for the determination of the ultimate bearing capacity of piles is based on empirical relations with the results of cone penetration tests (CPT's). The relations are tested and compared with the results of load tests over the past 60 years. If necessary they were modified.
This national report explains the philosophy of the Dutch codes concerning the design of foundation piles.

1 Geological setting of the Netherlands

1.1 Introduction

The stratigraphy of the natural sub soil underlying a structure is determining the foundation design. In order to obtain useful and reliable information of the subsoil both geological and geotechnical knowledge is required. The study of data in archives, geological maps, geotechnical profiles and literature can be performed in an early stage. Based on data from additional site investigation, a more detailed model of the stratigraphy can be made in a later stage. The subdivision of the sub soil in different layers is done using litho-stratigraphical characteristics such as composition of the layer and sequence of deposition. A secondary division is made on the basis of geotechnical parameters such as the cone resistance.

In general the influence on soil strata, induced by loaded area's will be of significant importance unto a depth of 1.5 times the width of the loaded area below foundation level. In practice the depth of the soil to be investigated will vary from 25 m for domestic structures to 75 m for large structures like high rise buildings or large tanks.

In this chapter, a short introduction to the geological setting of the Netherlands will be given. The reader may thus gain insight in the geological factors controlling a design.

1.2 Geology

The Netherlands are located at the boundary of the North Sea basin, which was filled up with sedimentation in the Pleistocene (Diluvium), and Holocene (Alluvium) periods. These predominantly fluvial and marine sediments were deposited by the rivers Rhine and Maas and the North Sea intruding from the west. The location of these rivers and the North Sea is shown in figure 1.

In the western part of the Netherlands the groundlevel is about or below sealevel. The groundwater table is kept at about 0.5 meter below the surface. Towards the east the land rises until approximately 100 meter above sealevel. The depth of the groundwater table increases towards the east until a maximum depth of several m below groundlevel.

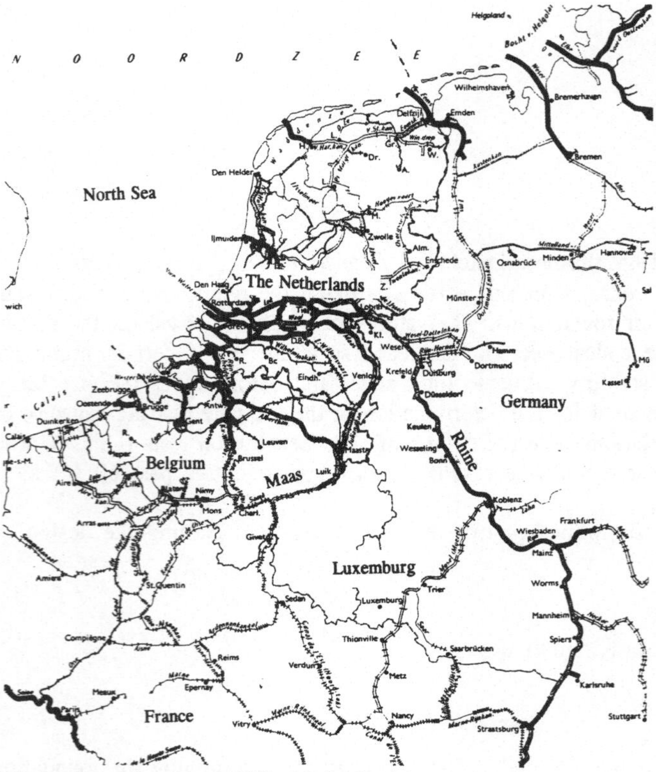

Figure 1. Location of The Netherlands

In general the typical geotechnical profiles in the Netherlands can be summarized in four different areas:

West part:

In the west part of the Netherlands the following layers have to be expected:

- from grade to NAP - 15 m (NAP equals to about mean sea level): very soft holocene clay and peat layers. The cone resistance is generally less than 1 MPa.
- from NAP -15 m to NAP -75 m: pleistocene fine to medium fine normally consolidated sandlayers. These sediments are deposited by the rivers Rhine and Maas. The cone resistance of the sand ranges from 5 MPa to 30 MPa. Locally shells and normally consolidated clay layers are present. The clay layers have a thickness of 1 to 5 meters and have a cone resistance of about 2 to 4 MPa. In the north area of the west part a marine clay layer of the Eem-Formation is present at NAP -25 m. The thickness of this layer varies from about 30 m near Amsterdam to a few meters towards the eastern part of the country. This layer is normally consolidated to slightly overconsolidated.

South west part:

In the south west part of the Netherlands the following layers have to be expected:

- from grade to NAP -5 m: very soft holocene clay and peat layers. The cone resistance is less than 1 MPa.
- from NAP -5 m to NAP -25 m: pleistocene fine to medium coarse sand with thin to very thin clay layers. The cone resistance of the sand varies from 5 MPa to 30 MPa. The sand is normally consolidated.
- from NAP -25 m to NAP -75 m: medium stiff clay layers. The clay layers belong to the fluviatile Formation of Kedichem (deposited 1.700.000 years - 800.000 years ago) and the marine Rupel Formation (tertiary period, age older than 2.500.000 years). These clay layers have a thickness from 5 to 20 m. The cone resistance is about 2 to 4 MPa. The clay layers are (slightly) overconsolidated due to aging or to a low water table in the past.

South east part:
In a south east, small part of the Netherlands the subsoil consists of rock formations consisting mostly of marl and limestones. These rock formations are eroded by the river Maas and sedimentation of sand and gravel can be found in the fossil and young channel beds of the river.
North east part:
The subsoil of the north east part consists of:
- from grade to NAP -10 m: very soft holocene clay and peat layers. The cone resistance is less than 1 MPa. Locally from groundlevel to NAP -10 m medium dense pleistocene sandlayers with a cone resistance of 10 MPa are present.
- from NAP -10 m to NAP -75 m: pleistocene fine to medium coarse sand with intersections of glacial till. Because this part of the Netherlands is at least twice covered with an icecap the sand is overconsolidated. The cone resistance may be up to 75 MPa. Two distinct till layers can be detected in this area, deposited during two glacial periods. The till layers are, due to the deposition below the ice cap, overconsolidated.

Seismic activities are of minor importance.

2 Soil investigation.

In general, for piled foundations, the soil investigation consists of a number of cone penetration tests (CPT's) with electrical registration of the cone resistance and the local friction. The soil type is derived from the calculated friction ratio (= local friction divided by the cone resistance at the same depth), combined with the results of one or more borings. Depending on the complexity of the structure a poor to more extended laboratory test program on undisturbed samples from borings will be executed. Exceptionally, for instance if the lateral behaviour of the piles is of importance, in situ tests like the dilatometer test or cone pressiometer test is executed.
The national code NEN 6740 *Geotechnics, Principals and loads*, prescribes the minimum extend of the soil investigation as a function of the complexity of the structure.

3 Piling technology.

In general all piles are end bearing piles. The tip elevation varies from 1 m to 10 m into a sand layer; commonly the Pleistocene sand.

If, induced by the environment, there are no restrictions with respect to vibrations and or noise, driven displacement piles are commonly used. Prestressed precast piles or cast in place piles type Vibro are most popular. The dimensions of precast piles vary from 180 mm x 180 mm to 500 mm x 600 mm, with a length of 10-20 m to 33 m respectively. The diameter of the Vibro piles ranges from about 300/350 (shaft-/bottom diameter) to 600/680 mm, with a length up to 35 m.

If there are restrictions with respect to vibrations, augered piles like the auger pile or screwed piles like the Fundex pile are commonly used. If there are restrictions with respect to vibrations, to a small spacing to other structures or piles, or to the acceptable settlements, a combination of augering and simultaneous grout injection is common practice.
In general the piles are driven with a diesel or hydraulic hammer with a blow energy of about 40 to 220 kNm.

4 National documents

Since october 1992 *het Bouwbesluit* became operative. This law prescribes that all structures

should satisfy minimum performance requirements. These requirements are related to safety, serviceability, health and economical use of energy. *Het Bouwbesluit* distinguishes three types of structures: domestic structures, non-domestic structures and structures not being buildings. The minimum requirements are laid down in TGB 1990. This TGB 1990 encompasses codes like NEN 6700 *General principals*, NEN 6702 *Loads and deformations*, NEN 6740 *Geotechnics; Principals and loads*, NEN 6743 *Piled foundations*.
If the principal demonstrates that his structure meets the minimum requirements according to *het Bouwbesluit,* local authorities may not impose extra requirements in excess of the national building code; they are obliged to provide a building permit.

5 NATIONAL CODES FOR PILE DESIGN

For the design of piled structures the codes NEN 6702 *Loads and deformations*, NEN 6740 *Geotechnics; Principals and loads* and NEN 6743 *Piled foundations* are relevant. These codes will be discussed in the following sections.
The code NEN 6743 contains a design method which is based on the use of in situ soil tests (CPT's) and empiric design rules. The execution of CPT's is an extremely cheap and quick method of soil investigation, especially in the soft soils in the Netherlands. The design rules prescribed in NEN 6743 are easy in practical application; especially after introduction of computer codes. As a consequence there is a wide extended acceptance and use of these codes. Other design methods are only tolerated if it is demonstrated that with those methods the same level of foundation performance will be guaranteed as in the case of the methods mentioned in the codes. In practice this demonstration will only be successful by means of static pile load tests. For that reason other methods of design like dynamic pile load tests, wave equation analysis or driving formulas are seldom used. These methods are not discussed in this report. It has to be stated that static pile load tests are always required and executed in the case of the introduction of new pile types.

6 GENERAL PHILOSOPHY DUTCH CODES FOR PILE DESIGN

6.1 Introduction

In this article only requirements for structures concerning safety and serviceability (displacements and deformations) are discussed; requirements for durability and energy consumption are beyond of the scope of this report. The requirements for safety and serviceability as mentioned in *het Bouwbesluit* are laid down in the NEN-codes NEN 6702, NEN 6740 and NEN 6743. For structures not being buildings, the requirements concerning displacements and deformations are not prescribed. They have to be adopted by the structural engineer. The performance requirements are linked to limit states. Limit states are defined as failure of the structure or such deformations/displacements that the structure will not fulfil the requirements for normal use. The codes NEN 6740 and NEN 6743 provide tools to verify that those limit states will, with sufficient safety, not occur. These tools are based on a probabilistic approach. Uncertainties with respect to the loads on and the strength and stiffness of the structure are taken into account by means of partial factors. The basic philosophy is in line with Eurocode 7.

All structures are classified in geotechnical categories and safety classes. The geotechnical categories determine the extend of the required analysis. The safety classes determine the level of safety and durability.

6.2 *Limit states*

Two types of limit states, 1 and 2, are defined. State 1 is the ultimate limit state, when failure of the structure or failure in the interface soil-structure occurs. Limit state 2 deals with the serviceability and is defined as an occurrence of deformations or displacement which lead to a minor use of or damage to the structure or extra maintenance costs. The TGB 1990 requires that no limit state will occur. Limit state 1 is divided in type 1A and type 1B. Type 1A means failure in the soil around the structure or in the interface between soil and structure. Type 1B means deformations of the structure itself in such an extend that the requirements for structural safety are not fulfilled. Type 1A deals with failure of the soil, while type 1B deals with failure of the structure due to large deformations.
The difference between type 1A and 1B has been defined for reasons that large deformations will not necessarily lead to consequences for the structure, if these settlements are more or less uniform (tilting). The check for fulfilling limit state 1A and 1B are tasks of the geotechnical and structural engineer respectively. Normally a maximum relative rotation ß of 1: 100 is adopted (see figure 2). The relative rotation ß is defined as the absolute value of the difference between Θ and ω.

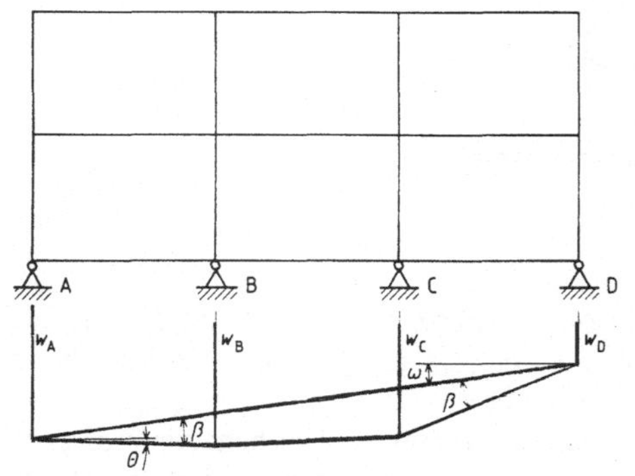

Figure 2. Rotation and tilting

For domestic structures limit state 2 has to be fulfilled. This means that the settlements of the structure, calculated for load and material factors of 1.0, should not exceed 0.15 m. The rotation Θ and relative rotation ß (see figure 2) should not exceed 1:300.
If for other structures no requirements for settlement and rotation are specified, the requirements for domestic structures, should be fulfilled.

6.3 *Safety*

The method of partial factors must ensure that the difference between calculated load and strength is so large that it is sufficiently certain that the real load is less than the real strength. Deformations are approached similarly. The partial factors create the required margins between the actual and the design values. The magnitude of load- and material factor is based on the reliability index. This index is a measure for the chance of reaching a limit state. Depending on the consequences of failure of the structure, each structure has to be classified in safety classes. NEN 6700 prescribes for each safety class the required reliability index (table 1). For each structure it should be proven that the reliability index (and the associated chance on failure) is smaller than given in table 1. It is possible to prove this with a full statistical analysis. For an average structure however, it is much easier to use the simplified model with load and material factors, which should ensure that the requirements with respect to the reliability index will be fulfilled.

The receipt for using the simplified model is the following:

- select characteristic values for loads and material parameters (typically a 5% upper or lower bound, as appropriate);

- determine the design values from the characteristic values using the partial factors;
- verify that the design value of the load is smaller than the design value of the strength.

Table 1 Reliability index

safety class	consequences of failure		reliability indices		
	chance on personal damage	chance on economical damage	limit state 1		limit state 2
			if wind decisive	if other loads decisive	
1	negligible small	small	2.3	3.2	1.8
2	small	large	2.4	3.4	1.8
3	large	large	2.6	3.6	1.8

7 Methodology of NEN6740 "Principals and loads"

7.1 Geotechnical categories

In accordance with Eurocode 7 three categories of geotechnical structures are distinguished. Each category contains structures with similar circumstances. Before starting the soil investigation the structure has to be placed in it's category:

- geotechnical category 1: light and non complicated structures. The behaviour of similar foundation types must be known from other structures in the same conditions. The load per m wall should not exceed 100 kN. Concentrated loads should not exceed 250 kN. The load on floors should be less than 20 kPa. The soil conditions and water table are known from investigations at adjacent area's;
- geotechnical category 2: structures which are not belonging to geotechnical category 1 are placed at least in geotechnical category 2;
- geotechnical category 3: complicated, large structures, with complicated soil and/or loads. Examples of these structures are: underground structures, structures with dynamic loads like machine foundations, chemical installations, high rise buildings, offshore structures, etc.

7.2 Soil investigation

For Geotechnical Category 1 structures no additional soil investigation is required when results of adequate soil investigation is available for at least 2 locations within a distance of 50 m to the new structure. The foundation type should be similar to adjacent structures. Only driven piles are allowed and negative skin friction may not expected. The level of the water table should be known.

For Geotechnical Category 2 structures the distance between the investigated locations should be in balance with the geology of the area, the soil conditions and the extend of the building area. For piled structures at least 2 CPT's are required, each CPT covering an area not greater than 25 x 25 m^2. The depth of the CPT's should be at least 5 m below the final tip elevation of the piles. At least one CPT should reach to a depth of 10 times the smallest width of the pile. If compressible layers are expected below tip elevation or when hydrological problems are to be solved, the investigation should reach to a greater depth, to be determined by a sensibility analysis.

The investigation for structures of Geotechnical Category 3 should fulfil at least the requirements for Geotechnical Category 2. Additional requirements can be prescribed on basis of the specific characteristics of the project.

7.3 *Interpretation and reporting*

After evaluation of the results of the soil investigation it has to be checked that the requirements for the geotechnical category are still fulfilled; if not, the structure has to be placed in the right, more severe category.

NEN 6740 prescribes that the factual report of the soil investigation contains at least the method of investigation, the results of all tests, and specifics like the geology of the area, water table, etc, and an evaluation of all geotechnical data related to the structure.

It also prescribes the contents of the design report.

7.4 *Analysis of limit states*

The requirements for the limit states are fulfilled if:

- for limit state 1A:
  $F_{s;d} <= F_{r;d}$

- for limit state 1B and 2:
  $w_d\ (F_{s;d}, F_{s;rep;nk}, \gamma_{f;nk}, E_{rep}, \gamma_{m;E}) <= w_{req}$

  where,

$F_{s;d}$	is the design value of the force on the structure;
$F_{r;d}$	is the design value of the strength of the structure;
$w_{;d}$	is the design value of the settlement of the pile head;
$w_{;req}$	is the maximum tolerated accepted settlement of pile head;
$F_{s;rep;nk}$	is the representative value of the force due to negative skin friction;
$\gamma_{f;nk}, \gamma_{m;E}$	is the partial load factors for negative skin friction and the stiffness;
E_{rep}	is the representative value of the stiffness of the soil below tip elevation.

7.5 *Determination of $F_{s;d}$*

Loads on structures may be induced by loads or deformations. Only those forces, of which the magnitude will not be influenced by the interaction of the structure with the sub soil will be considered as loads according to NEN 6702. The external loads should be determined according to NEN 6702 *Loads and deformations*. This code specifies the material factors to be applied on static loads. Three types of load are distinguished: permanent loads, live loads and special loads; each with an own load factor. The total load will consist of a combination of the different types of loads. NEN 6702 specifies which combinations have to be considered. The magnitude of the load factor depends not only on the type of load but also on

the safety class of the structure and the limit state considered. Common, non industrial buildings, are classified in the highest safety class 3 (reference period of the structures is 50 years). The load factor for structures belonging to safety class 3, for the loading combination without live loads, and the ultimate limit state, is set to 1.35. In the combination of the permanent and live load, the load factor for the permanent load is 1.2 and for the live load 1.5. In practice the permanent load will be 50 to 80% of the total load. In that case the mean load factor will range from 1.35 to 1.25. In the case of limit state 2 all load factors are set equal to unity (1.0).
In the loading combinations with special loads all load factors are also equal 1.0.

7.6 *Determination of $F_{r;d}$*

In the determination of the design value of the strength of the foundation the following items have to be considered:
- the overall stability of the foundation mass;
- failure of the soil around the pile or group of piles due to tension forces, compression forces or lateral forces should be analyzed;
- the design value of the capacity of the piles may be determined by means of pile load tests;
- the design value of the strength of the piles may be determined on basis of the results of CPT's according to NEN 6743;
- the resistance of pile groups should not be taken larger than the sum of the resistance of the individual piles, and determined according to NEN 6743.

7.7 *Determination of $w_{;d}$*

In the analysis of the settlement of the foundation the following items have to be considered:
- the settlement of the head of an individual pile. This settlement should be analyzed according to NEN 6743, taking into account the effects of negative skin friction. The load-displacement behaviour of a pile is based on empirical relations. If these relations are not available, load tests are required;
- the settlement of a pile group. The settlement of a pile group should be determined taking into account the extension of the area with increased stresses below tip elevation.
- if the design value of the maximum force due to negative skin friction is not included in the total load on the pile, the settlement of head of the pile should be increased with 50% of the settlement of ground level after pile installation, with a maximum of 0,05 m.

7.8 *Determination of relative rotation β and rotation Θ*

The relative rotation and rotation according to figure 2 should be determined based on the settlement $w_{;d}$. In the case of flexible foundations due to a possible heterogeneous subsoil and an imperfect installation of the piles an additional differential settlement has to be taken into account between two adjacent single piles or single pile groups. This additional settlement should be 1/3 of the average settlement of the head of the single pile or the single pile groups.
In the case of a stiff foundation this differential settlement may be set to zero.

7.9 *Material factors*

The material factors to be used are summarized in Table 2.

Table 2. Material factors

pile type	partial factor (γ_m)		limit state		
			limit state 1		limit state 2
			1A	1B	2
compression piles	$\gamma_{m;b1}$	without investigation[1)]	1.4	1.4	1
	$\gamma_{m;b2}$	with load test	1.25	1.25	1
	$\gamma_{m;b3}$	for preloaded piles	1.15	1.15	1
	$\gamma_{m;b4}$	design with CPT	1.25	1.25	1
displacements	$\gamma_{m;E}$		1.3	1.3	1

1) If during pile installation blow counts are registered; if not $\gamma_{m;b1}$ = 1.8.

8 METHODOLOGY OF NEN6743 "PILED FOUNDATIONS"

8.1 *Introduction*

NEN 6743 presents a method for the evaluation of the ultimate and serviceability limit states for piled foundations. Empirical load-settlement relationships are used, where the load is expressed as a percentage of the maximum bearing capacity.

For the evaluation of the serviceability limit state, total and differential settlements have also to be established. For the determination of these settlements, the partial load and material factors are set equal to unity (as indicated in chapter 7.5 and Table 2).

NEN 6743 further takes into account the effect of redistribution of the load between the foundation points in relation to the stiffness of the structure and the number of cone penetration tests carried out on the site.

With respect to the determination of the negative skin friction, NEN 6743 provides recommendations for calculating the maximum value which can be expected.

8.2 *Terms and definitions*

The following terms and definitions are applicable:

- *pile*: Element for which the length is at least five times the smallest dimension of the cross-section of the pile point.
- *pile base*: Geometric shape of the bottom part of the pile that could either be enlarged or not.
- *pile point*: Lowermost full cross-section of the pile base.
- *pile point level*: The level in the soil at which the pile is placed in relation to a reference level.
- *maximum point resistance*: The maximum pile point resistance in the soil when the pile penetrates the soil.

- *maximum shaft resistance*: The maximum friction between the pile shaft and the soil when the pile penetrates the soil.
- *pile shaft*: The part of the pile between the pile base and the pile head.
- *equivalent pile point diameter* (D_{eq}) *or pile shaft diameter* (d_{eq}): The value for the pile point diameter or the pile shaft diameter respectively to be used in calculations, for round pile points or pile shafts respectively, equal to the largest outer diameter; for rectangular pile points or pile shafts respectively defined by:

$$D_{eq} resp. d_{eq} = 1.13\, a \sqrt{\frac{b}{a}} \quad ,where$$

D_{eq} is the equivalent point diameter;
d_{eq} is the equivalent shaft diameter;
a is the smallest dimension of the largest cross-section of the pile point or the pile shaft respectively;
b is the largest dimension of the largest cross-section of the pile point or the pile shaft respectively, with $b \leq 1.5\, a$.

- *overconsolidation ratio* (*OCR*): The ratio of the vertical effective stress to which the soil at a certain level has been subjected and the present vertical effective stress at the same level.
- *maximum bearing capacity*: if determined from the results of load tests the force applied by the load test to the pile head, whereby the settlement is equal to $0.1D_{eq}$ for displacement piles and $0.2D_{eq}$ for augered or bored piles.

8.3 *Determination of the maximum bearing capacity*

8.3.1 *Determination of the design value of the maximum bearing capacity*

The design value of the maximum bearing capacity of the foundation under a structure or parts of it must be determined as follows:

$$F_{r;found;\max;d} = \frac{F_{r;found;\max;rep}}{\gamma_{m;b}} \quad ,where:$$

$F_{r;found;max;d}$ is the design value of the maximum bearing capacity of the foundation;
$F_{r;found;max;rep}$ is the representative value of the maximum bearing capacity of the foundation of the relevant part of the structure, determined in accordance with 8.4;
$\gamma_{m;b}$ is the material factor in accordance with Table 2.

8.4 *Determination of the representative value of the maximum bearing capacity of the foundation*

The representative value of the maximum bearing capacity of the foundation of a rigid structure or a rigid part thereof is determined by:

$$F_{r;found;max;rep} = M \bullet F_{r;max;rep}$$

where:
M is the number of piles under the relevant part of the structure;
$F_{r;max;rep}$ is the representative value of the maximum bearing capacity of a pile under the relevant part of the structure.

Representative values for the maximum bearing capacity of piles placed under a rigid structure or a rigid part of the structure

A structure or a part thereof may be considered to be rigid if the condition as stated below has been met.
If a foundation element is removed it must be apparent that the settlement of the structure at the location of the removed foundation element (see figure 3) is less than or equal to 5 mm, under the influence of the load for the serviceability limit state.

The representative value of the maximum bearing capacity of a pile in the foundation of a rigid structure or a part thereof shall have been determined using the following formula:

$$F_{r;\max;rep} = \xi \cdot F_{r;\max;mean}$$

whereby:

$$F_{r;\max;mean} = \frac{1}{N} \sum_{i=1}^{i=N} F_{r;\max;i}$$

where:

$F_{r;max;rep}$	is the representative value of the maximum bearing capacity of a pile;
$F_{r;max;mean}$	is the mean value of the maximum bearing capacity of a pile;
$F_{r;max;i}$	is the maximum bearing capacity for the selected pile point level at CPT i;
N	is the number of cone penetration tests in that part of the site;
ξ	is a factor that depends on the number of piles M and on N. For 1 pile and 1 CPT $\xi=0.75$; for M and N >10 $\xi=0.92$

The factor ξ serves to take account of:

- the capacity of the structure to transfer forces from a point with a "weak" foundation element to points with a "strong" foundation element and,
- the better knowledge of the variability and the quality of the soil which is obtained if more cone penetration tests are carried out.

The value of the factor ξ has been derived from a statistical analysis and calibration based on current practice. If the results of the soil investigation indicate that locally the elevation of the top of the foundation layer and the strength of it differs significantly it is recommended to treat the CPT's in that area as a different group.

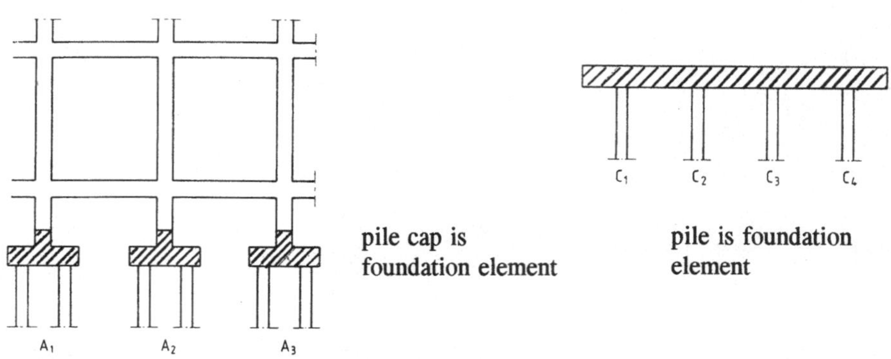

Figure 3 Type of foundation elements

Representative values for the maximum bearing capacity of piles placed under a non-rigid structure or a part thereof

The representative value of the maximum bearing capacity of a pile in the foundation of a non-rigid structure or a part thereof shall have been determined using the following formulas:

- For $N \leq 3$: $F_{r;max;rep} = \xi_{1;N} \; F_{r;max;min}$

- For $N > 3$: $F_{r;max;rep} = \xi_{1;N} \; F_{r;max;mean}$

where:

$F_{r;max;rep}$ is the representative value of the maximum bearing capacity of a pile;
$F_{r;max;min}$ is the lowest value of the maximum bearing capacity of a pile from the results of the cone penetration tests carried out in the considered part of the site;
$F_{r;max;mean}$ is the mean value of the maximum bearing capacity of a pile;
$\xi_{1;N}$ is the factor ξ for $M = 1$;
N is the number of considered cone penetration tests;

In the case of driven piles, the following formula is acceptable:

$$F_{r;max;rep;i} = 0.75 \; F_{r;max;i}$$

whereby the following conditions must be met:

1. For the purpose of the design of the pile foundation, the site is subdivided into adjoining areas surrounding each cone penetration test *i*.
2. The blow count diagram over the bottom metre of each driven pile in an area may not deviate more than 35% from the blow count diagram over the bottom metre of a pile driven at a distance of not more than 1 m from the cone penetration test location of that area.
3. If condition 2 is not met (deviation > 35%), then $F_{r;max;i}$ must be determined from the results of a cone penetration test carried out at a maximum distance of 1 m from the pile.

8.5 *Determination of the maximum bearing capacity of a single pile based on the results of a cone penetration test*

The maximum bearing capacity of the pile at the location of cone penetration test *i* shall have been determined as follows:

$$F_{r;max;i} = F_{r;max;point;i} + F_{r;max;shaft;i}$$

where:

$$F_{r;max;point;i} = A_{point} \cdot p_{r;max;point;i}$$

and

$$F_{r;max;shaft;i} = O_{p;mean} \int_0^{\Delta L} p_{r;max;shaft;i} \, dz \quad ,where$$

$F_{r;max;point;i}$ is the maximum point resistance of the pile determined from the results of cone penetration test i;

$F_{r;max;shaft;i}$ is the maximum shaft resistance determined from the results of cone penetration test i;

A_{point} is the cross-sectional area of the pile point;

$p_{r;max;point;i}$ is the maximum unit point resistance from the results of cone penetration test i;

$O_{p;mean}$ is the mean circumference of the part of the pile shaft in the layer in which the pile base has been positioned;

ΔL is the length of the part of the pile on which the pile shaft friction has been assumed, whereby:

- ΔL is equal to the entire pile length if the soil layers above the pile point level mainly consist of sand, and further only of clay and silt layers in which cone resistances greater than 2 MN/m² have been recorded;
- ΔL is not greater than the length of the enlarged part of the base, in the case of preformed piles with an enlarged base;

$p_{r;max;shaft;i}$ is the maximum unit shaft resistance determined from the results of cone penetration test i;

z designates the vertical direction.

The bearing capacity of an open pipe pile shall have been determined by taking the sum of the maximum shaft resistance of the outer and inner walls and the maximum point resistance of the bottom edge of the pipe pile. The adopted maximum shaft resistance of the inner wall must not exceed the maximum point resistance, which is mobilised if the soil in the pile forms a fixed plug.

For cast-in-place piles, including those with an enlarged base or baseplate, whereby during pile installation concrete or grout is placed in contact with the soil, shaft resistance may be included over the previously indicated length ΔL.

8.5.1 *Determination of the maximum point resistance*

The following determinations are applicable for a pile placed at the location of cone penetration test i. The index i has been further omitted.

The maximum point resistance $p_{r;max;point}$ must have been determined with the following formula, whereby the obtained value for sand and gravel must be reduced according to figure 7.

$$p_{r;\max;point} = 1/2\ \alpha_p\ \beta\ s\ \left(\frac{q_{c;I;mean} + q_{c;II;mean}}{2} + q_{c;III;mean} \right)$$

where:

$p_{r;max;point}$ is the maximum point resistance;

$q_{c;I;mean}$ is the mean value of the cone resistance $q_{c;z;corr}$ in trajectory I that runs from the pile point level to a level that is at least 0.7 times and at most 4 times the equivalent diameter (D_{eq}) deeper. The bottom of trajectory I must be selected within the above-mentioned limits in such a way that $p_{r;max;point}$ is minimal;

$q_{c;II;mean}$ is the mean value of the cone resistance $q_{c;z;corr}$ in trajectory II that runs from the bottom of trajectory I to the pile point level, whereby the value used for the cone resistance must never be higher than the previous value in the trajectory;

$q_{c;III;mean}$ is the mean value of the cone resistance $q_{c;z;corr}$ in trajectory III that runs from the pile point level to a level that is 8 times the equivalent diameter (D_{eq}) higher,

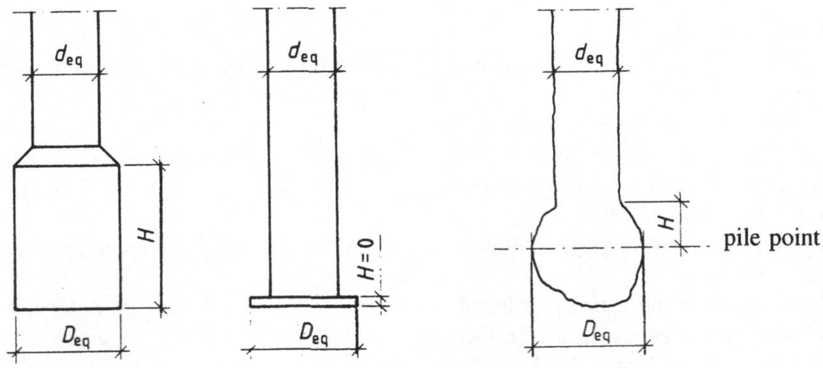

Figure 4 Shape of the pile base

whereby in the same way as for trajectory II the value used for the cone resistance must never be higher than the previous value in the trajectory. The starting value of the used cone resistance in trajectory III is the lowest value used for the cone resistance in trajectory II. For continuous flight auger piles this trajectory must start with a cone resistance less than or equal to 2 MN/m^2, unless the results of a cone penetration test are used, which has been carried out at a distance of 1 m from the pile after the pile installation;

α_p is the pile class factor, determined in accordance with table 3;

β is the factor that takes into account the influence of the shape of the pile base (figure 4);

s is the factor that accounts for the influence of the shape of the cross-section of the pile base, determined in accordance with figure 6.

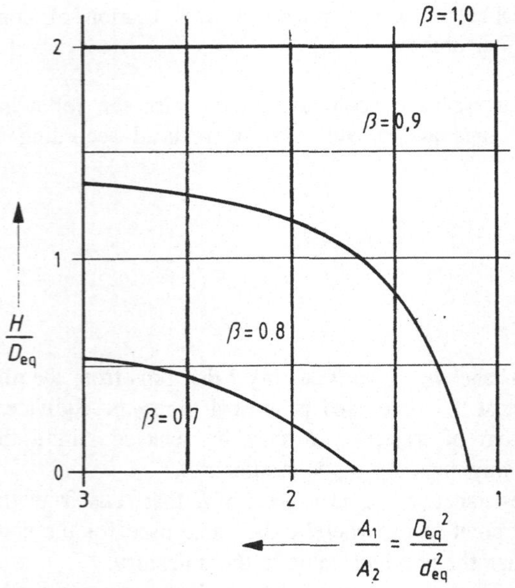

Figure 5 Pile base shape factor β

The maximum point resistance is determined as the weighted mean of the cone resistances below and above the pile point, where the failure of the soil occurs. By giving more weight to the low cone resistances in the area below the tip, the effect of for example punching is taken into account. The value of 2 MN/m^2 takes account of the disturbance that can occur, at the moment that the extraction of the continuous flight auger starts after reaching the required level. Before concreting or grouting can begin, space must be made to open the valve at the bottom end of the central tube. This allows relaxation or even flow of the sand at the location of the valve, whereby the originally recorded cone resistance can be reduced locally. The results of cone penetration tests made after the installation of continuous flight auger piles have shown that cone resistances can drop to values that are

Table 3 Values for the pile tip factor

pile class / type	α_p
displacement piles:	
- driven piles	1.0
- driven cast-in-place piles	1.0
- screw cast-in-place piles	0.9
- screw prefabricated piles	0.8
small displacement piles:	
- steel sections and steel open pipe piles	1.0
replacement piles:	
- continuous flight auger piles	0.8
- bored piles (formed with the use of drilling mud in an uncased borehole)	0.5
- bored piles (formed by use of shelling techniques and permanent casing)	0.5

less than 2 MN/m². The pile tip factor α_p is taken according to table 3, however it is allowed to use higher pile class factors, if certain (in this article not specified) conditions are satisfied.

The pile base shape factor β shall have been established from figure 5 after the effective height of the pile base (H), its equivalent diameter (D_{eq}), and the equivalent diameter of the pile shaft (d_{eq}) have been determined in accordance with figure 4, with the provision that $\beta = 1$ if the two following conditions have been met:

- β is determined for driven cast-in-place piles without permanent casing, whereby the full grout or concrete pressure acts against the soil during pile installation;
- $D_{eq}^{\ 2} / d_{eq}^{\ 2}$ is less than or equal to 1.5

  where:

  D_{eq} is the equivalent diameter, which in this case is the diameter of the steel base plate;

  d_{eq} is the equivalent diameter of the pile shaft (inner diameter of the driving tube).

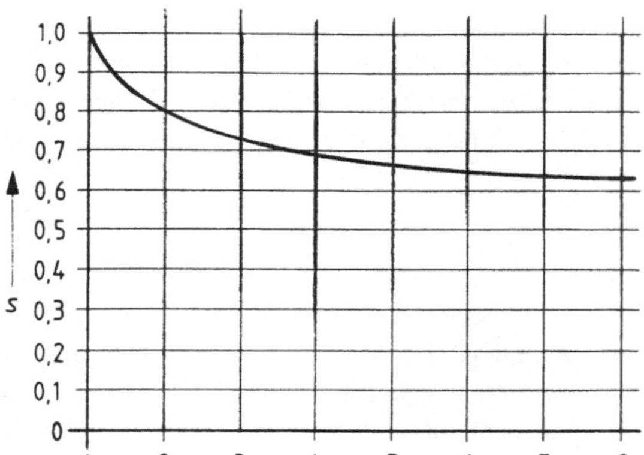

Figure 6 Values of s

The factor s shall have been established from figure 6, with the relevant ratio of the largest and the smallest dimensions of the cross-section of the pile point.

Value of the maximum point resistance in sand and gravel.

For the maximum point resistance $p_{r;max;point}$ in sand and gravel, a reduced $p_{r;max;point;red}$ value must have been determined using figure 7.

The overconsolidation ratio (*OCR* value) must at least have been generally established on the basis of geological data concerning the magnitude of previous loading by ice or soil.

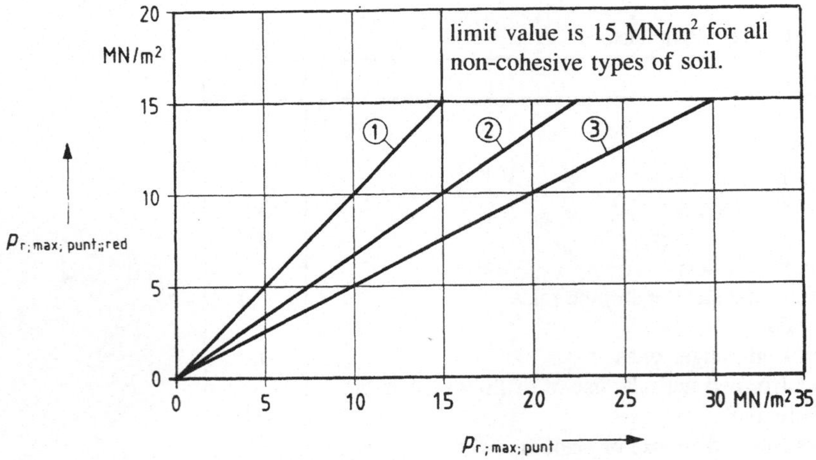

Figure 7 Value for the maximum point resistance in sand and gravel

1	sand and gravel	$OCR \leq 2$
2	sand and gravel	$2 \leq OCR \leq 4$
3	sand and gravel	$OCR > 4$

OCR = overconsolidation ratio

8.5.2 *Determination of the maximum shaft resistance*

The maximum shaft resistance $p_{r;max;shaft}$ must have been determined using the formula:

$$p_{r;max;shaft;z} = \alpha_s \, q_{c;z;a}$$

where

$p_{r;max;shaft;z}$ is the maximum shaft resistance at depth z;

α_s is the factor, according to tables 4 and 5, that takes into account the influence of pile installation, whereby:
- M_s is the sand median;
- M_g is the gravel median;
- z is the depth;

$q_{c;z;a}$ is the cone resistance whereby values higher than 15 MN/m² that occur over a depth range of more than 1 m have been limited to 15 MN/m² and values higher than 12 MN/m² that occur over a depth range of less than 1 m have been limited to 12 MN/m².

Requirements with respect to the cone penetration test

The cone resistance $q_{c;z}$ shall have been determined by means of cone penetration tests carried out at the building site in accordance with chapter 5 of NEN 3680:1982. Where excavation occurs after the execution of cone penetration tests, then $q_{c;z}$ shall have been corrected in the following manner:

$$q_{c;z;corr} = \frac{\sigma'_{v;z}}{\sigma'_{v;z;i}} \, q_{c;z}$$

Table 4 Maximum values of α_s in sand and gravelly sand

pile class / type	α_s [1)]
displacement piles:	
- driven straight-sided precast concrete pile and closed-ended steel pipe pile	0.010
- cast-in-place pile, whereby the driving tube is raised by upward blows of the hammer after the concrete has been placed	0.014
- cast-in-place pile, whereby the driving tube is raised by a vibrating unit after the concrete has been placed	0.012
- tapered timber pile	0.012
- screw piles:	
with grout injection or grout mixing	0.009
without grout	0.006
small displacement piles:	
- steel sections	0.0075
replacement piles:	
- continuous flight auger piles	0.006 [2)]
- bored piles (formed with the use of drilling mud in an uncased borehole)	0.006
- bored piles (formed by use of shelling techniques and permanent casing)	0.005

[1)] The values are valid for very fine to coarse sand, whereby in accordance with NEN 5104:1989, the following must have been satisfied: 105 μm < M_s < 600 μm.
For very coarse sand with M_s > 600 μm and gravel with $M_g \geq$ 2 mm, reduction factors must have been applied to α_s of 0.75 and 0.5, respectively.

[2)] This maximum value of α_s shall be applied if q_c values have been used from cone penetration tests that have been carried out before pile installation. If q_c values are used from cone penetration tests that have been made near the piles after pile installation, the maximum value of α_s = 0.01.

Table 5 Maximum values for α_s in clay, silt, and peat

type of soil	relative depth z/d	α_s
clay / silt $q_c \leq 1$ Mpa	$5 < z/d < 20$	0.025
clay / silt $q_c \leq 1$ Mpa	$z/d >= 20$	0.055
clay / silt $q_c > 1$ Mpa	-	0.035
peat	-	0

where

$q_{c;z;corr}$ is the corrected cone resistance at depth z below the bottom level of the excavation;
$\sigma'_{v;z}$ is the vertical effective stress at depth z below the bottom level of the excavation;
$\sigma'_{v;z;i}$ is the initial vertical effective stress, before excavation, at depth z below the bottom level of the excavation;
$q_{c;z}$ is the measured cone resistance at depth z.

8.5.3 *Load Tests*

If load tests have been conducted on the site as a check on the calculated maximum bearing capacity the maximum bearing capacity shall have been determined with,
for displacement piles:
$F_{r;max} \leq F_{0.1Deq} - F_{shaft;l}$
for augered piles:
$F_{r;max} \leq F_{0.2Deq} - F_{shaft;l}$
where:

$F_{0.1Deq}/F_{0.2Deq}$ is the force applied by the load test to the pile head, whereby the settlement is equal to $0.1D_{eq}$ *or* $0.2D_{eq}$ respectively.

$F_{shaft;l}$ is the force due to soil friction acting on that part of the pile shaft ($l = L - \Delta L$) that has not been included in the calculation of the maximum bearing capacity:

$$F_{shaft;l} = O_p \int_0^{\Delta L} p_{r;\max;shaft}\, dz$$

D_{eq} is the diameter or the smallest cross-sectional area of the pile point;
O_p is the circumference of the pile shaft;
L is the length between the pile point and pile head;
ΔL is the length of the part of the pile on which the pile shaft friction has been assumed; $\Delta L = L - l$;
l is the length between the top point of the part $L - l$ of the pile on which pile shaft friction has been assumed, and the pile head;
$p_{r;max;shaft}$ is the maximum unit shaft resistance whereby the following applies:

- $\alpha_s \geq 0.07$ for clay and peat;
- $\alpha_s \geq 0.015$ for sand.

9 Determination of the design value of the settlement of the top of the foundation

The design value of the settlement of the top of a foundation element shall have been determined as a function of the pile load, using the following formula:
$w_d = w_{1;d} + w_{2;d}$
where:
w_d is the design value of the settlement of the top of a foundation element;
$w_{1;d}$ is the design value of the settlement of pile head;
$w_{2;d}$ is the settlement due to compression of the soil layers situated below the pile point level.

Determination of the design value of the settlement $w_{1;d}$

The design value of the settlement $w_{1;d}$ shall have been determined with the following formula:
$w_{1;d} = w_{point;d} + w_{el;d}$
where:
$w_{1;d}$ is the design value of the settlement of the pile head;

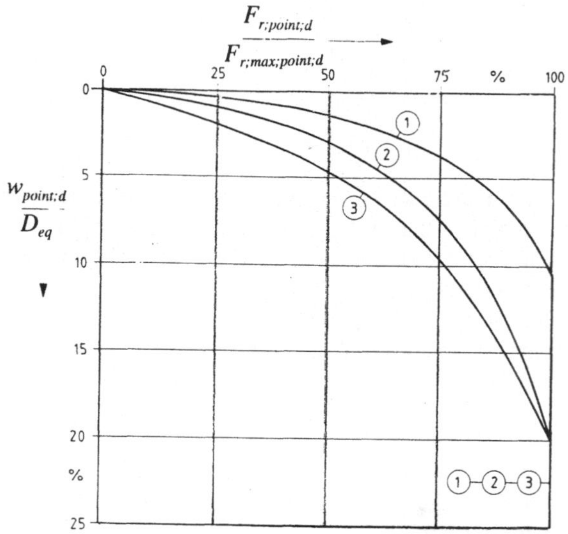

1 displacement pile
2 continuous flight auger pile
3 bored pile

Figure 8 Relationship between $F_{r;point;d}$ expressed as a percentage of $F_{r;max;point;d}$ and $w_{point;d}$ expressed as a percentage of D_{eq}

$w_{point;d}$ is the design value of the settlement of the pile point due to the load on the pile;
$w_{el;d}$ is the design value of the settlement of the pile head pile relative to the pile point, due to elasticity of the pile.

Determination of $w_{point;d}$

Determine the design value of the maximum point resistance ($F_{r;max;point;d}$) and the maximum

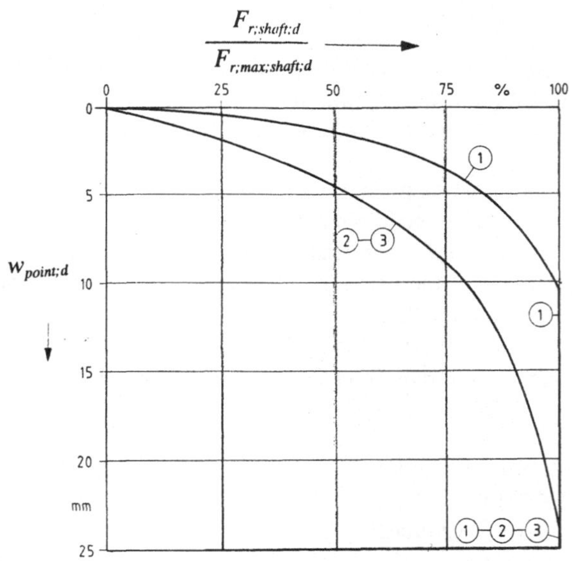

1 displacement pile
2 continuous flight auger pile
3 bored pile

Figure 9 Relationship between $F_{r;shaft;d}$ expressed as a percentage of $F_{r;max;shaft;d}$ and $w_{point;d}$ expressed in millimetres shaft resistance ($F_{r;max;shaft;d}$) in accordance with:

$$F_{r;max;point;d} = \xi \, (F_{r;max;punt} / \gamma_{m;b4})$$

$$F_{r;max;shaft;d} = \xi \, (F_{r;max;shaft} / \gamma_{m;b4})$$

Make a diagram with:

- values of $w_{point;d}$, on the vertical axis, in millimetres ranging from zero to at least the value whereby $F_{r;max;point;d}$ according to figure 8 and $F_{r;max;shaft;d}$ according to figure 9 is reached (in figure 8, the value on the vertical axis of $(w_{point;d} \,/\, D_{eq})$ x 100% must be converted to $w_{point;d}$ by multiplying with 0.01 D_{eq});
- on the horizontal axis to the left, values in kN of $F_{r;shaft;d}$ ranging from zero to $F_{r;max;shaft;d}$, and to the right, values of $F_{r;point;d}$ ranging from zero to $F_{r;max;point;d}$.

Determination of $w_{el;d}$

Determine the design value of the settlement of the pile head relative to the pile point due to the elasticity of the pile using the following formula:

$$w_{el;d} = \frac{L \cdot F_{mean;d}}{A_{shaft;d} \cdot E_{p;mat;d}}$$

where:

$$F_{mean;d} = \frac{l \, F_{s;tot;d} + 0.5 \, (L - l) \, (F_{s;tot;d} + F_{R;point;d})}{L}$$

and:

$w_{el;d}$ is the design value of the settlement of the pile head in relation to the pile point due to the elasticity of the pile;

$F_{mean;d}$ is the design value of the mean normal force in the pile shaft, determined from $F_{s;tot;d}$ and $F_{R;point;d}$ for the distribution of the pile shaft friction, which has been adopted for the determination of the bearing capacity over the part ($L - l$) of the pile in the soil layer supporting the pile;

$F_{s;tot;d}$ is the design value of the total load on the pile head;

$F_{r;point;d}$ is the design value of the actual force in the pile point;

L is the length between the pile point and the pile head;

l is the length between the top point of the part ($L - l$) of the pile on which pile shaft friction has been assumed, and the pile head;

$A_{shaft;d}$ is the design value of the cross-sectional area of the pile shaft;

$E_{p;mat;d}$ is the design value of the modulus of elasticity of the pile shaft material in accordance with:

- $E_{p;mat;d} = E_{p;b;d} = 20 \times 10^9$ N/m² for concrete;
- $E_{p;mat;d} = E_{p;a;d} = 200 \times 10^9$ N/m² for steel;
- $E_{p;mat;d} = E_{p;h;d} = 15 \times 10^9$ N/m² for timber.

Determination of the design value of the settlement $w_{2;d}$

For centre-to-centre pile spacings of more than ten times the smallest dimension of the cross-section of the pile base, $w_{2;d} = 0$ may be used. If the pile spacing does not comply with this requirement, then the design value of the settlement $w_{2;d}$ must be determined. The prescribed method is beyond of the scope of this article.

10 CALCULATION OF THE REPRESENTATIVE VALUE OF THE MAXIMUM NEGATIVE SKIN FRICTION FORCE AND SWELLING FORCES

NEN 6743 specifies a calculation method for the determination of the magnitude of the negative skin friction force and swelling forces if excavations are applied. These methods are not discussed in this paper.

11 EXAMPLE

Worked example of the determination of the maximum bearing capacity and the design value of the settlement of the pile head ($w_{1;d}$) for the serviceability limit state.

Data for calculation in the serviceability limit state

Pile: $a = b = 0.4$ m; pile is driven and made of concrete (curves (1) in figures 8 and 9).
The c.t.c. spacing with other piles is more then 10 times the diameter with other piles.
Design value of the modulus of elasticity of the concrete pile shaft: $E_{p;d} = 20 \times 10^6$ kN/m².
Negative skin friction: $F_{s;nk;d} = 500$ kN, whereby $F_{s;nk;d}$ must be considered as a part of the design value of the total load on the pile head, due to the fact that the ground surface settles more than 0.1 m after installation of the pile.
Design value of the load on the pile head: $F_{s;d} = 700$ kN.
Cone penetration test: according to figure 10, whereby shaft resistance may only be taken into account in the layer from $l = 15$ m to $L = 19$ m (the displacement of the pile shaft in relation to the soil is, in this layer, equal to the settlement of the pile point).
Design value of the cross-sectional area of the pile point:
$A_{point} = a^2 = 0.16$ m².
Design value of the effective shaft area on which the shaft resistance acts:
$A_{eff;shaft} = (4\,a) \bullet (L - l) = 4 \bullet 0.4\,(19 - 15) = 6.4$ m².

Determination of design value of maximum bearing capacity $F_{r;max}$

According to 8.4 and 8.4.1:
$F_{r;max} = F_{r;max;point} + F_{r;max;shaft}$
$F_{r;max;point} = A_{point} \cdot p_{r;max;point} = 0.16 \,.\, (1/2\,(1.1.1\,(13 + 11)/2 + 8) = 1.6 \text{ MN} = 1.600 \text{ kN}$
$F_{r;max;point;rep} = \xi \,.\, F_{r;max;point} = 0.75 \,.\, 1.600 = 1.200 \text{ kN}$

$F_{r;max;point;d} = 1.200$ kN (for limit state 2 $\gamma_{m;b4} = 1.0$)

According to 8.4 and 8.4.2:
$F_{r;max;shaft} = A_{eff} \cdot p_{r;max;shaft} = A_{eff}\,\alpha_s\,q_{c;z;a} = 6.4 \,.\, 0.01 \,.\, 10 = 0.64 \text{ MN} = 640 \text{ kN}$
$F_{r;max;shaft;rep} = \xi \,.\, F_{r;max;shaft} = 0.75 \,.\, 640 = 480 \text{ kN}$
$F_{r;max;shaft;d} = 480$ kN (for limit state 2 $\gamma_{m;b4} = 1.0$)

$F_{r;max} = 1.200 + 480 = 1.680$ kN

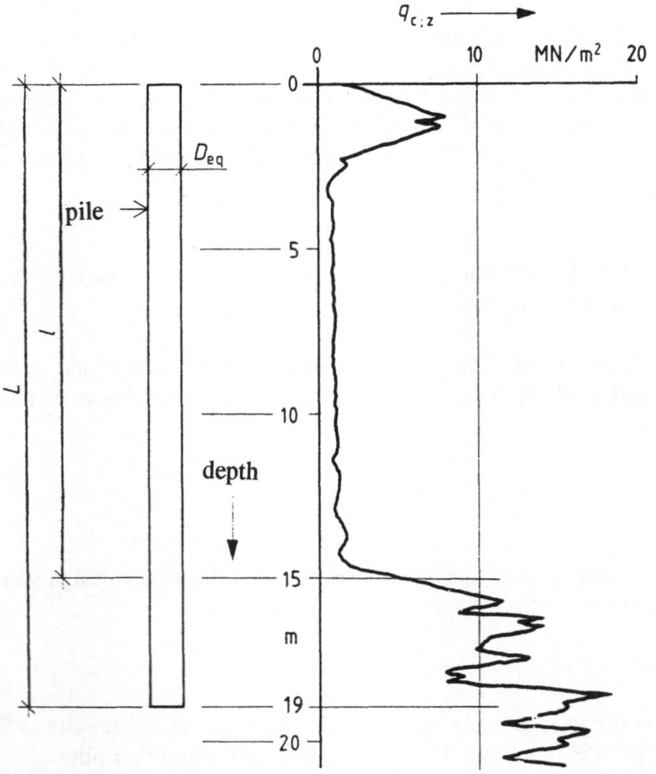

Figure 10 - Record of cone penetration test

Determination of $w_{1;d}$

According to chapter 9: $w_d = w_{1;d} + w_{2;d}$. Since the spacing between the piles is more then 10 times the diameter of the piles $w_{2;d} = 0$.

$w_{1;d} = w_{point;d} + w_{el;d}$

Using figure 8, on a graphical way or by trial and error:
$w_{point;d}/D_{eq} = 2\%$. Then $F_{r;point;d}/F_{r;max;point;d} = 62.5\%$. So $F_{r;point;d} = 0.625 \ . \ 1.200 = 750$ kN.
$w_{point;d}/D_{eq} = 2\%$, $w_{point;d} = 0.02 \ . \ 1.13 \ . \ 400 \approx 9$ mm. Using figure 9 then
$F_{r;shaft;d}/F_{r;max;shaft;d} = 95\%$. So $F_{r;shaft;d} = 0.95 \ . \ 480 = 456$ kN.
$F_{r;max} = 750 + 456 = 1.206$ kN ≈ 1.200 kN, so $w_{point;d} \approx 9$ mm.

According to chapter 9:
$w_{el;d} = (L \bullet F_{mean;d}) / (A_{shaft;d} \bullet E_{p;mat;d})$
whereby the following steps are taken:

Determination of $F_{mean;d}$:
Determine the magnitude of $F_{r;point;d}$ using figure 11.
This gives: $F_{r;point;d} = 1.200 - 456 = 744$ kN; then:
$F_{mean;d} = ((15)\ 1.200 + (19\text{-}15) \times 0.5\ (1.200 + 744)) / 19 \approx 1.152$ kN

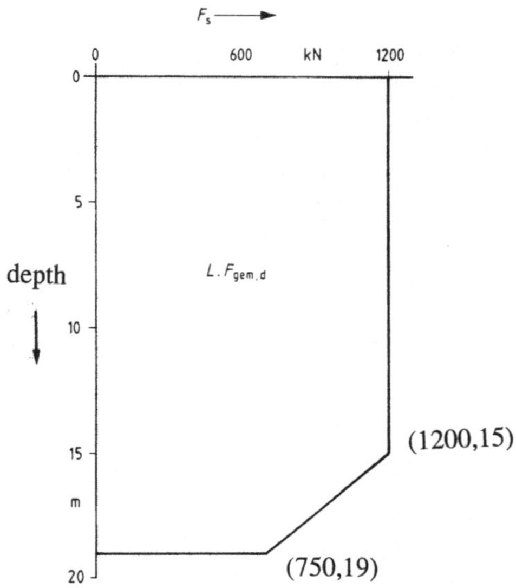

Figure 11 Determination of $L \bullet F_{mean;d}$ from the schematized relationship between the design value of the normal force in the pile shaft and depth

Determination of $L \bullet F_{mean;d}$

$L \bullet F_{mean;d} = 19 \times 1152 = 21888$ kNm

Determination of $A_{shaft;d} \bullet E_{p;mat;d}$

- $A_{shaft;d}$ is equal here to A_{point}: $0.16\ m^2$
- $E_{p;mat;d}$ is equal here to $E_{p;d}$: $20 \times 10^6\ kN/m^2$

Therefore: $A_{shaft;d} \bullet E_{p;mat;d} = 0.16 \times 20 \times 10^6 = 3.2 \times 10^6$ kN

It now follows for $w_{el;d}$:

$w_{el;d} = 21888 / (3.2 \times 10^6) \approx 6.8 \times 10^{-3}$ m ($w_{el;d} \approx 6.8$ mm)

$w_{1;d} = w_{point;d} + w_{el;d} = 9 + 6.8 \approx 16$ mm.

REFERENCES

KIVI syllabus: Ervaringen met NEN-Geotechniek normen. *Filosofie achter de normen*; L. de Quelerij; juni 1993.

NEN 6702: *Loads and deformations*. First print december 1991; NNI.

NEN 6740: *Geotechnics: principals and loads*. First print, december 1991 with corrections dated september 1993; NNI

NEN 6743: *Piled foundations*. First print, december 1991 with corrections dated september 1993; NNI

Design of Axially Loaded Piles – European Practice, De Cock & Legrand (eds)
© 1997 Balkema, Rotterdam, ISBN 90 5410 873 8

Design of axially loaded piles – Norwegian practice

Arne Schram Simonsen & Corneliu Athanasiu
NOTEBY A/S Consulting Engineers, Oslo

1. REGIONAL GEOLOGY

1.1 General Background

In total, only about 25% of the land area in Norway is covered by soil deposits. The major part is dominated by bedrock, or just a very thin cover of morainic materials, normally not representing serious geotechnical challenges. Areas covered by deep soil deposits are particularly concentrated in the lowlands, in the bottom of the valleys and in a narrow strip along parts of the coastline. The geological origin of these deposits is mostly related to the Quaternary period, covering the last 2 million years of the geological history. However, recent processes like landslides, erosion and weathering have many places influenced the local geology considerably.

During the Quaternary period, the climate was generally very cold and glaciers covered large land areas. This last ice-age represented several sequences where the Scandinavian peninsula was covered by a massive ice-cap. However, a number of warm interglacial periods caused the ice-cap to retreat temporarily. This period of time, rich of climatic fluctuations, represents the most significant influence on the Norwegian geology, both with respect to the type and structure of deposited materials.

The ice-cap was at its maximum extension about 20 000 bp, compressing the land masses to a level several hundred meters below the present. The movement of the glaciers scoured the rock surface, and transported eroded material in front, beneath and incorporated in the ice mass.

About 10 000 years ago, a warmer period set in and the glaciers started to retreat permanently. During deglaciation, large rivers of meltwater from the retreating ice cap deposited the glacially transported material into the sea. The sea level in the deglaciation period was significantly higher than present, reaching about 180-220 masl in some regions in the eastern and central parts of Norway. At the same time, the land masses started to heave, explaining the presence of marine sediments on dry land today. Land areas below this marine limit comprise large parts of the populated areas in Norway.

1.2 General geology below the marine limit

The geology below the marine limit is generally dominated by the sedimentological processes taking place in a marine environment. At some locations, glacifluvial deltas have been formed, mainly containing gravel and sand, but also strata of finer material. Clays and silts were deposited in the sea at some distance from the retreating glacier, and these soils dominate large areas below the marine limit. Glacial tills of variable thickness usually cover the bedrock, containing all soil fractions from clay to block.

1.3 General geology above the marine limit

The thickness of the soil deposits above the marine limit is generally less and more patchy than in the lowlands. The geology in these areas is dominated by glacial tills, glacifluvial sediments or dead-ice deposits from a stagnating glacier. Lacustrine clays and silts are scarce, but can be found in or near existing lakes or previous glacial lakes.

A typical soil profile above the marine limit may be described as follows:

- Top layer of organic matter, slide debris or other mixed materials.
- Alluvial gravels and sands.
- Lacustrine or sediments deposited in glacial lakes, mainly consisting of silts and sands.
- Moraines and/or glacifluvial sediments

1.4 Geotechnical properties

Coarse grained soils

Norwegian sand and gravel deposits usually contain equal portions of quartz and feldspar, sometimes representing as much as 80 to 90 % of free mineral grains. Typical geotechnical properties of our quartz-rich sands are summarised in the following:

unit weight:	γ	18-21 kN/m^3
unit weight of solids:	γ_s	26,0-27,0 kN/m^3
max. porosity:	n_{max}	45-50 %
min. porosity:	n_{min}	30-38 %
cohesion:	c	0-10 kPa
friction:	$\tan\varphi$	0.6-1.1
modulus number:	m	150-850

Fine-grained soils

Norwegian clays generally consist of low-activity clay minerals such as vermiculite and chlorite, in addition to minerals like feldspar and muscovite.

Some of the marine, fine-grained deposits have been overconsolidated due to ice loading, removal of previous overburden, capillary effects or a lowered groundwater table. The soft clays and silts show a small apparent overconsolidation, probably due to ageing or creep effects.

Typical ranges in properties for Norwegian clays and fine silts are shown below:

clay fraction:	< 2 mm	25-60 %
void ratio:	e	0,6-1,5
water content:	w	25-55 %
liquid limit:	w_l	35-45 %
plastic limit:	w_p	17-27 %
plasticity index:	I_p	10-35 %
unit weight:	γ	16-20 kN/m^3
unit weight of solids:	γ_s	26,5-28,5 kN/m^3
sensitivity:	S_t	> 5
undrained shear strength:	s_u	10-150 kPa
cohesion:	c	5-50 kPa
friction:	$\tan\phi$	0.4-0.7
modulus number:	m	15-50
coefficient of consolidation	c_v	5 - 100 m^2/yr

The detection of possible *quick clay* deposits is usually on the agenda when performing site investigations in fine-grained soils. This clay liquefy completely when remoulded, and the debris from a quick clay slide may overflood large areas. The undisturbed strength may however be considerable, which means that the sensitivity may be very high (100 - 1000). This behaviour can be explained by the open grain structure in a marine clay, established during sedimentation. The structure turns into a labile "cardhouse-structure" if salt ions are removed from the pore water by long-term leaching processes or groundwater transport.

With soil conditions ranging from soft quick clays to very stiff moraines, it is necessary to apply different geotechnical field equipment to remedy an optimal site investigation. The variations in geological conditions also represent given restrictions for general use of some types of field equipment.

2. COMMON PRACTICE FOR SOIL INVESTIGATION

2.1 Introduction

The soil investigation shall form the basis for the design of technical and economical good pile solutions.

The soil stratification, geotechnical properties, depth bedrock or firm ground and ground water level are important parameters for the selection of pile type and pile design. The soil investigation shall also comprise of sufficient data for the evaluation of the overall stability and settlements in the piled area.

The soil investigation is normally specified or reviewed by the geotechnical consultant.

Friction piles require soils data to sufficient depth below the pile tip. End bearing piles to bedrock calls for percussion drillings or total soundings. The various investigation methods are specified in guidelines from the Norwegian Geotechnical Society.

Friction piles require soils data to sufficient depth below the pile tip .

2.2 Rotary weight sounding

In the test, the rod system with a screw point is rotated by hand or machine, carrying a weight varying from 100 kg (starting value) to 25 kg (soft conditions with the rod system sinking). The number of half turns per 0,2 m penetration is recorded, and expresses the variation in soil stiffness in the profile. This method is now mostly used on small projects to assess the relative density of the soils strata.

2.3 Rotary pressure sounding

The equipment consists of a drill bit extended by rods with flush couplings. It is forced into the ground at a constant rate of penetration (3m/min) and a constant speed of rotation (25 rpm). The thrust necessary to maintain the penetration according to these requirements is measured and plotted versus depth.

The method may be used in preliminary investigations, where a grid of rotary-pressure soundings usually are the first investigations to be carried out. Based on results from these tests, the optimal location of sample bore holes and in situ tests can be determined.

2.4 Total sounding

Where the rotary-pressure sounding equipment cannot penetrate due to very hard layers, total sounding can be used. This sounding method combines the principles of rotary-pressure sounding and rock drilling into an effective method for mapping of soil conditions and determination of depth to bedrock.

When performing total sounding, the rig must be equipped with a hydraulic drilling system with sufficient capacity to supply a penetration force of 30 kN on the rod system. In addition, a percussion hammer and a flushing system may be activated for penetrating hard soil strata or boulders and rock.

The f45 mm rod system is equipped with a f57 mm drill bit with a spring loaded valve, which prevents clogging of the flushing system during penetration. When the flushing fluid is pressurised, the valve opens and drilling with flushing is continued, with or without additional activation of the percussion hammer. The drilling operations may be reverted to rotary pressure sounding procedures at any time during penetration. This equipment is especially suited for end bearing piles to firm layers or bedrock.

Mapping of the bedrock surface is also carried out by percussion drilling. A minimum of 2 m is normally drilled into bedrock. High capacity piles/bored piles or uncertain rock formation may call for core drilling.

2.5 Cone penetration test

In Norway, CPT/CPTUs are done from hydraulic drill rigs since the use of special CPT trucks are inconvenient with our topography. These rigs push about 1 m in each stroke before a new rod has to be added..

Two types of Swedish cone penetrometer systems dominate the Norwegian market:

ENVI Memocone
This equipment consists of the Memocone penetrometer itself and the corresponding Geoprinter data aquisition unit. The system is cableless, with the data acquisition based on synchronised clocks within the Geoprinter and the Memocone.

GEOTECH cordless cone penetrometer.
With this system, data measured at the probe is transformed into a sound signal via a microprocessor. This signal is transmitted through the CPT rods to a microphone which is located between the top rod and the drill rig. The data is then transferred through a cable from the microphone to the Geologg CPT interface. The same interface is also fed with the exact penetration depth from a sensor. Measured CPT data are displayed real time during the test, which is a large advantage compared to the memory type cone penetrometers..

The Norwegian Geotechnical Society has issued guidelines for use of CPT/CPTU.
CPT data may form the basis the design of friction piles.

Assessing values of the undrained shear strength fro CPT's is not a simple task since this parameter is influenced by factors such as anisotropy, stress history, rate of straining and mode of failure.

In the interpretation of s_u from CPTU records, empirical correlations are still the most popular approach, see for example Lunne et al (1985, 1989). These correlations are carried out by first assessing the empirical cone factor N_{KT} from the expression:

$$N_{kK} = (q_t - \gamma \cdot z) / s_u \qquad (2.5.1)$$

A reference s_u- value should then be found from other relevant in situ or laboratory tests in selected profiles at the site. After assessing a local value of N_{KT}, this value is used in subsequent interpretation of the CPTU results. If no reference values for s_u exist, N_{KT} = 15 is usually recommended.

2.6 Vane tests - undisturbed sampling

In situ vane tests and undisturbed sampling are typically used to calibrate CPT tests, establish geotechnical parameters for the design of friction piles or design of horizontal loaded piles or piles subjected to cyclic loads.

Special laboratory test for determination of geotechnical design parameters usually involve consolidation tests, static and or cyclic compression or extension triaxial , direct static and or cyclic shear tests and shear box tests.

3. PILING TECHNOLOGY

3.1 General

The most common pile type are friction or end bearing displacement piles of concrete, steel or wood. Bored piles to bedrock is also used.

Large areas in the southern Norway are covered by soft to medium stiff clay to bed rock. The depth to bed rock is typically less than 30 m and driven piles to bed rock is the most common pile type.

3.2 Prefabricated driven piles

3.2.1 Concrete

The dimensions are normally 230x230 mm, 270x270 mm and 345x345 mm. The concrete quality is typically of grade C50 and reinforcement 2 %. The piles are equipped with pile shoe for the purpose of chiselling the pile into bed rock..

The piles are driven with hydraulic impact hammer with 30-60 kN drop weight. End bearing piles are chiselled carefully into bedrock rock, typically 200 blows with drop height 100 mm and then driven until a specified driving criteria based on driving formula. The bearing resistance is occasionally verified by stress wave measurements. The piles shall also fulfil the driving criteria on restriking.

The characteristic resistance or installed capacity in ULS is in the order of 1000-2200 kN.

3.2.2 Steel

Steel piles, H profiles etc. are occasionally used when displacements are of concern or for special foundation purposes like underpinning of buildings. The piles are installed in a similar way as described above for concrete piles.

Steel pipe piles are used for foundation of structures with large concentrated loads such as bridges and piled harbour structures. They are normally 500-1500 mm diameter with wall thickness 8-20 mm. Steel pipe piles are normally driven to end bearing in firm layer or bed rock using large pile shoe . The pile are typically reinforced and filled with concrete after having obtained the driving criteria. The steel mantel is either considered as corrosion protection or partly as reinforcement. The pile shoe may be hollow allowing to drill through the pile tip and into boulders, bedrock etc. Insert of rock dowel is also used in difficult rock formations.

Typical characteristic resistance or installed capacity is in the range of 2500-8000 kN.

The piles are driven with diesel hammers and or hydraulic impact hammers with required net energy in the order of 30-150 kN.

3.2.3 Timber

Timber piles are normally friction piles in clays. The minimum pile tip diameter shall be greater than 125 mm. The piles may be spliced or combined with a top pile of concrete.

The piles are typically used as relief piles in bridge abutments and harbour structures, roads and foundation of building, especially to reduce possible differential settlements.

The characteristic resistance or installed capacity is in the range of 80-200 kN.

3.3 Bored Piles

The diameter ranges from approx. 900 mm to 1500 mm. The piles are basically used for foundations with large concentrated loads.

The piles are socketed into sound bedrock by chiselling/drilling or in combination of preblasting.

They are reinforced and concreted with typically grade C 45. Bored piles are installed with water filled casing during excavation. The casing is normally pulled in steps during concreting.

The characteristic resistance or installed capacity is in the range of 2500-20000 kN. The capacity and quality is documented by core drilling, ultrasonic testing or dynamic testing.

4 RELEVANT NATIONAL DOCUMENTS WITH REGARD TO PILE DESIGN

The design philosophy is described in «Safety principles in Geotechnique» and is common for all type of geotechnical engineering design.

The design of timber, steel and concrete structures are covered by Norwegian Standards, NS 3470, 3472 and 3473 respectively. General rules for pile design and execution together with examples is covered by Guidelines for Piles, issued by the Norwegian Council for Building Standardisation. The publication is partly translated to English. Specification for fabrication of concrete piles is given in Norwegian Standard NS 3046 (Norwegian and English issue).

General specification texts for tenders covering steel, concrete and timber piles are given in Norwegian Standard NS 3420.

Other guidelines, handbooks and papers etc. have also been published by the Norwegian Road Directorate, University of Trondheim, NGI and geotechnical consultants.

The loads and load actions to be use din the design are described in Norwegian Standard (NS 3479, 1991).

5. DETAILED DESCRIPTION OF THE NATIONAL DESIGN METHODS

5.1 General philosophy. Definitions

The design should ensure that the piles and the pile foundation is -with acceptable probability - able to support the loads without failure or excessive movements.

The design of piles is based on limit state method. The limit states are defined as most adverse situations in which the stresses - in the pile or in the ground - (or the deformations or movements) due to loads combined in unfavourable way can be greater than the strength of the pile material or of the soil (or greater than the movements causing structural damages or loss of functionality).

The engineer should document that none of the possible limit states will occur. The following classes of limit states have to be considered:

-ultimate limit states, in which the stresses induced in the pile material or in the surrounding soil exceed the material strength causing failure

-fatigue limit states, in which the repeated actions (such as vibrations or traffic loads) can reduce the material strength below the applied stresses causing failure

-accidental limit state, in which possible but little probable actions of high intensity such as earthquakes, flood, landslides, etc. can induce failure of the piles or of their foundation soil .

-serviceability limit states, in which the induced deformations or movements can cause damage or loss of functionality for the structure.

The loads used in the design are **design loads** (defined as the most unfavourable values which could occur in extreme situations). They are calculated by multiplying the **characteristic loads** (the unfavourable values which could occur in normal circumstances) by **load factors** (accounting for the uncertainties in the characteristic loads).

The **characteristic strength** of the material is the value of strength for which there is little probability that a lower value can occur. The **design strength** is obtained by dividing the characteristic value by a **material coefficient** , accounting for the uncertainty in strength determination

The pile **design structural capacity** is determined using the design strength of the pile material. The pile **installed capacity** is determined by multiplying the design structural capacity by a reduction factor accounting for installation conditions at the site.

The pile **design geotechnical capacity** is however obtained by dividing the **characteristic geotechnical capacity** (calculated using the characteristic strength) by an equivalent factor, γ_e , accounting for uncertainties in determining the characteristic bearing capacity.

The pile **design capacity** is the minimum of design geotechnical capacity and installed capacity

The design procedure is summarised by the flow chart (Fig.5.1.1)

5.2 Design on basis of static load tests

The most reliable method to determining pile capacity for most sites is the load test. The load test procedure is described in Guide Lines for Pile Design (1991) and will not be reproduced here. The following issues are however pointed out as they are very important for a successful pile load test:

-Except for pile driven to bed rock, the bearing capacity of all piles does not reach the ultimate until after a period of rest. Load tests are not reliable unless made after this period of adjustment . Table 5.2.1 shows the minimum period of adjustment. For piles in clayey/silty soils it is recommended that loading test should be repeated after a period longer than that indicated in Table 5.2.1 to see whether the bearing capacity increases with time.

-To begin with, the load is applied in increments of one-tenth the expected bearing capacity. Each increment of load is maintained constant and the settlement is measured at 1, 2, 5 and every 5 minutes interval, until the rate of movement is less than 3 mm /hour. After the load has reached 60-79% of the expected bearing capacity, or the pile settlement is 3-4 % of the pile diameter/side, the pile is completely unloaded in decrements of load. Each decrement is hold at least 5 minutes and readings are taken at at least 1, 2 and 5 minutes. Thereafter the pile is reloaded in the same manner as unloading up the maximum previous load. The pile is further loaded at a constant settlement rate of 0.5 mm/minute while load and settlement are recorded every minute. The test is stooped when the load is constant or starts to decrease (as for clays) or when it increases linearly with settlement (as for sands).

-If the pile is going to be tested again, the constant settlement rate loading should be stopped as soon there is sign of incipient failure and the unloading should be recorded as previously.

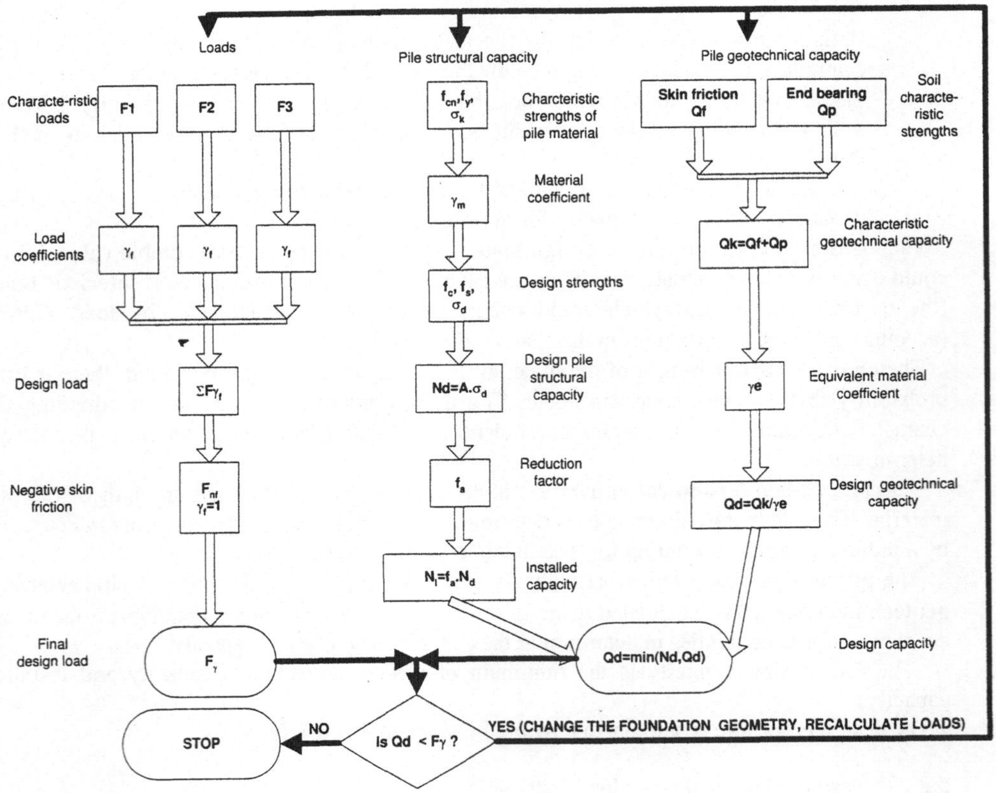

Fig.5.1.1

Table 5.2.1

Soil type	Pile material	Minimum time after driving (weeks)
Clay	Wood	4
	Concrete	4-12
Silt	Wood	2
	Concrete	4-6
Sand/Gravel	All	0-1

The results of loading test are summarised in a load-total settlement curve.

The characteristic geotechnical bearing capacity, Q_k , is determined by one of the following methods:

-«90 % rule» (Brinch Hansen) : Q_k is the minimum load that induces a settlement twice as large as the settlement corresponding to 90 % of Q_k.

-Davisson method (Davisson, 1972): Qk is the load corresponding to a settlement, s_k, defined as:

$$s_k = k + \frac{Q_k \cdot L}{A \cdot E} \tag{5.2.1}$$

where:

$$k = 4 \text{ mm} + D/120 \tag{5.2.2}$$

D - diameter/side of the pile in mm.

-«Pile Commission» method : the same as Davisson's but with

$$k = 20 \text{ mm} + D/20 \tag{5.2.3}$$

-«ISSMFE»'s method : the same as Davisson's but with

$$s_k = D/10 \tag{5.2.4}$$

-«Creep bearing capacity» method: the settlement records for the last part of each load increment are used to plot the settlement rate as a function of load. The «creep bearing capacity», Qk, is determined as the load where a sharp change in settlement rate occurs.

The 90% rule is the most common method used in Norway. The designer should always document that the chosen method is appropriate to local experience and site conditions.

The design geotechnical capacity is then calculated by dividing the characteristic capacity by an equivalent material coefficient , γ_e, that accounts for uncertainties in determining the characteristic capacity. It is establish based on local and regional experience and engineering judgement in interpreting the results of loading test. Recommended values are described in Section 5.6.

5.3 Design based on in situ and in laboratory test results (static geotechnical capacity)

The characteristic geotechnical pile capacity is determined by summing up the ultimate values of the components by which a pile transfers the load to the soil, i.e. **skin friction** (although true friction does not develop in all cases) and **end bearing**:

$$Q_k = Q_f + Q_p \tag{5.3.1}$$

The skin friction and end bearing in cohesionless soils are calculated as:

$$Q_f = \sum \tau_{fi} \cdot A_{fi} \tag{5.3.2}$$

$$Q_p = q_n \cdot A_p \tag{5.3.3}$$

where τ_{fi} - average skin friction in the «i»-th soil layer along the pile shaft; q_n - specific end bearing capacity; $A_{f\,i}$- the lateral area of the pile shaft in the «i»-th soil layer; A_p - the cross section area of the pile tip.

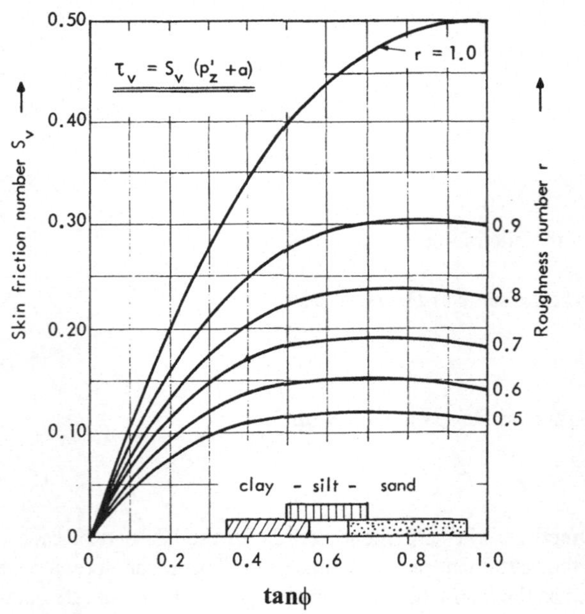

Fig. 5.3.1

5.3.1 Static formulae based on laboratory test results

Cohesionless soils

The average skin friction and the specific end bearing are determined using Janbu's method (Janbu,1976):

$$\tau_{fi} = S_v \cdot (p'_{zi} + a_i) \qquad (5.3.4)$$

where S_v - skin friction number (Fig. 5.3.1) depending on roughness number, r, defined as the ratio between mobilised shear at pile soil interface and the shear strength in the soil on a vertical plane - and on angle of internal friction, tan φ.; p'_{zi} - effective overburden at the average depth of layer «i»; a_i - attraction of the layer «i».

Alternatively, for a homogeneous layer, the average skin friction can be calculated as :

$$\tau_f = \beta \cdot \bar{p}'_o \qquad (5.3.5)$$

with β - an empirical skin friction factor depending on the length of pile in the soil (Fig.5.3.2); p'_o - the average overburden pressure in the soil along the part of pile for which skin friction is calculated.

$$q_n = N_q \cdot p'_p \qquad (5.3.6)$$

with p'_p - the overburden effective pressure at pile tip; Nq - the bearing capacity factor (Fig.5.3.3)

Cohesive soils

Undrained total stress analysis

For cohesive soils in undrained conditions the average skin friction and the specific end bearing capacity are calculated as:

$$\tau_{fi} = \alpha_i \cdot s_{ui} \qquad (5.3.7)$$

with α_i - empirical skin friction factor depending on clay type, average un-drained shear strength ratio (s_u/p'_o) of the soil layer and the ratio depth /diameter (width) as shown in Fig. 5.3. 4; s_{ui} - the average undrained shear strength of the clay layer «i»

$$q_n = 9 \cdot s_{up} \qquad (5.3.8)$$

with s_{up} - the undrained shear strength at the pile tip.

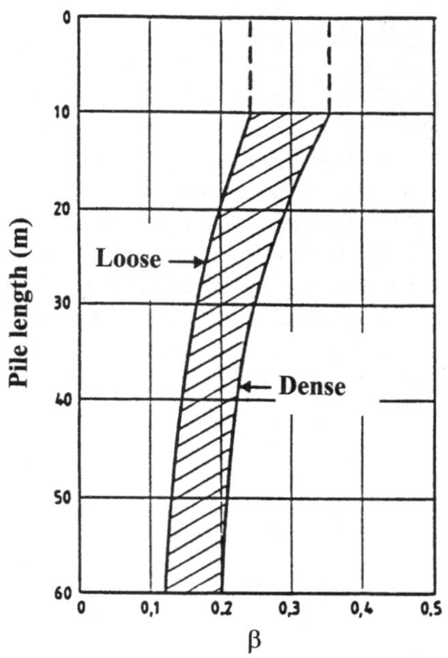

Fig. 5.3.2

Undrained, effective stress analysis of skin friction

The skin friction for clayey soils in undrained conditions can alternatively be calculated as:

$$\tau_f = \beta \cdot \bar{p}_o' \cdot (OCR)^{0.5} \tag{5.3.9}$$

with β - skin friction factor:

$$\beta = 0.4 \cdot (L + L_O) \cdot (2 \cdot L + L_o) \tag{5.3.10}$$

L - pile length; L_o - empirical constant (L_o = 20 m); OCR - overconsolidation ratio of the clay.

Piles in silt or layered soils

For piles in silt the geotechnical bearing capacity is taken as the minimum of the capacities calculated for cohensionless and cohesive soils.

A special attention should be paid to the cases where the layers have different stiffnesses such as a soft clay layer trapped in between sand layers. The soft clay will settle and will induce additional loads as negative skin friction (see Section 5.3.4). The designer should d ocument that the underlaying sand layer has sufficient capacity to carry the increased loads.

5.3.2 Design based on in situ test results

Rotary weight drilling and SPT

The characteristic geotechnical bearing capacity in sands can be derived from empirical correlations as shown in Fig. 5.3.5.

The correlations imply however so many uncertainties that this type of in situ tests are used only for preliminary calculation of geotechnical capacity

Cone Penetration Test

The characteristic geotechnical capacity of the pile can be calculated as:

$$Q_k = f_p \cdot q_{tp} \cdot A_p + \sum f_{fi} \cdot q_{tfi} \cdot A_{fi} \tag{5.3.11}$$

where q_{tfi} - cone tip resistance in the layer «i»; q_{tp} - cone tip resistance at the pile tip
f_f and f_p - empirical skin friction factor and end bearing factor respectively (Table 5.3.1)

5.3.3 Negative skin friction

Negative skin friction occurs if the soil around the pile settles more than the pile itself as is the case of soft normally consolidated clays. The settlements can be induced by lowering of the ground water table, filling, or by soil creep under constant overburden.

The negative skin friction can be calculated as described in Sections 5.3.1 - 5.3.3. As shown in Section 5.1, the negative skin friction is considered an additional load for the pile.

The design negative skin friction is equal to the characteristic negative skin friction:

$$F_{dnsf} = F_{knsf} \cdot \gamma_f \tag{5.3.12}$$

$$\gamma_f = 1.0 \tag{5.3.13}$$

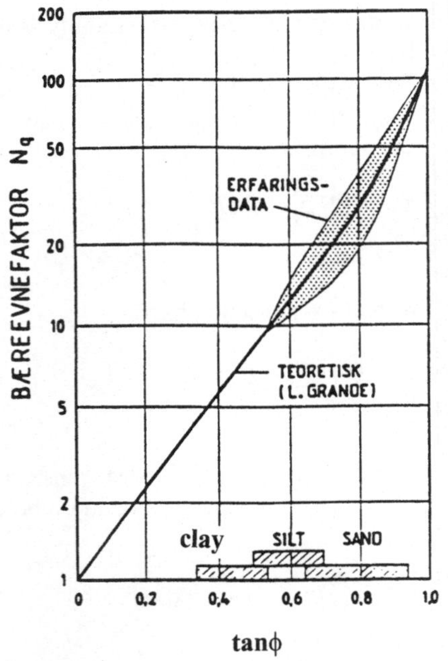

Fig.5.3.3

Table 5.3.1

Cone tip resistance, qc (MPa)	Skin friction factor, f_f	End bearing factor, f_p
1 - 10	0.010 - 0.007	0.5 - 0.7
10 - 20	0.007 - 0.005	0.7 - 1.0

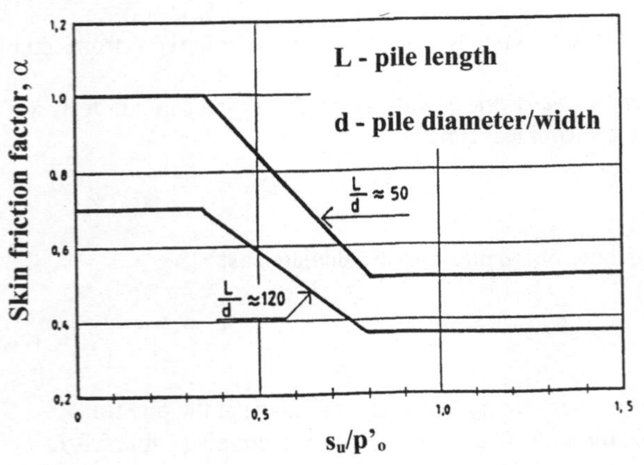

Fig.5.3.4

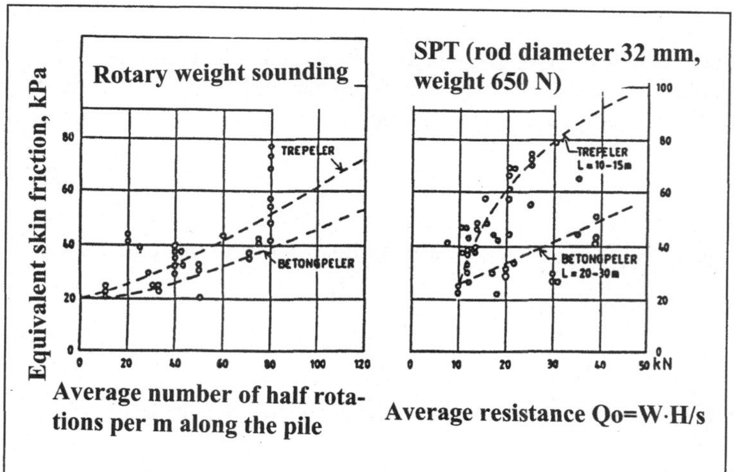

Fig.5.3.5

For pile groups the characteristic negative skin friction is reduced as explained in Section 5.6

5.3.4 Tension capacity of pile

For the pile loaded in tension the charac-teristic capacity is the skin friction plus the own weight of the pile. In cohesionless soils and in cohesive soils under long term loads the skin friction is lower in tension than it is in compression. The skin friction number, Sv, in tension and the empirical skin friction factor, β, in tension are presented in Fig. 5.3.6 and 5.3.7 respectively.

The equivalent material coefficient is $\gamma_e = 1.5$ for short term loading and $\gamma_e = 3.0$ for long term loading.

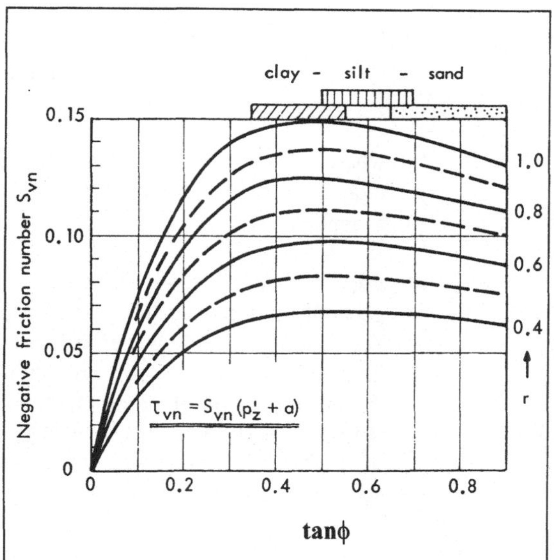

Fig.5.3.6

5.3.5. Buckling capacity of axially loaded piles

The pile shaft is a structural column that is fixed at the tip and restrained at the top. The elastic stability of a pile depends on its straightness, length, moment of inertia, and modulus of elasticity and the lateral resistance of the soil that supports it.

For many site conditions in Norway where deep deposits of very soft, sensitive silty clays are underlained by moraine and bed rock the buckling ultimate limit state is a critical one. The buckling capacity of a pile is calculated as:

$$Q_{kc} = \cdot\pi \cdot E \cdot I / L^2 + C \cdot L^2 / \pi^2 \qquad (5.3.14)$$

$C = k * D$ (5.3.15)

where E,I, L - are pile modulus of elasticity, moment of inertia and length, respectively; C - soil lateral reaction modulus; D - pile diameter/side; k -coefficient of lateral subgrade reaction .

The half of the elastic wave length should be calculated :

$$L_k = \pi \cdot \sqrt[4]{E \cdot I / C} \tag{5.3.16}$$

If the pile length is greater than L_k, L_k should be used in Eq.(5.3.14). In this case the equation becomes:

$$Q_{kc} = 2 \cdot \sqrt{E \cdot I \cdot C} \tag{5.3.17}$$

Recommended values for the coefficient of subgrade reaction are presented in Table 5.3.2

The design buckling capacity is calculated as

$$Q_{dc} = Q_{kc} / \gamma_e \tag{5.3.18}$$

The design buckling capacity is calculated as

$$Q_{dc} = Q_{kc} / \gamma_e \tag{5.3.18}$$

using an equivalent material coefficient of $\gamma_e = 1.5$.

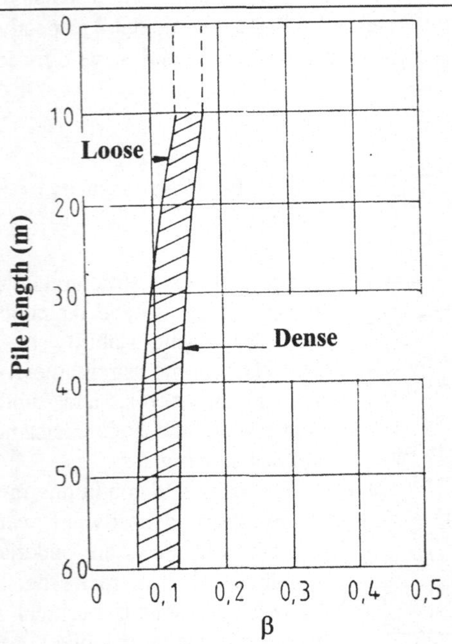

Fig.5.3.7

An initial deflection of the pile (due to, for instance, difficulties in installation or lateral movements of the soil) will induce a further deflection as the axial load increase:

$$\delta = \delta_o \cdot (1 - N / Q_{kc}) \tag{5.3.19}$$

where: δ_o - is the initial deflection; δ - is total deflection after the axial load, N, is applied.

For steel piles total deflection will generally not reduce the pile stiffness. For concrete piles, however, as the total deflection increases it will induce fissures at tension fibre and the bending stiffness (EI) will consequently be reduced. This should be taken into account in determination of buckling capacity.

Table 5.3.2

Soil type	k (kN/m²/m)
Clay (undrained condition)	$30*s_u$
Loose silt, sand, organic silt	2500
Medium dense sand	5000
Dens sand	10000

5.4 Design on basis of driving formulae

The driving formulae are based on the transfer of kinetic energy of the falling pile hammer to the pile and the soil.

For the pile driving with movement measurement, the characteristic geotechnical capacity is calculated as (Janbu's formula):

$$Q_k = \frac{\eta \cdot W \cdot H}{(s + \delta_e / 2)} \tag{5.4.1}$$

where η - the driving efficiency factor; W - the hammer weight; H - the hammer drop; s - the plastic (net) settlement of the pile per hammer blow; δe - the elastic (reversible) settlement of the pile for a hammer blow.

The design geotechnical capacity is then obtained from the characteristic capacity using an equivalent material factor of 1.6.

For pile driving without movement measurement, the Janbu's formula becomes:

$$Q_k = \frac{2 \cdot \eta \cdot W \cdot H}{s + \sqrt{s^2 + 2 \cdot \eta \cdot \alpha \cdot W \cdot H \cdot L / (A \cdot E)}} \tag{5.4.2}$$

where α - is force distribution factor; for short (10 to 15 m) end bearing piles the axial force is distributed over the most of the pile length and $\alpha = 0.9$; for long, friction piles only a fraction of total pile weight is accelerated at one time and $\alpha = 0.5$; s - is the average pile set measured in the last ten blows; L, A and E - the length, cross section area and Young modulus of pile material; η - the driving efficiency is taken as 0.5 for falling (simply acting) hammer, 0.6-0.9 for hydraulic hammers and 0.35-0.5 for Diesel hammers.

When the characteristic geotechnical capacity is determined by equation (5.4.2), an equivalent material coefficient , $\gamma_e = 1.8$ should be used for the design capacity.

The eqs. (5.4.1 and 5.4.2) can also be used with the net energy (ηWH) from PDA measurements giving a more reliable characteristic capacity.

5.5 Design on basis of wave equation analysis

The wave equation analysis is used to study the movement of the wave of compression in the pile due to hammer drop. The force variation in time at pile head depends on the pile elastic properties and dimensions as well as on soil properties. The total pile resistance can be determined from the record of force and velocity as a function of time at one point of the pile during driving. The total pile resistance can be determined as:

$$R_T = \frac{F(t1) + F(t1 + 2 \cdot L / c)}{2} + \frac{M \cdot c}{2 \cdot L}[v(t1) - v(t1 + 2 \cdot L / c)] \tag{5.5.1}$$

where F - the force; v - is the particle velocity of the pile ; t1 is the time of the first maximum force, c - is the stress wave velocity in the pile; L is the distance between the measuring point and the pile tip and M is the pile mass

The total pile resistance consists of the static resistance and the dynamic resistance which is can be calculated if the damping ratio of the soil is assumed. The geotechnical capacity can be thus determined as:

$$R_s = R_T - J_c \cdot [F(t1) + zv(t1) - R_T] \qquad (5.5.2)$$

where Jc - is the damping ratio of the soil; z- is the soil impedance.

If the wave stress measurements is combined with a wave equation analysis, using for instance the programme CAPWAP (Goble Rausche Linkis, 1996), the soil parameters can be found by matching the measured and computed force-time relation and thus a more reliable geotechnical capacity can be found.

An equivalent material coefficient, $\gamma_e = 1.5$ is used to obtain the design geotechnical capacity of the pile.

5.6 Prescribed global and partial factors. Other aspects of axially loaded pile design

5.6.1 Prescribed global and partial factors

The structural pile capacity is a matter of structural design and will not be treated here. The installed structural capacity however depends on pile initial deflection, driving conditions, corrosion, pile geometry, pile defects non-homogeneous soil conditions, and on control during construction phase. All these factors should be analysed and an overall rating mark (**favourable, middle** and **unfavourable**) is given for the particular conditions which could be met at the site.

The reduction factor used to obtain the installed capacity from design capacity of the pile is then chosen as shown in Table 5.6.1.

The conditions are rated as shown in Table 5.6.2

The following recommended equivalent material coefficients are used to determine the design geotechnical pile capacity, Table 5.6.3.

Table 5.6.1

Construction condition rating	**Reduction factor , fa**
Favourable	0.90
Middle	0.75
Unfavourable	0.60

5.6.2 Design of pile groups

The behaviour of piles in a pile group is different than that of a single pile due to such factors as change in initial stresses and density in the ground due to driving of piles close to each other, different negative skin friction acting on centre piles and on periphery piles, different skin friction and end bearing between centre piles and periphery piles and so on.

The minimum centre to centre pile spacing in a group of vertical driven piles is indicated in Table 5.6.4

The negative skin friction acting on a pile group can be calculated as:

$$Fnf = 2*(B+L)*D*su \qquad (5.6.1)$$

$$Fnf = 2*(B+L)*D*\beta*\bar{p}_o' \qquad (5.6.2)$$

Table 5.6.2

	Favourable conditions	Unfavourable conditions
Soil conditions	Homogeneous, no blocks and stones Increasing strength with depth Even bed rock of good quality rock type for fixing the pile tip	Variable soil conditions Soil type susceptible to induce corrosion Stones and blocks that can cause pile deflection Sloping, uneven, bed rock
Soil investigations	Thorough	Sparse
Constructive conditions	More than 5 pile in pile group Small variations in pile length Stiff, tension resistant joints	Less than 3 pile in pile group Large variations in pile length Bad joints, loss of driving energy
Driving equipment and technology	Adequate hammer weight, good helmet and cushions, good hammer guidance	Too light hammer, inadequate helmet and cushions, large loss of driving energy Inexperienced entrepreneur
Quality control of execution	Good control of the piles and pile driving Good concreteing control for cast-in-situ piles Complete pile protocol Specification followed during execution	Sparse control of pile execution Specifications not followed

Table 5.6.3

Basis for determination of characteristic pile capacity	Equivalent material coefficient γ_e
Static geotechnical capacity (laboratory soil testing)	1.6
In situ tests	2.0
Driving formulae	1.6-1.8
Wave equation analysis	1.5
Loading test (« 90% » method)	1.3
Loading test (Davisson's method and variants)	1.4
Loading test («creep» method)	1.2

Table 5.6.4

Pile length (m)	Friction piles sand	Friction piles clay	End bearing piles
< 12	3D	4D	3D
12 to 24	4D	5D	4D
> 24	5D	6D	5D

D - pile diameter/width

for piles in sand and in clays respectively, where B,L - the width and length (Fig.5.6.1) of the pile group; D - the depth at which the relative pile-soil movement is negligible; su - mean undrained shear strength within depth D; p_o'- the average overburden pressure within the depth D.

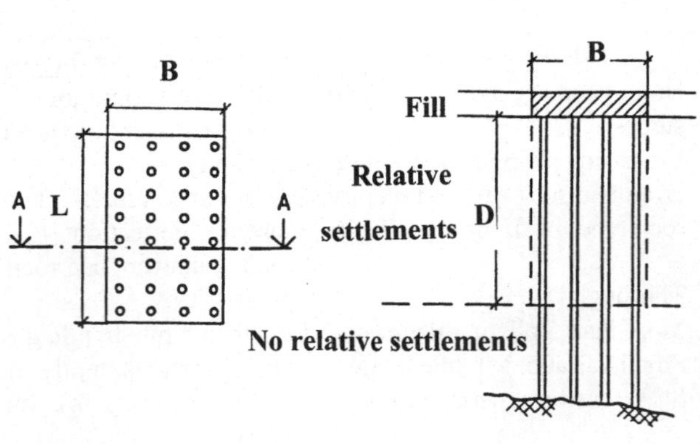

Fig.5.6.1

The average geotechnical capacity of a pile in a pile group is higher than that of a single pile in sands and less than that of a single pile in clays. The difference depends on pile spacing and installation. As a simple rule the pile geotechnical capacity of a pile group is the minimum between the sum of single pile capacities and the bearing capacity of an equivalent large foundation. More accurate evaluation can be obtained by use of other methods as explain in Section 5.6.2.

The settlement of a pile group can be approximated by the settlement of an equivalent foundation placed at a depth equal to 2/3 of pile length and having the dimensions of the pile group (Fig. 5.6.2).

5.6.3. Use of other methods for pile design

The design rules described in this report are regarded as guiding lines for design specifying the general frame, design philosophy and simple rules to design pile foundations.

In many situations the designer will meet conditions that are different and often more complex than described in this report. In such situations the design practice in Norway is to use more elaborate design methods and computer programmes in order to obtain solutions for both force distribution between the piles of a pile group and along individual pile and of pile capacity, deflections, settlements and capacity mobilisation degree, provi-ded the designer can docu-ment that adequate safety is achieved.

The force distribution be-tween piles can for example be obtained by use of com-puter programme SPLICE (Clausen, 1993) which is based on API (API, 1993) and DNV (DNV, 1977) codes or of programme GROUP (Lieng, 1989) based on effective stress analysis developed at Norwegian Technical University in Trondheim, Grande (1977), (Janbu,1976). The geotech-nical capacity of a single, axially loaded pile can be computed based on calculation of static bearing also using computer programme SPLICE, PIA1 (Nordal, 1985). The pile buckling capacity and deflections can be computed using the computer programmes PILESTAB (NOTEBY, 1996) .

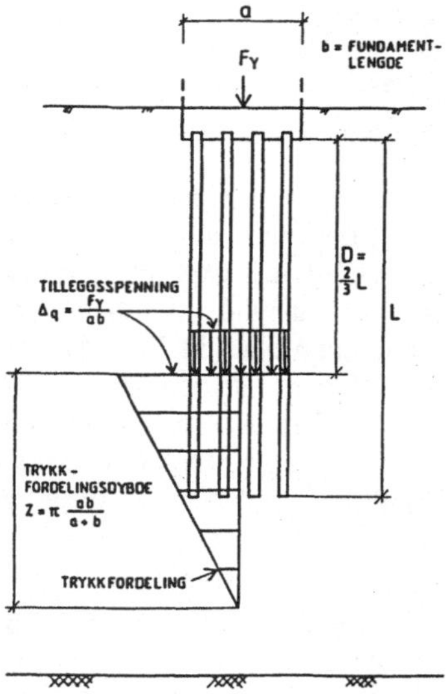

Fig.5.6.2

More complex problems such as axially loaded piles in a pile group subjected to soil lateral movements due to an adjacent excavation can be analysed using finite element codes such as, for example, PLAXIS (Vermeer and Brinkgreve, 1995) or ABAQUS (1989).

5.7 Rules for serviceability.

The settlements and lateral deformations of pile foundations are limited by the condition that the structure should not suffer damage or loss of functionality. Thus the serviceability rules are merely a matter of structural design than of foundation design.

The specific serviceability rules are determined for each project according to structural solution and functionality.

6. Some particular example to demonstrate the methods.

6.1 Friction pile in sand

Calculate the design geotechnical capacity (compression) of a concrete pile 22 m long in sand (Fig.6.1.1).

Average effective overburden is $p_o' = \gamma' * z_{av}$ = (19-9)*11 =99 kN/m^2 ;β=0.24 (Fig.5.3.2), $\tau_f = \beta * p_o'$=0.24*99=23.8 kN/m^2 ; Cha-racteristic skin friction: Qf = π* 0.277 * 22 * 23.8 = 455.6 kN; Effective overburden at pile tip : p'_p=22 *(19-10)=198 kN/m^2 ; End bearing $q_p = N_q * p_p'$ = 21*198=4158 kN/m2 (Fig.5.3.3). Characteristic end bearing: Qp=π*0.277^2/4*4158= 250.6 kN
Characteristic geotechnical capacity: Qk=Qf+Qp=455.6+250.6=706 kN
Design geotechnical capacity: Qd=Qk/γe=706/1.6=441 kN

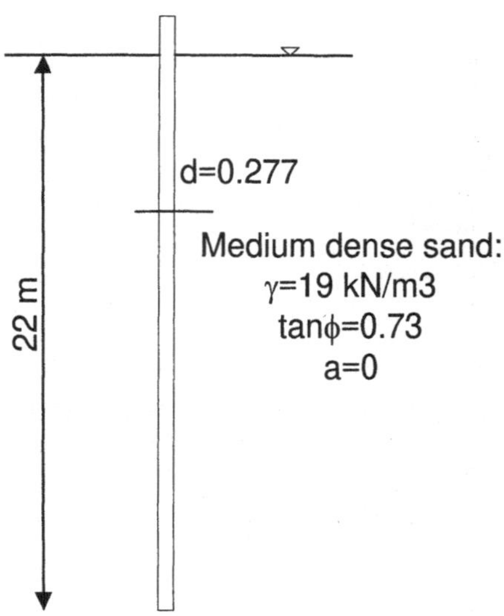

Fig.6.1.1

6.2 Friction pile in clay

Calculate the design geotechnical capacity (compression) of a concrete pile 20 m long in clay (Fig.6.1.2).

The lateral area of the pile is A_f=17.5 m2; the tip area is A_p=0.05 m^2. The average undrained shear strength is s_u=30 kN/m2; the average effective overburden pressure , p'_o= (19-10)*10 =90 kN/m^2; s_u/p'_o=30/90=0.33; L/d=20/0.277=72.

From Fig. 5.3.4, α=0.9; The characteristic end bearing is Q_p= 9 ·50·0.06=27 kN and the characteristic skin friction is Q_p=0.9·30·17.5=472.5 kN. The characteristic geotechnical capacity is Q_k=27+472.5=500 kN. The

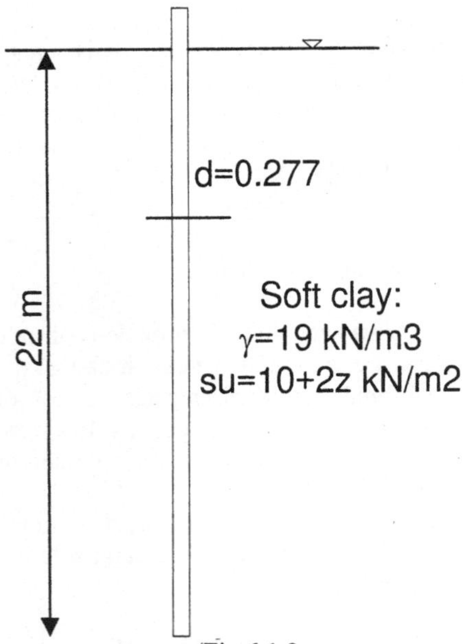

Fig.6.1.2

design geo-technical capacity is $Q_d=Q_k/1.6=312$ kN.

6.3 Use of pile load test and PDA measurements to determine geotechnical capacity

Extensive pile load testing and PDA measurements were undertaken by NOTEBY A/S Consulting Engineers in connection with construction of a new aluminium plant in Tyssedal, west coast of Norway (Simonsen and Jensen, 1983). The site conditions consist of glacial and glacifluvial deposits of loose sands and gravels with isolated boulders. The depth of bedrock varies between 10 and 70 m. The friction angle is $\varphi' = 35°$ and $\gamma = 20$ kN/m^3. Full scale load tests on close end steel pipe pile (800 mm diameter, 20 mm wall thickness) were performed at three levels (25, 30 and 35 m).

The characteristic geotechnical capacities for the three levels were determined from load test results as illustrated in Fig.6.3.1 using the «90%» rule (Section 5.2) and using PDA measurements with assumed Jc = 0.1 (eq. 5.5.2) . The characteristic capacities were also determined using soil laboratory test results (Section) and by CAPWAP analyses. A comparison of geotechnical characteristic capacities is presented in Table 6.3.1.

Table 6.3.1

Pile	CAPWAP (kN)	CASE, (Jc=0.1) (kN)	Static Formulae (kN)	Static load test (kN)
Test pile Level 1(25 m)	3700	4820	4300-5600	5700
Test pile Level 2(30 m)	-	6800	5600-7000	6100
Test pile Level 3 (35 m)	-	7800	7000-8800	7030
Anchor pile C40 (37.5 m)	6770-9000	6980-7380	7700-9500	-

The axial force - displacement relations as obtained by analyses with PIA1 (Section 5.6.2) for the three levels are compared to load test measurements in Fig. 6.3.3. The load test results, dynamic measurements and static formulae show that the differences in capacities determined from load test (taken as reference) and other methods show fairly good agreement.

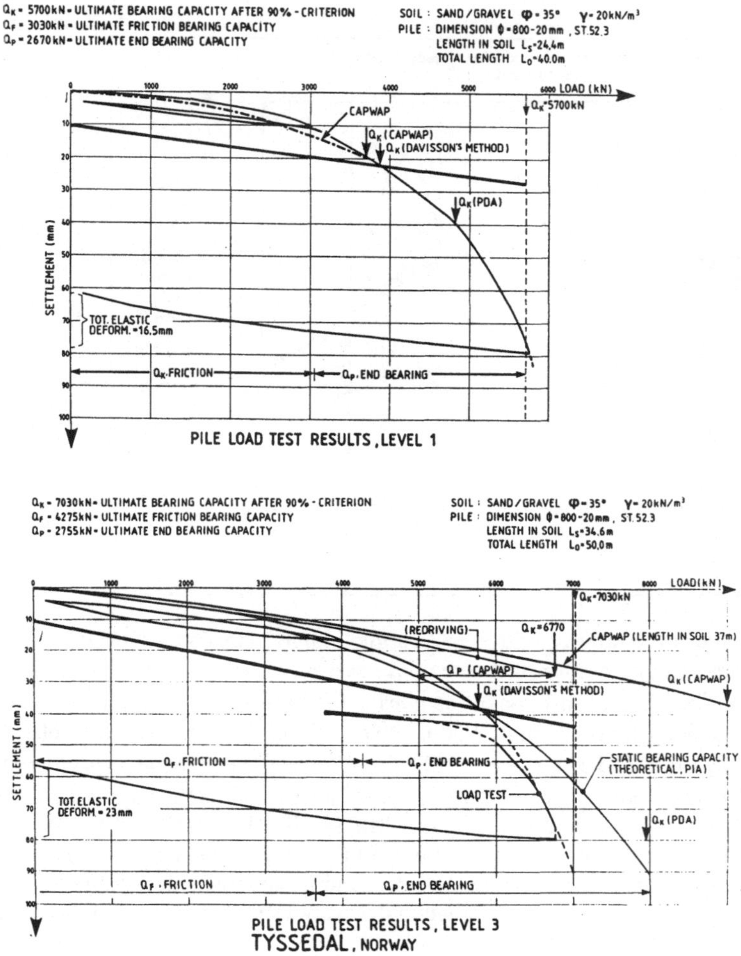

Fig.6.3.1

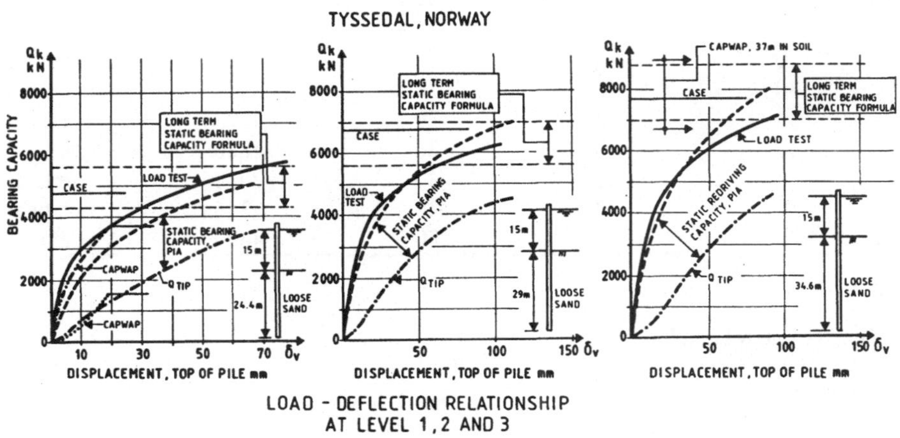

Fig.6.3.2

6.4 Buckling capacity of piles in soft clay

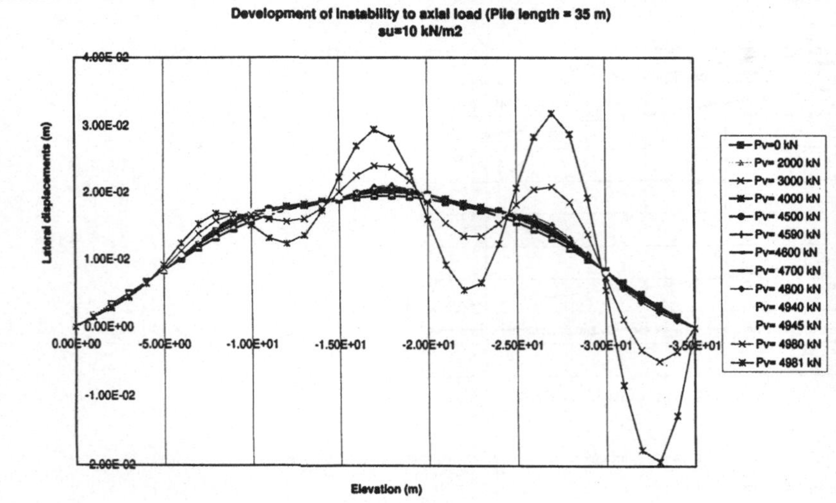

Fig.6.4.1

In connection with design of a railway bridge in Vestfold county, Norway, NOTEBY A/S carried out buckling capacity analyses for steel core piles in sensitive, soft clay, 25 - 60 m down to bed rock (Fig.6.4.1). The bending stiffness of the piles is $EI=5850$ kN·m^2, the pile diameter is 150 mm and outer pipe diameter is 193 mm. With $C=30 \cdot s_u= 300$ kN/m2, the half of the elastic wave length is according to eq. (5.3.16), $L_k= \pi \cdot (EI/C)^{0.25} = 4.2$ m and the buckling capacity (eq. 5.3.17) for piles of any length greater than L_k, is $Q_{kc}= 2 \cdot (\sqrt{EI \cdot C})= 2650$ kN. A more detailed analysis were undertaken using the programme PILESTAB. The lateral soil reaction to buckling is modelled by p-y curves as recommended by API Code (API 1993a and b). An initial deflection is assumed due to fabrication /installation errors and to soil creep along batter piles . The development of instability to axial load modelled in the analyses is shown in Fig. 6.4.2. The results of a parameter study are presented in Table 6.4.1

Table 6.4.1

Pile length(m)	**su Layer 1 (kN/m²)**	**su Layer 2 (kN/m²)**	**Initial deflection (mm)**	**Characteristic buckling capacity (kN)**	**Pile type**
10	30	10	10	6048	vertical
25	30	10	20	5375	vertical
35	30	10	20	4990	vertical
60	30	10	30	4880	vertical
60	30	20	30	7080	vertical
60	30	30	30	9345	vertical
60	30	10	30	4600	batter

7. Quality control and monitoring

7.1 Quality Control

A suitable qualified and experienced person shall be in charge of the execution of the work. He shall be responsible for that the work is carried out in according to the specifications and relevant standards, the monitoring of pile construction and keeping of the pile records and updated pile drawings.

7.2 Monitoring

The pile installation shall be supervised and all relevant data shall be recorded in a pile record for each pile. The monitoring should follow the European standards for displacement piles and bored piles, PR EN 288005 and EN 1536 .

The monitoring may also include among others measurements of noise, vibration, pore pressure, deformations, inclinations heave and corrosion protection.

8. Particular national experiences.

Particular norwegian experience is related to our soil condition with soft clays to bedrock.

Special experience has been developed in installing driven precast concrete piles and steel pipe pile to end bearing with high capacity. Piles driven to hard or weak rock with erratic rock surface combined with boulders or soft clays are situations met and solved on several projects. Many of these cases are reported.

9. References.

ABAQUS - 1989 - ABAQUS , Standard User's Manual, version 4.8, Hibbitt, Karlsson and Sorensen , Inc. 1989.

API - 1993 a- Recommended Practice 2A-LRFD (RP 2A-LRFD), July 1993

API - 1993 b - Recommended Practice 2A-WSD (RP 2A-WSD), July 1993.

Davisson, M. T. - 1972 - High capacity piles. Proc. of Lecture Series Innovation in Foundation Construction. ASCE Illinois Section.

Det Norske Veritas - 1977 - Rules for Design, Construction and inspection of offshore structures. Høvik.

Clausen C.J.F, -1993- SPLICE - Documentation of Computer program package for Structure/Pile/Soil Interaction Analysis, Rev. 3, July 1994.

Goble Raushe Likins and Associates, Inc. -1996 - CAPWAP. CASE Pile Wave Analysis Program. User's Manual, Cleveland, Ohio, 1996.

Grande,L.O. - 1977 - Pile -soil interaction. Geotechnical Division, NTH, Dec. 1977

Janbu, N. - 1976 - Static bearing capacity of friction piles. Proc. of the Sixth European Conference on SMFE, Wien, 1976.

Lunne, T. and R. Sandven - 1995 - Cone Penetration Testing - CPT'95. National report for Norway. Proc. of the International Symposium on Cone Penetration Testing. Lindköping, Sweden. Swedish Geotechnical Society, 1995.

Nordal, S. - 1985 - PIA1 - Version 02 , SINTEF Report no: STF69 - F855004.

NOTEBY - 1996 - PILESTAB- User's Manual Report no: 39119-1, Sept. 1996.

Norwegian Pile Commetee - 1991- Guide lines for pile design.(in norwegian). Norwegian Council for Standardisation.

Norwegian Road Research Laboratory - 1991 - Handbook No. 16 . Geotechnical design of highways structures (in norwegian).

Simonsen, A.S. and H.P. Jensen - 1983 - Bearing capacity of steel pipe piles and hammer performance. Proc. of Symposium Dynamic measurements of piles and piers, ASCE, Philadelphia, Pennsylvania, 1983.

Vermeer,P.A. and R.B.J. Brinkgreve - 1995 - PLAXIS, Finite element code for soil and rock Analyses , Balkema, Rotterdam, 1995.

Design of Axially Loaded Piles – European Practice, De Cock & Legrand (eds)
© 1997 Balkema, Rotterdam, ISBN 90 5410 873 8

Polish design methods for single axially loaded piles

K.Gwizdała
Geotechnical Department, Technical University of Gdańsk

ABSTRACT: The calculation procedure of piles, based on Polish Code No. PN-83/B-02482, "*Foundations. Bearing capacity of piles and pile foundations*" is presented. The attention has been focused on the bearing capacity of single pile calculation. The results of comparative analysis for a determination of complete load-settlement relation with application of load transfer functions are briefly shown. Values of unit resistance under a base and along a shaft of pile, for load transfer functions were assumed on the basis of Polish Code. The results obtained seem to be satisfactory from the engineering point of view.

1 INTRODUCTION

Piles and pile foundations are mostly applied in the cases when part of a subsoil is built of weak soils of small bearing capacity such as: loose non-cohesive soils, soft cohesive or liquid soils and peat. Additionally, piles are used in the case of soils with good bearing capacity when there is a need to transmit very high, vertical and horizontal concentrated loads and theirs combinations.

In Poland piles are widely applied in industrial building, bridges, civil engineering and housing.

Typical kinds of a subsoil for structures founded on piles can be grouped as follows:

a) layered non-cohesive soils occurring along entire length of pile and under its base,
b) layered cohesive soils occurring along entire length of pile and under its base,
c) very weak soils occurring directly under the ground level such as weak cohesive soils or peat, bottom part of pile and its base are embed in good capacity non-cohesive or cohesive soils,
d) cohesive and non-cohesive soils of highly differentiated shear strength occurring alternately along pile shaft; in the base region occur non-cohesive and cohesive soils of good bearing capacity,
e) cohesive and non-cohesive soils of highly differentiated shear resistance occurring alternately along pile shaft; in the base region exist high strength non-cohesive soils, under the base at the depth equal to several pile diameters are again weak cohesive soils.

Foundation on piles is then somewhat affected by subsoil conditions together with continuously increasing vertical and horizontal loads and new areas of application of piles.

New kinds of piles, technology of their construction and calculation methods are strongly related to the introduction of new technique achievements. It regards both new technologies of piles as well as methods of subsoil investigations.

Calculation methods of pile foundations regard the analysis of an individual pile or a group of piles. For both cases two main states are considered: ultimate bearing capacity state and serviceability limit states.

2 METHODS OF A SUBSOIL INVESTIGATION

An exploration and classification of a subsoil should provide with any possible information for safe and economically effective choice and design of structures' foundations.

In many cases the archive data are very helpful in planning of *in situ* investigations. Theirs scope should consider the conditions related to design of deep foundations.

The conclusions summarising geotechnical report include usually some indications and recommendations referred to a way of founding together with necessary geotechnical parameters.

Geotechnical report consists mostly of three main parts:

- *in situ* investigations, borings, cuts, excavations, sampling and *in situ* testing,
- laboratory investigations,
- analytical elaboration of investigations and tests, formulation of final conclusions.

Nowadays in Poland, *in situ* investigations are mainly based on borings together with sampling which provides with undisturbed samples of soils of natural moisture content and natural particle-size gradation.

Among various *in situ* investigations most common are the following methods:

- standard penetration tests (SPT), light and heavy dynamic penetrometer,
- cone penetration tests (CPT and CPTU),
- Vane shear tests,
- presiometer tests (PMT) and dilatometer tests (DMT).

The results of *in situ* and laboratory investigations are a basis for establishing the profiles of general subsoil conditions with stratification, groundwater conditions and physical and mechanical parameters of individual soils.

According to the calculation method contained in Polish Code No. PN-83/B-02482 for pile foundations, for calculations of bearing capacity of piles it is sufficient to know a general geotechnical profile with specification of soils including either density or liquidity indexes, I_D and I_L respectively (see chapter 5). For cohesive soils the calculations may be performed using undrained shear strength S_u (c_u).

Presently, the code does not yet contain direct methods of application of *in situ* investigations for calculations of piles.

3 COMMON PILE TYPES

Kinds of piles applied are basically related to geotechnical conditions (see chapter 1), a potential to be used and technical possibilities. Additionally, they are always connected with a kind of building e.g. bridges, civil engineering, industrial building etc.

During many years in Poland classical kinds of piles have been used such as (see also Table 4):

- precast reinforced concrete piles,
- Franki piles,
- Vibro, Vibro-Fundex piles, recently also Vibrex piles,
- bored Wolfsholz piles with placing the concrete under pressure,
- large diameter bored piles performed in casing pipe and in bentonite,
- driven and vibrated steel pipe or steel section piles either closed- or open-end.

Recently, CFA and jet-grounting piles are also being introduced to the engineering practice.

4 NATIONAL DOCUMENTS AND PUBLICATIONS ON PILES CALCULATIONS

Methodology of calculation of piles and pile foundations has been initiated in Poland in fifties. In 1955 first Polish Code No. PN-55/B-02482, *"Foundations. Bearing capacity of piles and piles foundations*" appeared. In subsequent issues of pile code, in 1958, 1969 and 1983 new kinds of piles were being introduced and calculation methods specified.

The introduction to engineering practice large diameter piles had required determination of common standards for theirs design. Thus, in turn in 1975 and 1993 were issued "Design and application recommendations of large diameter piles in bridge objects".

In post-war period a lot of papers referred to the piles and pile foundation problems have been published. Among others, the most important are works of Hueckel (1963), Tejchman (1966, 1971), Mazurkiewicz (1968), Jarominiak et al (1976), Kłosiński (1976), Kłos (1977), Gwizdała (1977, 1984, 1996).

At present, a basic report which regulates the principals of pile calculations and loading tests of piles is code No. PN-83/B-02482, "Foundations. Bearing capacity of piles and pile foundations".

Investigation works and implementations of new technologies in the nearest future should be focused on the analytical determination of complete load-settlement curve, application of direct methods with the use of *in situ* investigations (mainly cone penetration tests) and the determination of a behaviour of large groups of piles with field measurements.

5 CAPACITY CALCULATIONS OF PILES ACCORDING TO POLISH CODE PN-83/B-02482

5.1 *Bearing capacity of piles*

In Polish practice so-called „geostatical" calculations and loading tests are recommended (see PN-83/B-02482 and Gwizdała, Jacobsen 1992).

Bearing resistance from pile driving formulae is applied additionally for non-cohesive soils mainly. It is required to determine calibration factors using the results of loading tests (see Tejchman, Gwizdała 1993).

For the ultimate bearing capacity state the following condition has to be fulfilled:

$$Q_r \leq mN, \qquad (1)$$

where: $Q_r = \gamma_f Q_n$,

Q_r - calculated load,

γ_f - load coefficient,

Q_n - characteristic load,

m - assumed correction factor equal to 0.9 for pile foundations; for single piles and group of two piles $m = 0.7$ and 0.8 are assumed, respectively,

N - bearing capacity of single pile.

From Eq. (1) it results that for calculations a method of partial coefficients is applied. Coefficient γ_f ($\gamma_f > 1$) includes different influence of loadings, whereas correction coefficient m refers to an accuracy of the method and a number of piles considered.

Non-uniformity of a subsoil has been taken into account by a material coefficient γ_m, see equation (2).

Bearing capacity N of a single pile is a sum of base N_p and shaft N_s resistances:

– in compression:

$$N = N_p + N_s = S_p q^{(r)} A_p + \sum S_{si} t_i^{(r)} A_{si} \qquad (2)$$

– in uplift:

$$N^w = \sum S_i^w t_i^{(r)} A_{si} \tag{3}$$

where: $q^{(r)} = \gamma_m q$,

$t^{(r)} = \gamma_m t$,

i - index refers to a number of soil layers,

q - is unit base resistance,

t - is unit shaft resistance,

S_p, S_s, S_i^w - are engineering factors.

γ_m - material coefficient, $\gamma_m < 1$, depending on ether density or liquidity indexes, I_D and I_L respectively.

Values of bearing capacity calculated on the basis of the above formulae should be compared to measurements from loading tests for a structure analysed. A number of loading tests depends on local subsoil conditions and number of piles performed.

Base resistance

The unit base resistance q of non-cohesive soils at the pile base is specified in Table 1, depending on the type of soil and density index I_D or liquidity index I_L. The influence of pile base diameter on q (and h_c) should be taken into account for medium dense and dense non-cohesive soils (Figure 1a and 1b).

Table 1. Unit base resistance q (kPa).

Soil type	Density index			
	$I_D = 1.00$	$I_D = 0.67$	$I_D = 0.33$	$I_D = 0.20$
Gravel, sand-gravel mix	7750	5100	3000	1950
Coarse and medium sand	5850	3600	2150	1450
Fine sand	4100	2700	1650	1050
Silty sand	3350	2100	1150	700
Soil type	**Liquidity index**			
	$I_L < 0, w \cong 0$	$I_L = 0, w = w_p$	$I_L = 0.50$	$I_L = 0.75$
Loamy gravel, gravel-sand-clay mix	4150	2750	1650	850
Loamy sand, sandy loam, loam, silty loam	2750	1950	850	450
Firm sandy loam, firm loam, firm silty loam, sandy clay, clay, silty clay	2800	1950	800	400
Sandy silt, silt	1850	1250	500	250

Unit base resistance q is specific for depths, equal to or exceeding the critical depth $h_{ci} = h_c^0 \sqrt{D_i / D_0}$, where: $h_c^0 = 10$ m, $D_0 = 0.4$ m and D_i is the actual base diameter. Linear interpolation should be adopted to determine values of q for depths less than h_c with zero taken as the value of q for the initial or equivalent ground level, Fig. 1.

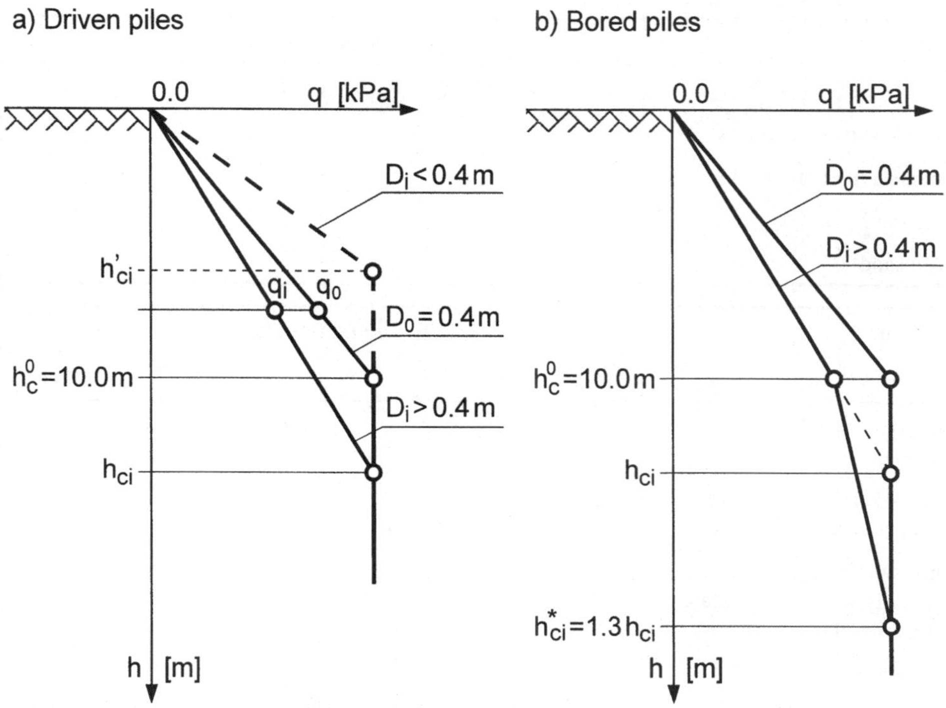

Figure 1. Interpolation of unit base resistance (non-cohesive soils).

The critical depth should be increased by 30% in case of bored piles, $h_{ci}^* = 1.3\ h_{ci}$. The ultimate resistance of soil under the pile base should be interpolated according to Fig. 1b.

For other soils (as specified in Table 1) values of q do not depend on the pile diameter and become constant and independent of the depth when the critical value $h_c^0 = 10$ m is exceeded. For cohesive soils (clays, $\phi_u \approx 0$) the following formula may be adopted:

$$q = 9\,S_u \tag{4}$$

where S_u is undrained shear strength.

Shaft resistance

The value of the unit shaft resistance t, is specified in Table 2. It depends on the type of soil and its density index I_D or liquidity index I_L.

Values of t, as shown in Table 2, should be used for depths equal to or exceeding 5 m below the ground level. For smaller depths the appropriate value of t ought to be determined by interpolation between the table entry and zero assumed for the initial or equivalent ground level as shown in Fig. 1.

For cohesive soils (clays, $\phi_u \approx 0$) values of t may be set depending on the undrained shear strength of soil, S_u, see Fig. 3.

Values of t do not depend on pile diameter.

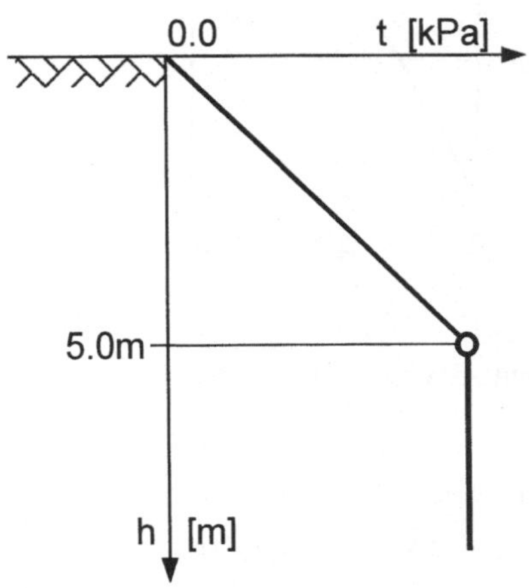

Figure 2. Interpolation of unit shaft resistance.

Table 2. Unit shaft resistance t (kPa).

Soil type	Density index			
	$I_D = 1.00$	$I_D = 0.67$	$I_D = 0.33$	$I_D = 0.20$
Gravel, sand-gravel mix	165	110	74	59
Coarse and medium sand	132	74	47	34
Fine sand	100	62	31	22
Silty sand	75	45	25	16
Soil type	**Liquidity index**			
	$I_L < 0, w \cong 0$	$I_L = 0, w = w_p$	$I_L = 0.50$	$I_L = 0.75$
Loamy gravel, gravel-sand-clay mix	134	95	67	44
Loamy sand, sandy loam, loam, silty loam	95	50	31	14
Firm sandy loam, firm loam, firm silty loam, sandy clay, clay, silty clay	95	50	25	11
Sandy silt, silt	65	30	16	7
Mud	48	18	0	0

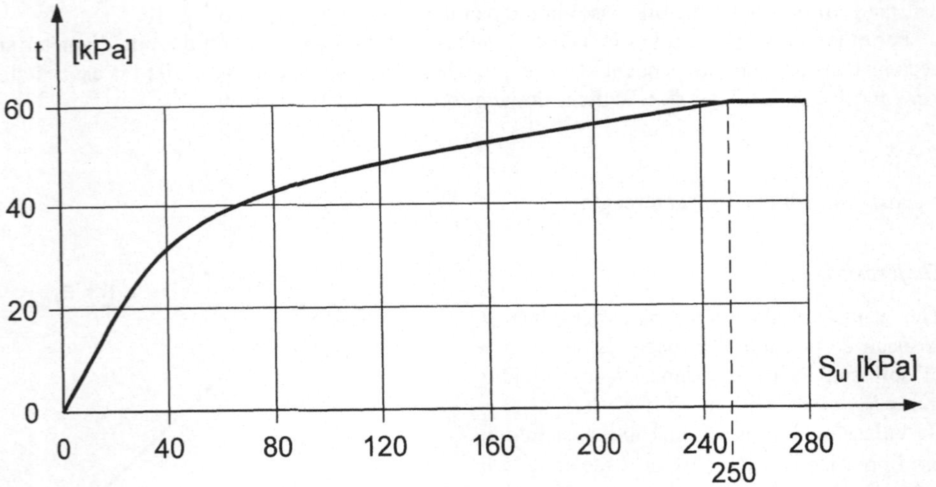

Figure 3. Unit shaft resistance t for cohesive soils.

Engineering factors

S_p, S_s, and S^w are dimensionless factors, specified in Table 3 and 4.

Table 3. Engineering factors S_p, S_s, and S^w, non-cohesive soils.

Type of pile and method of installing	Values of factors for soils non-cohesive					
	$I_D > 0.67$			$I_D = 0.67 \div 0.20$		
	downward movement of pile		uplift of pile	downward movement of pile		uplift of pile
	S_p	S_s	S^w *)	S_p	S_s	S^w *)
1. Precast reinf. concrete piles						
a) driven	1.0	1.0	0.6	1.1	1.1	0.6
b) installed by jetting (the last 1 m driven)	1.0	0.8	0.4	1.0	0.8	0.4
c) driven with vibr. equip.	—	—	—	1.0	0.8	0.5
2. Franki piles	1.3	1.1	1.0	1.8	1.6	1.0
3. Vibro piles	1.1	1.0	0.6	1.4	1.1	0.6
4. Piles bored in non-coh. soils **) (except of fine and silty sands)						
a) in temporary casing	1.0	0.8	0.7	1.0	0.9	0.7
b) cased	1.0	0.8	0.6	1.0	0.8	0.6
c) with casing sunk into the ground and lifted out by means of a rotary cap .	1.0	1.0	0.7	1.0	1.1	0.7
d) in drilled fluid	1.0	1.0	0.7	1.0	1.0	0.7
e) washbored	1.0	1.0	0.7	1.0	1.0	0.7
f) Wolfsholz piles	1.0	0.8	0.6	1.0	0.9	0.6
5. Piles bored in fine and silty sands **)						
a) in temporary casing	0.8	0.6	0.4	0.9	0.7	0.5
b) cased	0.8	0.6	0.4	0.9	0.7	0.5
c) with casing sunk into the ground and lifted out by means of a rotary cap	0.8	0.7	0.5	0.9	0.8	0.5
d) in drilling fluid	1.0	0.9	0.6	1.0	0.9	0.6
e) washbored	1.0	1.0	0.7	1.0	1.0	0.7
f) Wolfsholz piles	0.8	0.6	0.5	0.9	0.7	0.5
6. Closed-end pipe piles						
a) driven	1.0	1.0	0.5	1.1	1.0	0.5
b) installed by jetting (the final 1 m driven)	1.0	0.7	0.4	1.0	0.6	0.4
c) driven with vibr. equip.	—	—	—	1.0	0.8	0.5
7. Steel section piles						
a) driven	1.0	0.8	0.5	1.0	0.9	0.5
b) installed by jetting (the last 1 m driven)	1.0	0.5	0.3	1.0	0.6	0.3
c) driven with vibr. equip.	—	—	—	1.0	0.7	0.4

*) for anchor piles used only during pile loading tests values of S^w may be increased by 20%.

**) factors specified under No. 4 and 5 in the table do not cover cases of special treatment of subgrade improving its performance under the base or along the shaft. Values of the engineering factors for such cases should be based on results of site investigations.

Table 4. Engineering factors S_p, S_s, and S^w, cohesive soils.

Type of pile and method of installing	Values of factors for soils					
	cohesive					
	$I_L < 0$			$I_L = 0 \div 0.75$		
	downward movement of pile		uplift of pile	downward movement of pile		uplift of pile
	S_p	S_s	S^w *)	S_p	S_s	S^w *)
1. Precast reinf. concrete piles						
a) driven	1.0	1.0	0.7	1.0	0.9	0.6
b) installed by jetting (the last 1 m driven)	—	—	—	—	—	— —
c) driven with vibr. equip.	—	—	—	—	—	— —
2. Franki piles	1.2	1.1	0.8	1.1	1.0	0.7
3. Vibro piles	1.0	1.0	0.6	1.0	0.9	0.6
4. Piles bored in non-coh. soils **) (except of fine and silty sands)						
a) in temporary casing	1.0	0.9	0.6	1.0	0.9	0.6
b) cased	1.0	0.8	0.6	1.0	0.8	0.5
c) with casing sunk into the ground and lifted out by means of a rotary cap	1.0	1.0	0.7	1.0	1.0	0.6
d) in drilled fluid	1.0	0.9	0.5	1.0	0.9	0.5
e) washbored	—	—	—	—	—	—
f) Wolfsholz piles	1.0	0.9	0.6	1.0	0.8	0.5
5. Piles bored in fine and silty sands **)						
a) in temporary casing	1.0	0.9	0.6	1.0	0.9	0.6
b) cased	1.0	0.8	0.6	1.0	0.8	0.5
c) with casing sunk into the ground and lifted out by means of a rotary cap	1.0	1.0	0.7	1.0	1.0	0.6
d) in drilling fluid	1.0	0.9	0.5	1.0	0.9	0.5
e) washbored	—	—	—	—	—	—
f) Wolfsholz piles	1.0	0.9	0.6	1.0	0.9	0.5
6. Closed-end pipe piles						
a) driven	1.0	1.1	0.5	1.0	0.9	0.5
b) installed by jetting (the final 1 m driven)	—	—	—	—	—	— —
c) driven with vibr. equip.	—	—	—	—	—	— —
7. Steel section piles						
a) driven	1.0	1.0	0.5	1.0	0.9	0.5
b) installed by jetting (the last 1 m driven)	—	—	—	—	—	—
c) driven with vibr. equip.	—	—	—	—	—	—

*) for anchor piles used only during pile loading tests values of S^w may be increased by 20%.

**) factors specified under No. 4 and 5 in the table do not cover cases of special treatment of subgrade improving its performance under the base or along the shaft. Values of the engineering factors for such cases should be based on results of site investigations.

Open ended piles

The open ended pipe piles should be determined according to formula:

- in compression:

$$N = N'_p + N'_s = b_1 S_p q^{(r)} A_p + \sum b_2 S_{si} t_i^{(r)} A_{si} \tag{5}$$

- in uplift:

$$N^w = \sum b_2 S_i^w t_i^{(r)} A_{si} \tag{6}$$

where b_1 and b_2 are reduction factors as specified in Table 5 and Table 6.

In the case of non-cohesive soils of large grain size and higher degree of compaction, the actual bearing capacity of a pile will be larger than that calculated based on these values.

Use of open ended pipe piles in non-cohesive soils characterised by degree of compaction lower than 0.4 is not recommended.

Table 5. Values of b_1 and b_2 for non-cohesive soils.

No.	$I_D = 0.40$				$I_D = 0.70$			
	h/D		b_1	b_2	h/D		b_1	b_2
	moist	wet			moist	wet		
1					4.0	6.0	0.22	0.27
2	6.0	9.0	0.28	0.61	5.5	8.0	0.50	0.35
3	7.5	11.5	0.78	0.61	6.5	10.0	0.90	0.37
4	17.0	26.0	1.00	0.61	17.0	26.0	1.00	0.65

Table 6. Values of b_1 and b_2 for cohesive soils.

No.	h/D	Liquidity index $I_L \leq 0.50$ and $t \geq 20$ kPa		
		b_1	b_2	
			steel pipe	reinforced concrete pipe
1	6 -15	0.7	0.8	1.0
2	15	pipe empty inside = 0.9	0.8	1.0
		pipe filled with concrete = 1.0		

Notes to tables of b_1 and b_2:

a) small relative penetration depths h/D specified in the tables should be regarded as minimum required for driven open ended pipe piles.

b) factors b_1 and b_2 for intermediate values of relative penetration depths and for $0.4 < I_D < 0.7$ ought to be determined by linear interpolation. For $I_D \geq 0.7$, b_1 and b_2 are assumed to be constant.

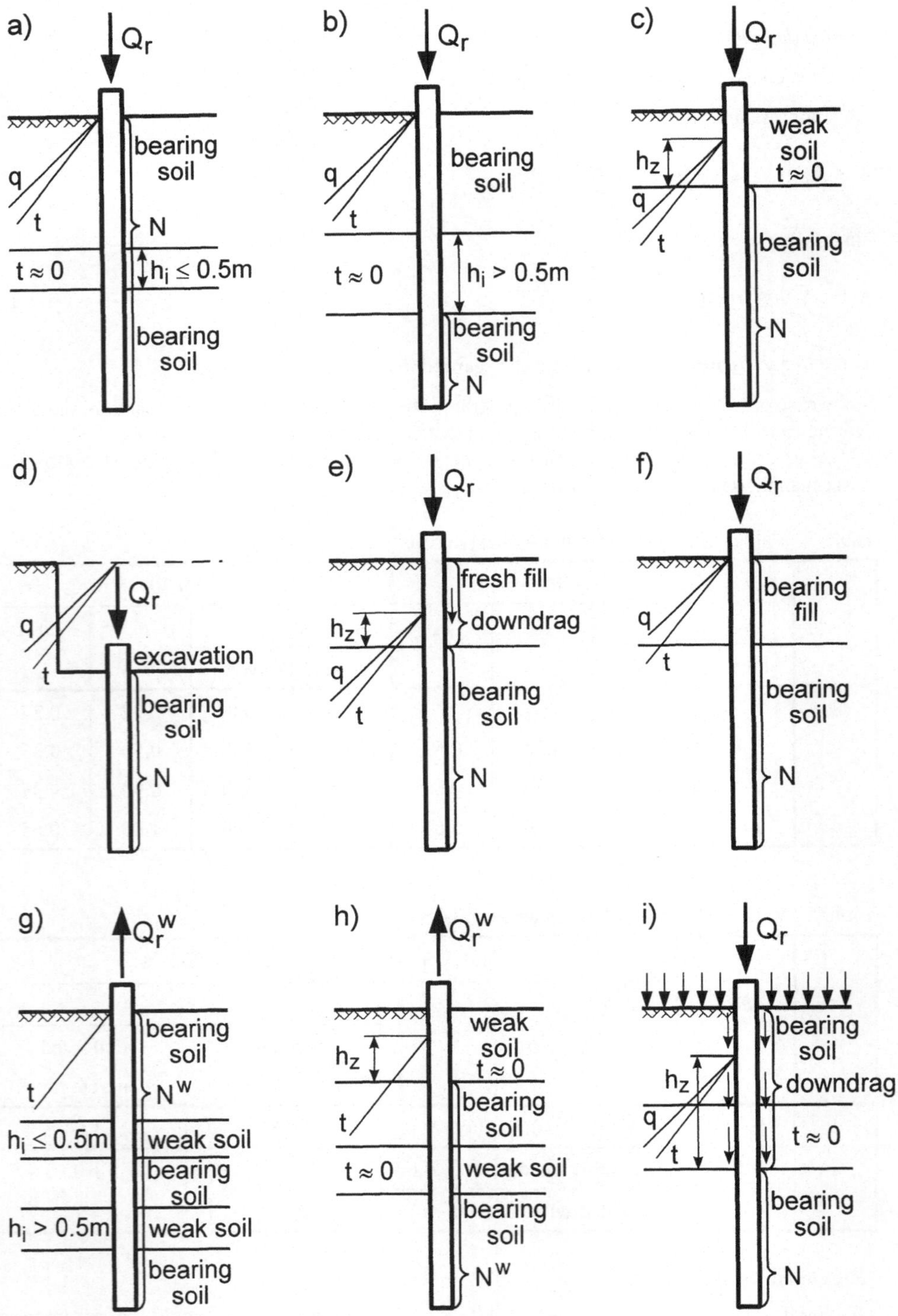

Figure 4. Interpretation of the q and t for layered soil according to the equivalent ground level.

For piles embed in layered soils it is suggested to make the interpolation of q and t from an equivalent level, based on equivalent height h_z calculated from the following relation (see Fig. 4):

$$h_z = \frac{0.65}{\gamma'} \left(\sum \gamma'_i \, h_i \right) \tag{7}$$

where: γ' - unit weight of bearing soil

γ'_i - unit weight of i - layer above the bearing soil

0.65 - correction factor reflecting non-uniformity of soils.

5.2 *Settlement curve of pile*

For the evaluation of real behaviour of pile in a subsoil it is necessary to know load-settlement curves for pile base and shaft resistances, loading distribution along length and shortening of pile.

Practically, the relations accurate enough from the engineering point of view can be obtained applying load transfer functions methods, see Gwizdała (1984), Gwizdała and Jacobsen (1992), Gwizdała (1996), Gwizdała and Tejchman (1997).

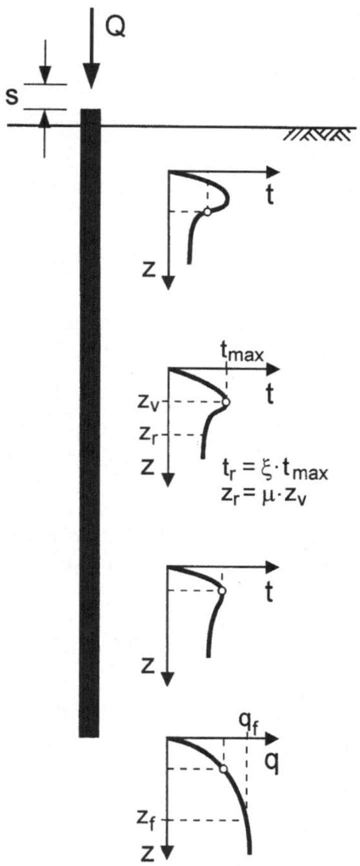

Figure 5. Calculation model of single elastic pile.

Many years experience of application of unit resistance values under a base and on a shaft of pile included in Polish Code PN-83/B-02482 has been used in the research. The method presented below of determination of complete load-settlement curve is not yet included in Polish Code.

A pile is divided into a defined number of elastic elements. Soil response on the contact with pile is expressed by non-linear functions *q-z* and *t-z*, describing relationships between base or shaft resistance and settlement of the considered pile point, respectively, see Fig. 5. A solution consists in the iterative search of the relation between axial force in a pile at any depth and settlement at this depth. The calculations are made using numerical code PALOS. The calculations procedure provides with a possibility of a choice of almost 20 types of *t-z* and 13 types of *q-z* functions. Numerical code has been verified on experimental data from piles' loading tests.

The accuracy of the results being calculated by PALOS depends mostly on the following factors:

- the correct prediction of ultimate unit shaft and base resistance t and q, respectively,
- a determination of reliable non-linear stiffness functions for particular depths and soil layers,
- calculation or assumption of limit settlements, at which both t and q resistance reach their maximum values, denoted as z_v for shaft (for given point of pile) and z_f for base.

In the verification calculations the values of resistances under a base q and on a shaft t and the way of theirs interpolation were used according to PN-83/B-02482 (see Fig. 4).

In order to receive better compatibility the values were increased of 30% for Franki piles and 20% for Vibro-Fundex piles in comparison to, standard values. The values of z_v and z_f were assumed equal to 1% and 5% of the pile diameter, respectively.

Generally, values of q and t can be determined on the basis of any arbitrary method using the results of loading tests, laboratory or *in situ* investigations

Summary results of the analysis containing totally 240 piles of different kinds are presented in Table 7.

As a criterion of the correctness of calculations a ratio of calculated to measured values (from load-settlement curve) has been assumed in terms of conformity coefficient:

$$\eta_i = \frac{P_i}{Q_i}, \tag{8}$$

where P and Q denote the values of calculated and measured load, respectively and index i refers to subsequent real values of settlement. The values of η coefficient less then unity show that the calculation results are on the safe side.

Examples of conformity coefficients distributions for 32 large diameter bored piles in non-cohesive soils are presented in Fig. 6.

In Fig. 6a are shown calculation results for all investigation points referred to complete load-settlement curves of 32 piles. The calculation results for chosen settlement $s = 10$ mm are presented in Fig. 6b. The results obtained in the histogram form were approximated in terms of log-normal distribution.

Table 7. Values of conformity coefficients η.

Type of pile	Measured points	Number of points	Mean value $\bar{\eta}$	Standard deviation σ	Coefficient of variation v
Precast reinf. concrete piles.	all points	484	0.933	0.282	0.302
	s = 3 mm	51	0.955	0.322	0.337
	s = 5 mm	37	1.014	0.304	0.300
	s = 10 mm	21	0.930	0.276	0.297
Wolfsholz piles	all points	391	0.925	0.340	0.368
	s = 3 mm	28	0.907	0.350	0.386
	s = 5 mm	28	1.042	0.374	0.359
	s = 25 mm	8	0.786	0.201	0.256
Large diameter bored piles non-coh. soils	all points	303	0.924	0.250	0.270
	s = 3 mm	30	0.951	0.264	0.278
	s = 10 mm	25	0.994	0.238	0.240
	s = 20 mm	19	0.960	0.200	0.208
Large diameter bored piles cohesive soils	all points	207	0.987	0.285	0.288
	s = 5 mm	25	0.976	0.278	0.284
	s = 10 mm	19	0.976	0.205	0.210
	s = 20 mm	10	0.935	0.233	0.238
Franki piles	all points	151	0.978	0.154	0.157
	s = 3 mm	13	0.980	0.154	0.157
	s = 5 mm	12	0.954	0.161	0.169
	s = 7 mm	11	0.961	0.151	0.157
Fundex piles non-cohesive soils	all points	429	0.974	0.269	0.276
	s = 3 mm	35	1.028	0.259	0.252
	s = 5 mm	18	1.084	0.200	0.185
Fundex piles cohesive soils	all points	362	0.989	0.240	0.267
	s = 3 mm	26	0.972	0.270	0.278
	s = 5 mm	23	1.042	0.278	0.267
	s = 7 mm	20	1.009	0.264	0.262

5.3 *Determination of pile bearing capacity from loading tests*

Bearing capacity of piles calculated by statical formulae should be examined in terms of loading test for a given structure. As a principle two loading tests at the first 100 piles and one test for the next 100 piles are assumed.

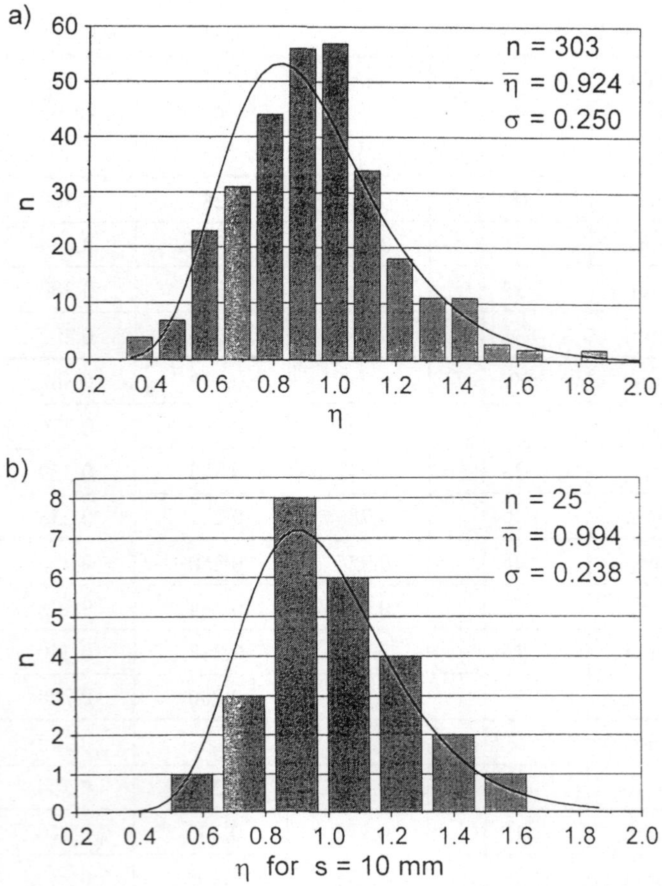

Figure 6. Histogram of conformity coefficients (non-cohesive soils).

The final number of tests should be determined taking into account local subsoil conditions, type of construction and previous practical experience resulting from the application of a given type of pile in similar subsoil conditions.
Taking into account the results of loading tests, the following condition has to be fulfilled:

$$Q_r \leq k N_c^0, \qquad (9)$$

where: k - is a correction coefficient, $k = 0.8 \div 1.0$ resulting from the analysis of real load-settlement curve,

N_c^0 - denotes bearing capacity of pile resulting from the analysis of load-settlement curve obtained on the basis of loading tests, according to PN-83/B-02482.

The above considerations show that in Poland as a principle for larger constructions being founded on piles the application of loading tests is assumed. The tests are usually carried out using anchoring piles or ballast.

A base of pile work evaluation is real relation of load-settlement (settlement curve). The load which can be safely transmitted by a pile is determined in terms of graphical differentiation of that curve.

The details of the problem can be found in „Dynamic and static analysis and in situ tests for evaluation of pile bearing capacity" (Tejchman, Gwizdała, Kłos 1985).

5.4 *Settlement of piles and pile foundations*

The problem of settlement of piles and pile foundations is very wide and will not be analysed in details in this paper. The attention will be only focused on some aspects of the problem with regard to Polish Code regulations.

In general, the calculation method according to Poulos and Davis (1974, 1980) has been accepted. Independently, some author's verifying calculations based on statical loading tests were being carried out together with establishing the principles of the modulus of determination. The method assumed allows for settlements' calculations of a single pile and a group of piles in the range of characteristic loads Q_n.

For the case when the highly deformable layer of a subsoil occurs below the pile base (see e) subsoil type in chapter 1) the settlement of pile foundation should be calculated as for equivalent deep foundation.

The basic method of the determination of complete load - settlement curve is statical loading test. Recent works show that there is a possibility of an analytical evaluation of load-settlement curve, see Gwizdała (1996).

6 SUMMARY

The evaluation of piles' bearing capacity contained in Polish Code No. PN-83/B-02482 relies on statical calculations and examination of the correctness of results obtained in terms of statical loading tests.

The calculations are performed for the specified subsoil conditions and determined parameters of soils from the subsoil profile. The values of unit base and shaft resistances are presented in the tabular form (see Tables 1 and 2). The technology of a given pile included in terms of technology coefficients (see Tables 3 and 4).

Polish Code requires the examination of calculations by loading tests performed on chosen piles in nature. The base of the proper evaluation of calculations is a real load-settlement curve obtained during such tests.

Application of load transfer functions of *q-z* and *t-z* type enables the analytical evaluation of complete load-settlement relation. In calculations the unit base and shaft resistenaces recommended by Polish Code are used. The accuracy of that method was shown in Table 7.

For the more accurate evaluation of piles bearing capacity and complete load-settlement curve it is suggested to apply direct methods. In the methods mentioned CPT tests are widely used. Such analyses can be found in Gwizdała (1984), Tejchman, Gwizdała (1988), Gwizdała, Tejchman (1995) and Gwizdała (1996).

REFERENCES

Gwizdała K. 1977. *Bearing capacity and behaviour of large diameter piles in non-cohesive soil.* (in Polish). Politechnika Gdańska, Wydział Hydrotechniki.

Gwizdała K. 1984. *Large diameter bored piles in non-cohesive soils. Determination of the bearing capacity and settlement from results of static penetration test: (CPT) and standard penetration test: (SPT).* Swedish Geotechnical Institute, Report No 26, 1-129.

Gwizdała K. 1993. The determination of load-settlement curve for large diameter piles. *Proc. 6 Colloque Franco-Polonais Mecanique des Sols Applique.* Douai, 20-23 Septembre 1993, 194-201.

Gwizdała K. 1996. *The analysis of piles settlements employing load-transfer functions.* (In Polish). Technical University of Gdańsk, Zesz. Nauk. PG Nr 532, Budownictwo Wodne Nr 41.

Gwizdała K., Jacobsen M. 1992. *Bearing capacity and settlements of piles.* University of Aalborg.

Gwizdała K., Kłosiński B. 1993. Nonlinear load-settlement behaviour of bored piles. *Proceedings of the 2nd International Geotechnical Seminar on Deep Foundation on Bored and Auger Piles.* Ghent, Belgium, 1-4 June 1993, 255-258.

Gwizdała K., Tejchman A. 1993. Evaluation of bearing capacity and settlement of Wolfsholz piles bored into layered subsoil. *Proceeding of the 2nd International Geotechnical Seminar on Deep Foundation on Bored and Auger Piles.* Ghent, Belgium, 1-4 June 1993, 285-288.

Gwizdała K., Tejchman A. 1995. Pile settlement analysis using CPT and load-transfer functions t-z and q-z. *Proceeding of the International Symposium on Cone Penetration Testing.* Linkoping, Sweden October 4-5, 1995.

Gwizdała K., Tejchman A. 1997. Numerical modelling of pile-subsoil interaction. *Proceeding of the XIV International Conference on Soil Mechanics and Found. Eng.* Hamburg, 6 - 12 September 1997.

Hueckel S. 1963. *Foundations.* (In Polish). Vol. IX. Rosiński B. Warszawa: Arkady.

Jarominiak A., et al. 1976. *Piles and pile foundations.* (In Polish). Warszawa: Arkady.

Kłos J. 1977. *Bearing capacity of open-end pipe piles.* (In Polish). Technical University of Gdańsk.

Kłosiński B. 1976. *Design of large diameter piles.* (In Polish). Warszawa: IBDiM 1976.

Mazurkiewicz B. 1968. *Negative skin friction on piles.* (In Polish). Technical University of Gdańsk, Zesz. Nauk. PG Nr 126.

Poulos H.G., Davis E.H. 1974. *Elastic solutions for soil and rock mechanics.* New York: John Wiley and Sons.

Poulos H.G., Davis E.H. 1980. *Pile foundation analysis and design.* New York: John Wiley and Sons.

Tejchman A. 1966. *Investigation of the skin friction and the point resistance on piles in non-cohesive soil.* (In Polish). Archiwum Hydrotechniki, vol. XIII, z.1, 53-120.

Tejchman A. 1970, 1971. *Bearing capacity of pile groups in cohesionless soil media.* (In Polish). Archiwum Hydrotechniki No 2/1970, 1/1971, s.83-99.

Tejchman A., Gwizdała K. 1988. Determination of the load settlement curve for large diameter piles based on CPT results. *Proc. I International Symposium on Penetration Testing.* Orlando, USA, 20-24 March 1988, 1015-1020.

Tejchman A., Gwizdała K. 1993. Dynamic and Static Analysis and in situ Investigation of Vibro-Fundex Piles. *Proceedings of the Third International Conference on Case Histories in Geotechnical Engineering.* St. Louis, Missouri, 1-4 June 1993, 197-199.

Tejchman A., Gwizdała K., Kłos J. 1985. Application of in situ tests for evaluation of pile bearing capacity. *Proc. of the 11 th Int. Conf. on Soil Mech. and Found. Eng.*, San Francisco, 1479-1482.

Tejchman A., Gwizdała K., Świniański J. 1988. Analysis of bearing capacity and settlement of piles driven into weak soils. *Proc. II Baltic Conference on Soil Mechanics and Foundation Engineering.* Tallin, 10-15 May 1988.

Polish Code PN-55/B-02482, 1955. *Foundations. Bearing capacity of piles and pile foundations.*

Polish Code PN-69/B-02482, 1969. *Foundations. Bearing capacity of piles and pile foundations.*

Polish Code PN-83/B-02482, 1983. *Foundations. Bearing capacity of piles and pile foundations.*

Design and application recommendations of large diameter piles in bridge objects. (In Polish). Warszawa: Instytut Badawczy Dróg i Mostów, 1975, 1993.

Design of Axially Loaded Piles – European Practice, De Cock & Legrand (eds)
© 1997 Balkema, Rotterdam, ISBN 90 5410 873 8

Design of axially loaded piles – Romanian practice

Iacint Manoliu
Technical University of Civil Engineering, Bucharest

ABSTRACT : Main types of piles have been used in the past decades in Romania as a result of the simultaneous recurence of soils of high compressibility and of constractions carrying heavy loads. The domain of piles is completely covered by technical regulations, among which two national standards (codes) dealing with design matters. A strong emphasis is put on the field tests, especially on static load tests on piles. The design of pile foundations in Romania is governed by the concept of limit state design.

1. REGIONAL GEOLOGY

1.1 Main Romania' s tectonic units

The territory of Romania has a very complex and varied geological structure (Marchidanu & Antonescu 1995). The main geological unit of Romania is represented by the Carpathian Geosyncline that makes up a segment of the great Alpine - Carpathian - Himalayan Geosyncline and includes the Eastern and Curvature Carpathians, the Southern Carpathians and the Western Mountains.

On the eastern and southern border the Carpathian Geosyncline is limited by the Pericarpathian Depression that continues toward the western extremity with the Getic Depression.

Between the Eastern Carpathians, Southern Carpathians and Western Mountains there is the Transilvanian Depression. Westward of Western Mountains stretches the eastern frame of the Pannonian Depression.

The territory comprised between the Pericarpathian Depression in the western side and Prut river on the eastern side belongs to the Moldovian Platform, while the territory situated southward of Pericarpathian Depression and Getic Depression belongs to the Moesian Platform.

The geological unit of Dobrudja, comprised between Danube river and the Black Sea, is considered as a distinct unit with an intricate geology.

The main tectonic units of Romania are tightly correlated with the structure of lithosphere that is intensely divided into tectonic plates and microplates. The great East - European plate stretches eastward of the Eastern Carpathians and has the Moldovian Platform as its western limit. Southward of Southern Carpathians there is the Moesian Microplate over which the Moesian Platform is disposed. Westward of the curvature of Carpathian Mountains there is the Intralpine Microplate. Between the Moesian Microplate and the East - European Plate, the Microplate of the Black Sea with a shape of a promontory is located.

The great dislocations that mark the limits of tectonic plates make their junction in the curvature zone of Eastern Carpathians, in the region of Vrancea. This area represents the most sensitive zone of Romania' s territory from the seismic point of view. The hypocentres of earthquakes in this region have depth usually comprised between 70 and 160 km. The maximum magnitude of earthquakes produced in the region has reached 7.2 - 7.3 degrees on the Richter' s scale, while the seismic intensity increased until 9-10 degrees on the Mercalli' s scale.

1.2 Geological features of the Romanian main units

The Carpathian Mountains, belonging to the Alpine - Carpathian - Himalayan chain, are young mountains, in full process of folding, intensely divided into fragments by the erosion processes due to the existing dense hydrographical network. The diluvial deposits on the mountain slopes are relatively thin due to strong denudation process favoured by the vigorous slope proving a high slope energy.

The Pericarpathian Depression forms an uninterrupted strip at the outside of the Eastern and Southern Carpathian Mountains. It is formed by the mio - pliocene molasse formations, in clay-sandy facies with intercalations of chemical precipitation rocks, in particular salts, generating numerous diapiric folds. The geological formations in this area are much folded and fractured, easy erodible and covered by shallow deposits, especially diluvial, colluvial and proluvial, affected by important erosional processes and landslides.

The Getic Depression represents a continuation westwards of the Pericarpathian Depression, covered in its upper part by mio - pliocene sedimentary deposits with intercalations of brown coal, in clay - sandy facies. All this area is covered by shallow deposits, especially by eluvial and diluvial materials. The hilly zones, with a sloped ground surface, are often subjected to active landslide phenomena.

The Transilvanian Depression is characterized by sedimentary deposits which predominantly develop in the clay - sandy facies with frequent intercalations of volcanic tuff and diapiric salt structures. The main shallow deposits are represented by clayey - sandy dilluvial materials. The geological formations of the depression are also affected by many landslides.

The Pannonian Depression. The pannonian (mio - pliocene) deposits are present, in the clayey - sandy facies, covered at the surface by shallow formations made by clayey and silty alluvial and aeolian deposits.

The Moldovian Platform is predominantly covered bt clayey, shallow, eluvial, diluvial and alluvial deposits and partially by loessial sediments. It is affected by numerous landslides and erosional processes.

The Moesian Platform has a sedimentary coverlet formed by miocene and pliocene clayey and sandy layers covered by the pleistocene clayey - sandy deposits placed, sometimes, on fluvio - lacustrine deposits predominantely formed by sand and gravel. The alluvial holocene deposits are mostly clayey and sandy while the aeolian deposits are represented by loessial collapsible soils.

The geological unit of Dobrudja . Various parts of the geological unit of Dobrudja have different geological features.

The north - west part of the Northern Dobrudja is formed by old magmatic intrusions, while the southern part is formed by carboniferous detrital sedimentary formations, triassic detrital and calcareous rocks and cretaceous flisch in marly - sandstony facies. At the surface a coverlet of loessial deposit is placed, having a thickness of more than 10-15 m.

The Central Dobrudja is formed by the discontinuous clayey - sandstony flisch formations, slightly affected by metamorphic processes, and much newer detrital and calcareous deposits. An important area of this zone is covered by loess deposits, too.

In the Southern Dobrudja cretaceous deposits are predominant in the flisch facies ; the sarmatian deposits are well represented by slightly inclined or horizontal calcareous deposits and sometimes pliocene deposits in clayey - sandy facies. All the surface of the Southern Dobrudja is covered by thick loessial deposits.

The Danube Delta is characterised by the presence of highly compressible deposits of unconsolidated silty - clayey formations.

The low plains bordering the course of the Danube or of other important rivers have also unconsolidated young sediments or loose sands.

2. COMMON PRACTICE FOR SOIL INVESTIGATIONS

Soil investigation is carried out by means of trial pits and trenches, boreholes, penetration tests.

Trail pits and trenches are used quite seldom, when undisturbed samples of very good quality, which cannot be obtained from boreholes, are required, like in the case of loessial soils. Their depth is limited by the presence of the ground water.

The most used method for soil investigation is the test boring. Drilling is done by augering, wash boring or rotary drilling with a revolving cutter and circulating liquid to remove the cuttings. In stiff soils, the borehole is unsupported. In soft clays and in sands below the water table a casing is used.

Two kinds of samplers are used.

The open drive sampler is a thick walled steel table of 140 mm diameter driven dynamically by means of a drop weight.

The thin walled sampler has an internal diameter of 70 mm allowing to obtain directly undisturbed samples to be used in oedometer and direct shear tests.

Among penetration tests, the Standard Penetration Test is the most widely used.

Dynamic penetration is also used, with a 50 kg hammer falling 0.50 m. From the static tests, the Dutch cone is the most popular (cone with a 60^0 point angle, a diameter of 36 mm, a projected area of 10 cm^2 and a penetration rate of over 2 cm/sec).

3. PILING TECHNOLOGY

During the regime which collapsed in December 1989, and particularly in the 60' s and 70' s when constructions for the heavy and petrochemical industries were booming, pile foundations were very much in favour of both designers and contractors in Romania. The highest authority of the country became so worried about a possible misuse of this foundation solution and the resulting waste of money and resources that in September 1976 issued an order stating that for the use of a pile foundation, regardless the type and the number of piles, a decree signed by the President of the country was mandatory!

In reality, there was no abuse of the pile foundations. Many of the huge metallurgical and petrochemical combinats ss well as other industrial plants were erected along the Danube or other rivers ; ground conditions on the sites were characterized by the presence of alluvial soils of high compressibility. Since loessial collapsible soils cover about 17% of the teritory of the country, many industries, like for instance the big metallurgical plant in Galaţi, were developed on such soils prone to large and uneven settlements. The use of deep foundations was the only rational solution for industrial buildings carrying heavy loades. The same applied for bridges and for Danubian harbours. Considerably less was the use of the piles for dwelling houses, even in the case of multi - storeys buildings.

Most of the known types of piles have been used : precast reinforced concrete piles, cast - in - situ driven piles using Franki technology, uncased bored piles in dry using Salzgitter and Calweld equipments, uncased bored piles under slurry using Romanian equipment FA - 12, bored piles with recoverable casing using Benoto technology. For exceptional works, like new bridges over Danube, bored piles with non - recoverable steel casing lowered by vibration have been used. Several quays in harbours on Danube have been founded on bored piles with non - recoverable reinforced concrete casing lowered by vibration. For a number of very high chimneys, of 250 m and over, rectangular piers (barrettes) with excavation and pouring of the concrete done under slurry, have been used.

4. NATIONAL RELEVANT DOCUMENTS WITH REGARD TO PILE DESIGN

For the time being, there are several levels at which technical regulations in civil engineering are enforced in Romania.

At the highest levels are situated the Romanian standards which represent documents of national importance and coverage.

Standards must be observed. However, according to a new rule introduced after 1990, some deviations are admitted provided they are very well documented and that the entire responsibility is assumed by those who propose the deviations .

At the following level come prescriptions called technical instructions or "norms", which are also documents of national coverage but regarded merely as guidelines.

The third level belongs to documents of sectorial (departmental) coverage. They are a synthesis of the experience accumulated in a certain sector (transportation, energy etc.) but which is worth to be known and applied by the whole technical community.

Altogether, the domain of piles is covered in Romania by seven documents which are in the reference list, from which five standards, one technical instruction and one norm.

The "package" of four standard in the serie 2561 is put at the base of the pile design in Romania. The first in serie, STAS 22061/1-83 refers to the classification and terminology of piles. The second one, STAS 2561/2-81 is devoted to the problems of pile loading tests. The most comprehensive standard is STAS 2561/3-90 which deals with design problems for all types of piles. The last in the serie, STAS 2561/4-90, refers only to the large diameter bored piles but is broader in scope, covering design, construction and reception matters.

Similar in content with STAS 2561/4-90 are the technical instructions P 106-85 dealing with piers (barrettes).

The norm C 160-75 has as objective the execution of all types of piles.

The standard 7484/74 defines constructive provisions and quality criteria for precast reinforced concrete and prestressed concrete piles.

5. DETAILED DESCRIPTION OF THE NATIONAL DESIGN METHODS

5.1 General philosophy

Pile design, like geotechnical design in general , is governed in Romania by the concept of limit state design. The general principles of the limit state design, to be applied to geotechnical problems, are defined in the Romanian standard 3300/1-85.

Ultimate limit state analysis is required for all categories of piles. Serviceability limit state analysis is required only for friction piles.

For the ultimate limit state, the following condition should be checked :

$$S \leq R \qquad (1)$$

Where S is the load on the pile of the foundation and R is the bearing capacity of the pile.

For the serviceability limit state, the following condition should be checked :

$$\Delta \leq \Delta \qquad (2)$$

where Δ is the computed probable deformation of the pile foundation and Δ in the allowable deformation.

The Romanian Standard 2561/3-90 has the following provisions concerning the ways of establishing the bearing capacity of piles.

a. In preliminary phases of the design, the bearing capacity of the piles, regardless the class of importance of the construction, can be determined by using empirical formulae given in the standard.

b. For regular constructions it is allowed to use the empirical formulae also in the final design phase if under the piles tips is encountered a practically incompressible ground and only if the total number of the piles on the site is less than 100.

c. In the final design phase, the bearing capacity of the piles is established on the basis of field loading on test piles constructed with the same technique and with the same equipment previewed in the project of the pile foundation.

d. For constructions of exceptional importance, and whenever possible, it is recommended to perform the field loading tests in phases of the design preceeding the final one.

e. The static loading test of piles is done in accordance with the Standard 2561/3-90. The number of piles to be tested under static load on a site is established by the designer and the geotechnical consultant in relation with the total number of piles, the area covered by the piles, the uniformity of the geological profile and the degree on knowledge about the site. When no other types of field tests are foreseen, the number of statically loaded piles (except large diameter bored piles) should be at least the one given in table 1.

Table 1

Number of piles according to the project	under 100	100...500	501...1000	1001...2000
Number of test piles	2	3	5	6

Note to Table 1.

In foundations with over 2000 piles, beside the six piles given for 2000 piles, one extra pile is tested for each 1000 or 2000 additional piles depending on the uniformity of the stratification..

f. For the large diameter bored piles, the minimum number of test piles, depending on the total number of piles and the type of loading is established by Standard 2561/4-90 as follows :

Number of the pile in the work or in the zone	Type of loading		
	Compression	Tension	Horizontal forces
1......20	1	1	1
21...100	2	2	2
101...200	3	2	2
≥ 201	3 + one pile for each hundred piles over 200	2	2

The same standard prescribes that at constructions at which is necessary to limit the deformation (settlements, horizontal displacements, rotations etc.) testing of piles is performed regardless the number of piles in the work.

In the case of precast driven piles, when along with static load tests are performed on the site other types of field tests, like for instance dynamic tests on piles (in agreement with Standard 2561/2-81) or static penetration tests (in agreement with standard 1242/6-76) whose results, reduced according to the Standard 2561/3-90 are close to those obtained by static load tests, the number of test piles can be reduced but without becoming less than half of the number shown in table 1 and no less than two. Also, the number of piles subjected to static load tests can be reduced to half of the number shown in table 1 (but no less than two) in zones characterized by uniform stratification in case when on neighbouring sites a corresponding number of static load tests on piles similar to those on the considered site have been performed.

At jobs with a small number of piles on a site, 20 or less, it is allowed to perform static load tests on piles which will be incorporated in the foundation. In this case, the maximum load applied to the pile during the test should not be greater than the load generated by the design loads in the most unfavourable combination.

5.2 Design on basis of field tests

According to the Romanian standard 2561/3-90, the bearing capacity of a compression pile is established with the relationship :

$R = km\, P_{cr}$

where k is a partial factor of safety taken as 0,7 ;

m is a partial factor of safety taken as 1 ;

P_{cr} is the critical load on pile, determined on the basis of field tests.

5.2.1 Static load tests

The static load tests of piles is regulated in Romania by the standard STAS 2561/2-81. A stress-controlled type of test is prescribed. The loading is performed in steps, each step being selected as 1/15...1/10 of the presumed critical load. Each load is maintained until the vertical displacement of the pile is stabilized. Stabilization is considered to be reached when difference between settlements recorded at four consecutive readings is less than 0,1 mm.

The ultimate load is considered the load at which one of the following conditions is fulfilled :

- average settlement is larger than 1/10 of the diameter (width) of the pile ;
- after 24 hrs since the application of the load the stabilization was not reached.

Critical load P_{cr} is the greatest load for which the condition of stabilization was reached.

When during the test critical load cannot be reached, the maximum load reached during the test is considered as critical load.

5.2.2 Dynamic tests

The total penetration of the pile under 10 blows with low frequency of hammer is recorded. The set is then computed as the average penetration under a single blow.

The critical load based on dynamic tests is computed with the relationship :

$$P_{cr} = -\frac{aA}{2} + \sqrt{(\frac{aA}{2})^2 + \frac{aA}{e} \cdot \frac{Q_0 + 0.2q}{Q_0 + q} Q_0 \cdot H_0} \qquad (3)$$

where P_{cr} is the critical load, in kN ;

a factor depending on the type of pile and driving conditions, given in table 2, in kP_a ;

A transverse area of the pile, in sqm ;

e the set of the pile, in cm ;

Q_0 the weight of the hammer, in kN ;

q the weight of the pile, including the protective cap and the stationary part of the hammer, in kN ;

H_0 falling height of the hammer (established according to table 3), in cm ;

H_1 the total movement of blowing part of the hammer, in cm ;

E_0 energy delivered by the hammer, in kJ .

Table 2

Type of pile and driving conditions	*a* kPa
Reinforced concrete pile (with protective cap)	1500
Word pile (without protective cap)	1000

Table 3

Type of hammer	Vertical piles	Battered piles 3 : 1
Free fall hammer or single action hammer	$H_0 = H_1$	$H_0 = 0,8\, H_1$
Diesel hammer or double action hammer	$H_0 = \frac{100E_0}{Q_0}$	$H_0 = \frac{80E_0}{Q_0}$

5.2.3 Static penetration test

Based on the results of a static penetration test, the critical load on precast driven piles can be computed with the relation :

$$P_{cr} = \frac{R_p}{2} A + F_e \frac{U}{u_p} \text{ (kN)} \tag{4}$$

where R_p is the cone resistance, calculated as :

$$R_p = \frac{R_{p1} + R_{p2}}{2} \quad (kP_a)$$

R_{p1} is the average of the cone resistances recorded in the layers situated below the tip of the pile down to a depth of 4 d, in kP_a ;

R_{p2} is the average of the cone resistances recorded from the level of the tip of the pile to a height βd above that level, in kP_a ;

d diameter or large side of the rectangular section of the pile, in cm ;

β coefficient, given in table 4.

Table 4

Layer in which penetration occurs	β
Cohesive soils, loose sands	3
Medium dense sands	8
Dense sands and sands with gravel	15

A area of the transverse section of the pile, in sqm ;

F_l friction force on the lateral surface of the penetrometer, lowered to the level of the pile tip,in kN ;

U perimeter of the transverse section of the pile, in m ;

u_p perimeter of the transverse section of the penetrometer casing, in m.

The above relationship can be used in the case of a static penetrometer having a constant rate of penetration on the whole depth of the testing and the following characteristics :

- diameter of the base of the cone d_c = 3,6 cm ;
- diameter of the casing d_{cas} = 3,6 cm ;
- rate of penetration $v \geq 3{,}3$ cm/s.

5.3 Design on basis of soil characteristics obtained by laboratory tests or by empirical methods

a. The bearing capacity in compression of an end bearing pile

The following relationship is used:

$$R = k\, m\, p_v\, A \quad \text{(kN)} \tag{5}$$

where k is a partial factor of safety taken as 0,7 ;

m is a partial factor of safety taken as 1 ;

p_v is a conventional resistance of the ground below the pile tip, in kP_a, which is taken as follows :

- for displacement piles with the base on hard or soft rocks or on macrogranular non-cohesive soils (blocks, boulders) p_v = 20,000 kP_a ;

- for displacement piles with the base on gravels, according to table 6 ;
- for non-displacement piles with the base on hard or soft rocks :

$$p_v = \sigma_{cs} \left(\frac{t}{d} + 1.5\right)$$

σ_{cs} is the average compression strength of the rock, established in accordance with the Standard 6200/5-71 on saturated samples, in kP_a ;
t depth penetration of the pile in rock, in m ;
d diameter of the pile at the level of the base ;
- for non-displacement piles with the base in macrogranular, non-cohesive soils (blocks, boulders, gravels) the relationship (5) is used.

A area of the maximum section at the level of the pile base, in sqm, which is taken as follows :
- for cast in situ piles :

$$A = \frac{\pi d^2}{4}$$ for piles with constant circular section, with diameter d ;

$$A = 0{,}9\,\frac{\pi d_b^2}{4}$$ for piles with enlarged base, when the diameter d_b of the enlarged base can be controlled.

- for tubular piles, A is taken equal to the total area of the section with external diameter d only if the void has been filled with concrete for a height of at least 3 d from the level of the tip ; in contrary case, A is taken as the netto area of the annular concrete section.

When in the bearing stratum below the pile tip there is a heavily fissured rock, or soil insertions, in all situations is compulsory to check the bearing capacity by static loads on test piles.

b. The bearing capacity in compression of a precast friction pile

The following relationship is used :

$$R = k\,(m_1\, p_v\, A + U\, \Sigma\, m_2\, f_i\, l_i) \quad \text{(kN)} \qquad (6)$$

where k is a partial factor of safety taken as 0,7

m_1 partial factor of safety given in table 5 ;
m_2 partial factor of safety given in table 5 ;
A has the same significance as in relationship (1) ;
U perimeter of the pile transverse section, in m ;
p_v conventional resistance of the soil below the pile tip, given in table 6, in kP_a ;
f_i conventional resistance on the lateral surface of the pile corresponding to the "i" layer, according to table 7, in kP_a ;
l_i length of the pile in contact with the "i" layer, in m.

c. The bearing capacity in compression of a friction cast in situ pile

The following relationship is used :

$$R = (m_3\, p_v\, A + U\, \Sigma\, m_4\, f_i\, l_i) \quad \text{(kN)} \qquad (7)$$

Table 5

Type of pile from the execution standpoint	m	m
Driven piles	1,0	1,0
Jetted piles in sandy soils, provided driving is performed on the last metter without jetting	1,0	0,6
Piles introduced by driving in :		
a) sandy saturated soils, medium dense :		
- large and medium	1,2	1,0
- fine	1,1	1,0
- silty	1,0	1,0
b) clayey soils having consistency $0,5 < I_c \leq 1$:		
- sandy silts	0,9	0,9
- sandy clays or silty clays	0,8	0,9
- clays	0,7	0,9
c) clays with consistency index $I_c > 1$	1,0	1,0

Table 6

Depth of penetration m	Noncohesive soils				Silty sand	Cohesive soils with I_c :						
	Gravel	Sands: Large	Sands: Medium	Sands: Fine		≥1,0	0,9	0,8	0,7	0,6	0,5	0,4
	p_v kPa											
3	7500	6500	2900	1800	1200	7000	4000	3000	2000	1200	1000	600
4	8300	6600	3000	1900	1250	8300	5100	3800	2500	1600	1200	700
5	8800	6700	3100	2000	1300	8800	6200	4000	2800	2000	1300	800
7	9700	6900	3300	2200	1400	9700	6900	4300	3300	2200	1400	850
10	10500	7300	3500	2400	1500	10500	7300	5000	3500	2400	1500	900
15	11700	7500	4000	2800	1600	11700	7500	5600	4000	2800	1600	1000
20	12600	8200	4500	3100	1700	12600	8200	6200	4500	3100	1700	1100
25	13400	8800	5000	3400	1800	13400	8800	6800	5000	3400	1800	1200
30	14200	9400	5500	3700	1900	14200	9400	7400	5500	3700	1900	1300
35	15000	10000	6000	4000	2000	15000	10000	8000	6000	4000	2000	1400

Notes to table 6

1. The penetration depth of the pile is measured from the level of the natural ground to the level of the pile tip, when the fill or the cuts to be made are not exceeding 3 m. When fills or cuts are larger than 3 m, the penetration depth is measured from a level above or, respectively below 3 m from the natural ground level.

2. Tabulated values of p_v can be used provided the pile penetrates in the stable soil (which is not ubjected to erosion or slides) at least 4 m in the case of the foundations of bridges and hydrotechnic works and at least 3 m in the case of other constructions.

3. In the case of non-cohesive soils, tabulated values of p_v are valid for dense or medium dense soils ($D_R > 0,33$).

4. In the case of large sands and gravels, tabulated values of p_v can be used only when the relative embedment of the tip of the pile in the layer is $t/d \geq 15$. For values of $t/d < 15$, a corrected design resistance should be computed using the relationship :

$p_{v\,corr} = p_v\,(0,7 + 0,02\ t/d)$ (kP_a)

where t is the embedment depth of the pile tip in the layer of large sands or gravels, in m ; d diameter of the pile in the base plane, in m.

5. In the case of sands (except large sands mentioned at p.4) and of cohesive soils, tabulated values of p_v can be used only when the relative penetration of the pile tip $t/d \geq 4$. For t/d values 4 less than the corrected resistance is computed with the relationship:

$$p_{v\,corr} = p_v\,(0.5 + 0.125\ t/d) \quad (kP_a)$$

6. For intermediate values of the depth or of the consistency index I_c, values of p_v should be obtained by linear interpolation.

Table 7

Depth to the middle of the layer m	Noncohesive soils			Cohesive soils with I_c :					
	large and medium	fine	silty	≥0,8	0,7	0,6	0,5	0,4	0,3
	f, kP_a								
1	35	23	15	35	23	15	12	5	2
2	42	30	20	42	30	20	17	7	3
3	48	35	25	48	35	25	20	8	4
4	53	38	27	53	38	27	22	9	5
5	56	40	29	56	40	29	24	10	6
7	60	43	32	60	43	32	25	11	7
10	65	46	34	65	46	34	26	12	8
15	72	51	38	72	51	38	28	14	10
20	79	56	41	79	56	41	30	16	12
25	86	61	44	86	61	44	32	18	-
30	93	66	47	93	66	47	34	20	-
35	100	70	50	100	71	50	36	22	-

Notes to table 7

1. Values of f are taken for the average depths, corresponding to the distance between the middle of the "i", layer to ground surface, taking into consideration the note 1 to the table 6. In case of layers with thicknesses greater than 2 m, the establishment of the f values is made by dividing in sub-layers of maximum 2 m.

2. For intermediate values of the depths or of the consistency index I_c , f is obtained by linear interpolation.

3. If within the limits of the pile penetration there is a layer of very compressible soil, of low consistency (peat, mud etc.) of at least 30 cm thickness and the ground surface is going to be loaded (because of the vertical planning or other reasons), values of f for the highly compressible layer and for the layers located above, are established as follows :

- when the surcharge does not exceed 30 kP_a, for all the layers located above the lower limit of the highly compressible layer (including fills) f is taken 0 ;

- when the surcharge is between 30 and 80 kP_a, for layers located above the highly compressible layer (including fills) the tabulated value f is multiplied by 0.4 and taken with negative sign while for the highly compressible layer f = - 5 kP_a ;

- when the surcharge exceeds 80 kP_a for layers located above the highly compressible layer the tabulated values of f are taken with negative sign, while for the highly compressible layer f = - 5 kP_a.

4. If the pile penetrates through recent fills, clay layers under consolidation or collapsible soils sensitive to wetting, with thicknesses exceeding 5 m, the tabulated values of f are taken with negative sign.

Table 8

Procedure used for concreting the pile	Type of soil at the pile base	
	cohesive	non-cohesive
	m_3	
Concreting in dry	1,0	1,0
Concreting under water :		
- with base grouting	0,9	1,0
- without base grouting	0,8	0,9
Concreting under slurry :		
- with base grouting	0,8	0,9
- without base grouting	0,6	0,8

Table 9

Procedure used for the pile construction	Type of soil around the pile	
	cohesive	non-cohesive
	m_4	
Pile with the casing introduced by driving and the concrete compacted by driving	1,0	1,0
Pile with the casing introduced by vibration and the concrete compacted by vibration (at the extraction of the casing)	0,7	0,6
Pile bored in dry, uncaused	0,6	0,7
Pile bored under slurry	0,5	0,6
Pile bored with recoverable casing	0,6	0,7
Pile bored with non-recoverable casing	0,6	0,8

where k, A, U, f_i, l_i have the same significance as in relationship (2) ;

m_3 partial factor of safety, depending on the procedure used for concreting the pile, given in table 8 ;

m_4 partial factor of safety, depending on the procedure for the construction of the pile, given in table 9 ;

p_v design resistance of the soil at the pile base, in kP_a.

For displacement piles (constructed by driving or by vibration) the value of p_v is taken from table 6.

For non-displacement piles with the base on cohesive soils, the value of p_v is computed with the relationship (4), provided the pile penetrates in the cohesive layer on a depth of minimum the diameter of the pile or of the base :

$$p_v = N_c\, c_u + \gamma_1 D \quad (kP_a) \qquad (8)$$

where N_c bearing capacity factor taken as 9 ;

c_u design value of the undrained cohesion, in kP_a ;

γ_1 the average weighted value, through the thicknesses of the layers, of the design values of the unit weights of the layers penetrated by the pile, in kN/cum ;

D the real embedment of the pile (depth at which the pile base is located, measured from the ground surface or, in case of bridge foundations, from the level of the river bed, considering the erosion depth) in m.

When data concerning the shear strength of the layer at the pile base are missing, it is allowed, for cohesive soils, to use the values p_v from table 10.

Table 10

Depth of the pile base, m	I_c						
	≥ 1	0,9	0,8	0,7	0,6	0,5	0,4
	p_v, kP_a						
3	850	750	650	500	400	300	250
5	1000	850	750	650	500	400	350
7	1150	1000	850	750	600	500	450
10	1350	1200	1050	950	800	700	600
12	1550	1400	1250	1100	950	800	700
15	1800	1650	1500	1300	1100	1000	800
18	2100	1900	1700	1500	1300	1150	950
20	2300	2100	1900	1650	1450	1250	1050
30	3300	3000	2600	2300	2000	-	-
40	4500	4000	3500	3000	2500	-	-

Notes to table 10

1. For intermediate values of consistency index I_c and of the depth, p_v values are established by linear interpolation .

2. In case of a cohesive soil with e > 0,5, corrected values of p_v have to be used, established with the relationship :

$$p_{v\,corr} = p_v \left(1 - 0{,}5\,\frac{e - 0{,}5}{0{,}6}\right)$$

For non-displacement piles with the base on non-cohesive soils, p_v is computed with the relationship :

$$p_v = \alpha\,(\gamma\, d_b\, N_\gamma + \gamma_1\, D_c\, N_q)\quad (kP_a) \tag{9}$$

where α coefficient given in table 11, depending on the relative density D_R of the soil at the pile base ;

γ design value of the unit weight of the soil below the pile base, in kN/cm;

γ_1 the average weighted value, by the thicknesses of layers penetrated by the pile, of the design values of the unit weight of the layers penetrated by the pile, in kN/cu m ;

D_c design embedment of the pile, in m, to be established as follows :

- if $D \geq \beta d$, then $D_c = \beta d_b$;
- if $D < \beta d_b$, then $D_c = D$.

D real embedment of the pile having the same significance as in relationship (4) ;

N_γ, N_q bearing capacity factors, given in table 12 in function of the design value of the angle of internal friction ϕ' of the soil at the pile base.

Note.

When above the non-cohesive layer in which the pile penetrates there is a recent, uncompacted fill, a layer of soft or very soft cohesive soil or a layer of peat, as real embedment D is considered the depth of penetration in the bearing layer, and to the expression p_v defined by (12) the term γ_2 h is added,where :

γ_2 design value of the unit weight of the weak layer, in kN/cu m ;

h thickness of the weak layer, in m.

Table 11

I_D	α	β
0,00...0,35	0,5	10
0,34...0,66	0,4	15
0,67...1,00	0,3	20

Table 12

ϕ'	26^0	28^0	30^0	32^0	34^0	36^0	38^0	40^0
N_γ	9,5	12,6	17,3	24,4	34,6	48,6	71,3	108,0
N_q	18,6	24,8	32,8	45,5	64,0	87,6	127,0	185,0

d. Bearing capacity of a compression pile in group

The following relationship is used :

$$R_g = m_u\ R\quad [kN] \tag{10}$$

where R bearing capacity of the single pile (outside the group), in kN ;

m_u partial safety factor to consider the group effect, which is established as follows :

- for end bearing piles $m_u = 1$;

- for foundations with embedded cap and friction piles $m_u = 1$;
- for foundations with cap above the ground level and friction piles, m_u is established according to the table 13, except the case of displacement piles fully embedded in non-cohesive soils for which $m_u = 1$.

Table 13

r/r_0	≥ 2	1,8	1,6	1,4	1,2	1,0	0,8
m_u	1,00	0,95	0,90	0,85	0,80	0,70	0,60

r minimum distance between neighbour sides of neighbour piles, in m ;

r_0 influence radius of the single pile, at the base level, in m :

$r_0 = \Sigma\, l_i \tan \varepsilon_i$

l_i thickness of the "i" layer penetrated by the pile, in m

$$\varepsilon_i = \frac{\phi_i}{4}$$

ϕ_i design value of the angle of internal friction of the "i" layer.

Notes

1. Values below unit from table 13 can be increased to a maximum of $m_u = 1$ when the computed probable settlement of the pile foundation is within allowable limits for the given construction

In layers where negative skin friction is considered (see notes 3 and 4 for table 7) is taken $\varepsilon = 0$.

e. Bearing capacity of a tension pile

The following relationship is used :

$$R_{sm} = 0{,}6\, k\, U\, \Sigma\, m\, f_i\, l_i \quad [\text{kN}] \tag{11}$$

where k, U, l_i meanings as in (2)

f_i according to table 7, which in all cases is taken with positive values ;

m partial factor of safety, equal to m_2 from relationship (6) in case of precast piles and to m_4 from relationship (7) in case of cast in situ piles.

5.4 Rules for serviceability

As shown, the condition (2) should be checked for the serviceability limit state.

For the orientation of the designers the Romanian Standard 3300/2-85 gives recommended values of the displacements or deformations allowable for structures which are not specially adapted for large or uneven settlements. These values can be considered regardless the type of foundation. Of course, is up to the designer to establish the value of the allowable deformation in order to avoid the occurrence of a limit state in the structure.

6. PARTICULAR NATIONAL EXPERIENCE

The most relevant experience related to the use of pile foundations in Romania is represented by the acquisition over the years of a large volume of data from static load tests. This is a direct consequence of the provision of the national code requiring the load test in the final phase of the design.

A very impressive test was performed on a 1.97 m diameter bored piles, with left in place steel casing, on the site of a new bridge over the Danube (Manoliu, Teodorescu, Tranca 1994). Under a maximum direct load of 41,000 kN the settlement was of only 21 mm. In the opinion of the author' s report, this was the largest axial load ever reached during the test of a pile.

7. REFERENCES

STAS 2561/1-83 Piloţi. Clasificare şi terminologie (in Romanian)
Piles. Classifications and terminology

STAS 2561/2-81 Fundaţii pe piloţi. Incercarea în teren a piloţilor de probă şi din fundaţii (in Romanian)
Pile foundations. Field loading on test piles and on piles in foundations

STAS 2561/3-90 Piloţi. Prescripţii generale de proiectare (in Romanian)
Piles. General design provisions

STAS 2561/4-90 Piloţi foraţi de diametru mare. Prescripţii generale de proiectare, execuţie şi recepţie (in Romanian)
Large diameter bored piles. General design, execution and reception provisions.

STAS 7484 - 74 Elemente prefabricate din beton armat şi beton precomprimat. Piloţi (in Romanian)
Precast reinforced concrete and pretensed concrete elements. Piles

P 106 - 85 Instrucţiuni tehnice pentru proiectarea şi executarea baretelor pentru fundarea construcţiilor (in Romanian)
Technical instructions for the design and execution of barettes for foundations

C 160 - 75 Normativ privind alcătuirea şi executarea piloţilor pentru fundaţii (in Rmanian)
Code for the construction and execution of piles for foundations

STAS 3300/1-85 Teren de fundare. Principii generale de calcul (in Romanian)
Foundation ground. General principles for design

STAS 3300/2-85 Teren de fundare. Calculul terenului de fundare în cazul fundării directe (in Romanian)
Foundation ground. Geotechnical design for shallow foundations.

Manoliu, I., Teodorescu, D., Trancă, M., 1994 Large bored piles for the new Danubian bridges in Romania Proceedings XIII ICSMFE, New Delhi

Marchidanu, E., Antonescu, I., 1995 Some Engineering Geology features of the Romanian territory, Geotechnical Engineering in Romania ed. Romanian Geotechnical Society

Design of Axially Loaded Piles – European Practice, De Cock & Legrand (eds)
© 1997 Balkema, Rotterdam, ISBN 90 5410 873 8

Calculation of pile foundations on limiting states – Russian practice

A.A. Bartolomey, I.M. Omeltchak & A.B. Ponomaryov
Perm State Technical University
B.V. Bakholdin
N.M. Gersevanov All-Union Scientific Research Institute of Foundations and Underground Constructions, Moscow

Abstract: here the methods of calculation of pile foundations bearing capacity and settlements are given. Methods of pile foundations projection proceeding from permitted settlements of buildings and constructions are described here.

1 CALCULATION OF THE PILE FOUNDATIONS BEARING CAPACITY

All-Russia standards of pile foundations projecting (Building Standards and Rules - BS&R - 2.02.03-85 (1)) currently in force nowadays, are the forth remade and expanded reissue of the projecting standards of 1962. Since that time they have being used in all regions of Russia. That is why these standards are considered as the well-tested by a 35-year practice of their successful use by all building organizations. Calculation of pile foundations on their bearing capacity is the important stage of projection. According to the requirements of BS&R 2.02.03-85 (1), the number of piles in foundation (and in the most cases their length) is defined on the basis of the piles bearing capacity.

Calculated load which may be applied to a pile, is defined (according to the standards) by dividing of the pile bearing capacity meaning *Fd* into the safety coefficient which is equal to 1,4 in calculations on the table standard calculated meanings of ground resistance, and 1,2-1,25 when calculating on the data of static sounding or static and dynamic pile testing carried out with measuring of residual and elastic refusal in the piles immersion.

According to the BS&R, ground carrying capacity is considered as calculated on the data of static testing conventional limiting resistance of the ground foundation of a pile. This resistance formally corresponds to the load when the pile settling (in case of its immersing until stabilization of the settling, equivalent to 0,1 mm per 1 hour) is equal to 0,2 of the meaning of permitted settling for a building or construction.

These methods of the piles carrying capacity calculation are accepted because settlements registered during static testing of the piles, in fact, are considered as the settlements caused by the ground deformations which are the consequence of development of the local but rather fast areas of limiting balance near the pile upper and lower parts. Because of the short-term tests, there are no any evident deformations of ground in other parts of pile foundation.

In N.M. Gersevanov Scientific Research Institute of Foundations and Underground Constructions on the basis of experimental researches results (some of them are given on the figures 1 and 2), the methods of the pile carrying capacity calculation were worked out. Development of the limiting balance areas in which a total pressing-out of the soil from under the piles may take place, was taken into account. These methods also imply acception of the thesis about substitution of the principle of indissolubility of the soil environment for the principle of the environment mass preservation, they

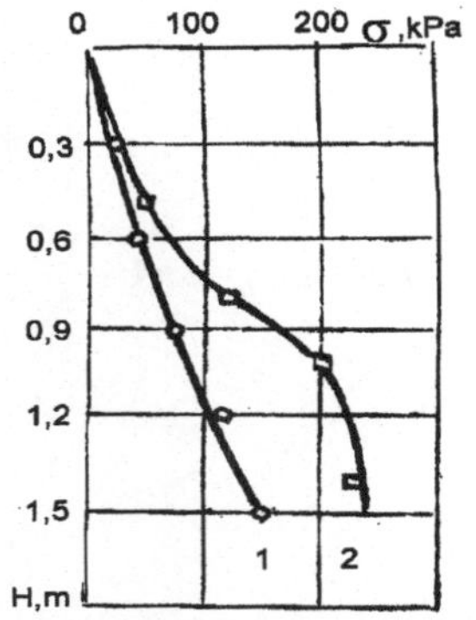

Figure1.Fragment of the results of the experimental research of the stress distribution on the surface of the driven pile, immersed by the depth of 1,7 m
1 - before the pile loading
2 - after the pile loading.

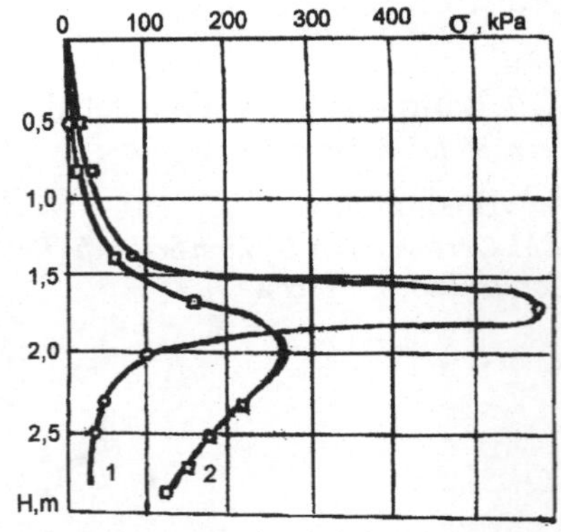

Figure 2.Fragment of the results of the experimental research of the stress distribuion in the soil around the experimental pile
1- vertical stress
2 - horizontal stress

assume the possibility of the soil compression with its internal out-pressing. Stated above methods of calculation make possible to do mathematical description of the soil limiting resistance under the lower part of the piles and on their side surface with dependence on durability and deformation characteristics of soil concerning the interpretation of resistance used in the BS&R 2.025.03-85 (1). Methods of calculation were elaborated for dusty-clayey and sandy soils as well. As for the sandy soil, these methods allow to get meanings of the soil limiting resistance under the lower parts of the driven piles R and on the side surface f in a quite simple form:

$$R=\frac{1}{\varsigma}\left\{\frac{9}{2}\cdot\frac{\gamma h}{\varsigma}\left[\frac{\xi Su\,k\frac{E}{1+\beta}}{\xi Su+2\left(1-\frac{E}{Eo}\right)ro}\right]^{2}\right\}^{\frac{1}{3}}+U\sum fi\,hi-\frac{\gamma h}{2\varsigma} \tag{1}$$

$$f=\frac{tg\varphi}{2}\left[\frac{9}{2}\left(\frac{kEp}{1+\beta}\right)^{2}\gamma h\right]^{\frac{1}{3}} \tag{2}$$

$$\varsigma=tg\left(45-\frac{\varphi}{2}\right) \tag{3}$$

where: $\gamma =$ the specific weight of soil (in case of its weighing under the level of underground waters);

h = the depth of the soil layer; Su = the limiting settling for a building or construction; ξ= the coefficient taking into account the degree of permissibility of the settling development by means of internal soil pressing-out from under the foot of limiting balance, it is equal to 0,2 (according to the BS&R 2.02.03-85 (1)); k = the coefficient taking into account the technology of the pile immersion (pressing, driving-in, vibroimmersion , etc.); β = the nondimensional coefficient, equal to 0,8 (according to the Building Standards and Rules 2.02.01-85 (1)); E = the modulus of the soil deformation; Eo = the instantaneous modulus of the soil deformation; ro = the radius (half of the side) of the cross section of a pile; U = the perimeter of the pile trunk; f_i is the resistance of the soil on the side surface of a pile; h_i = the thickness of the first soil layer within the limits of the side surface of a pile; φ = the angle of the internal friction of the soil; Ep = the unloading modulus of the sands deformation, for the sands of average and high density it is equal to 70 t/sq.m, for the sands of low density - 60 t/sq.m, for the dusty sands - 50 t/sq.m.

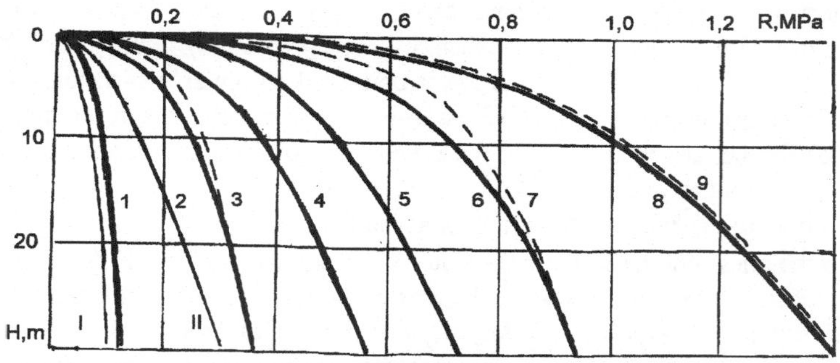

Figure 3.Calculated resistance of the soil under the lower part of the driven and bore piles, with dependence on the depth H of the pile immersion:
a) of the driven piles: - in dusty-clayey soil with the fluidity indicators I_L equal to: 1 - 0,6; 2 - 0,5; 3 - 0,4; 4 - 0,3; 5 - 0,2; 6 - 0,1; 8 - 0; - in the sands of the average density: 2 - fine sand, 7 - large sand, 9 - gravel sand ;
b) of the bore piles in dusty-clayey soil with the fluidity indicator I_L, equal to: I - 0,6; II - 0,0.

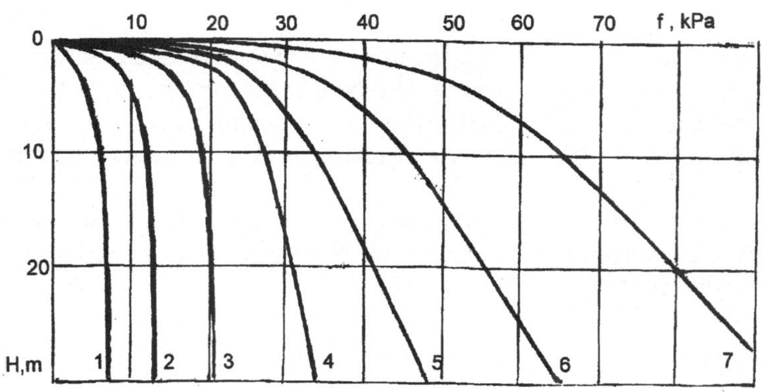

Figure 4.Calculated resistance of the soil on the side surface with the dependence on the depth H of the layer: - in dusty-clayey soil with the fluidity indicators I_L, equal to: 1 - 1,0; 2 - 0,7; 3 - 0,6; 4 -0,5; 5 - 0,4; 6 - 0,3; 7 - 0; - in the sands of the average density: 5 - dusty, 6 - fine, 7 - of the average size.

Although the dependence (1) and (2) coincide with the consequences observed in nature and accepted in the BS&R 2.02.03-85 (1), these consequences are not widely used in projecting practice as well as all other theoretical solutions of the considered problem. In fact, projection of pile foundations in Russia is carried out with the use of general experimental meanings of the piles foundations soil resistance. Fixed by BS&R 2.02.03 (1) calculated meanings of the ground resistance under the lower parts of the piles R and on the side surface f with account of granulometric structure of the soil, its consistence and depth of bedding are given on the figures 3 and 4, which are made according to the table data of these standards. When accepting durability and deformation characteristics of the soil, according to BS&R 2.02.01-83 (2), these calculated meanings R and f coincide with the meanings calculated on the formulas (1) and (2).

With use of the fixed meanings of the soil resistance, carrying capacity of the driven piles Fd is calculated (according to the Building Standards &Rules 2.02.03-85) on the following formula:

$$F_d = \gamma_c \left(\gamma_{cr} R A + U \sum \gamma_{cf} h_i f_i \right) \qquad (4)$$

where: $\gamma_c =$ the coefficient of the pile work conditions, equal to 1, with the exception of dusty-clayey soils with the degree of humidity $Sr < 0{,}9$, when $\gamma_c =$ equal to 0,8; γ_{cr} and γ_{cf} are the coefficients of the ground work conditions under the lower parts and on the side surface of the piles, taking into account the influence of the way of the driven piles immersion and the driven piles structure. These coefficients are equal to: - for the driven piles (in case of their immersing by hammers) $\gamma_{cr} = \gamma_{cf} = 1{,}0$; for vibro-immersed piles $\gamma_{cr} = 0{,}7{:}1{,}0$ and $\gamma_{cf} = 0{,}9{:}1{,}0$ depending on the kind of soil; - for the bore piles, concreted without water in the bore hole (dry way) $\gamma_{cr} = 1{,}0$ and $\gamma_{cf} = 0{,}6{:}0{,}7$, and for the piles concreted under the water $\gamma_{cr} = 0{,}9$ and $\gamma_{cf} = 0{,}8$; R and f is the calculated resistance of soil under the lower and side parts of the piles, equal to the fixed table meanings of the Russian Building Standards & Rules, given on the figures 3 and 4; $A =$ the area of leaning of the lower pile parts on the soil; $U =$ the perimeter of the cross section of the pile trunk; $h_i =$ the thickness of the i layer of the soil contacting with the sides of the pile.

During the last ten-year period for calculation of the driven piles carrying capacity, the results of static sounding of the soils are also used. In this case, according to the Russian standards (Building Standards & Rules 2.02.03-85 (1)), piles carrying capacity is recommended to calculate on the following formula:

$$F_d = \frac{\gamma_c \sum_{1}^{n} F_U}{n \gamma_g} \qquad (5)$$

where: $\gamma_c =$ the coefficient of the work conditions equal to1; $Fu =$ the special significance of the limiting resistance of the pile in the point of sounding, calculated on the formulas (6) or (7); $n =$ the number of the sounded points; $\gamma_g =$ the safety coefficient of the soil calculated in accordance with variability of specific meanings of limiting resistance of the pile Fu in the points of sounding with probability $\alpha = 0{,}95$.

Specific limiting resistance of the driven pile in the point of sounding Fu is calculated on the formulas: - with the use of the sounder of the first type (which measures a general resistance of the soil on the sides of the sounder)

$$F_u = \beta_1 q_s A + U \sum \beta_i f_{si} h_i \qquad (7)$$

-with the use of the sounder of the second type (which measures soil resistance of separate layers on the sides of the sounder)

$$F_u=\beta_1 q_s A+\beta_2 f_3 hU \qquad (6)$$

where: β_1, β_2, βi are the transitional coefficients defined on the table 1; q_s = the average meaning of the soil resistance under the tip of the sounder, one pile diameter higher and 4 diameter lower than the mark of the foot of the projected pile; A = the area of leaning of the lower part of the pile; f_3 = the average meaning of the soil resistance on the sides of a sounder; h = the depth of the pile immersion; U = the perimeter of the cross section of the pile trunk; fs_1 = the average resistance of the layer "i" on the sides of a sounder; hi = the thickness of the soil layer i.

Table 1.Determination coefficients : β_1, β_2, βi

	β_1 - coefficient of transition from q_s to R			fs,	β_2 - coefficient of transition from f_s to f for the sounder of the first type		β_i - coefficient of transition from f_{si} to f for the sounder of the second type	
q_s, kPa	for the driven piles	for the spiral piles with the load		fsi, kPa				
		compressing	pulling ant		in sandy soil	in dusty -clayey soil	in sandy soil	in dusty -clayey soil
<1000	0,90	0,50	0,40	>20	2,40	1,50	0,75	1,00
2500	0,80	0,45	0,38	40	1,65	1,00	0,60	0,75
5000	0,65	0,32	0,27	60	1,20	0,75	0,55	0,60
7500	0,55	0,26	0,22	80	1,00	0,60	0,50	0,45
10000	0,45	0,23	0,19	100	0,85	0,50	0,45	0,40
15000	0,35	-	-	>120	0,75	0,40	0,40	0,30
20000	0,30	-	-	-	-	-	-	-
>30000	0,20	-	-	-	-	-	-	-

When calculating the driven piles carrying capacity, it is recommended (according to the BS&R 2.02.03-85 (1)) to use the data of dynamic tests, for the bore piles - the data of static tests held in conditions of a building ground.

In dynamic testing, according to stated standards and registered data of immersion of the driven piles, the limiting resistance of the ground foundation Fu is estimated on the formula :

$$F_u=\frac{\eta A}{2}\left[\left(1+\frac{4E_d}{\eta A S_a}\cdot\frac{m_1+\varepsilon^2(m_2+m_3)}{m_1+m_2+m_3}\right)^{1/2}-1\right] \qquad (8)$$

where: η = the coefficient, in case of concrete piles driving it is equal to 1500 kN/sq.m, in case of the wooden piles driving it is equal to 1000 kN/sq.m; A = the area, limited by the outer contour of the solid or hollow cross section of the pile; Ed = the actual calculated energy of the hammer shock or calculated energy of a vibroimmerser calculated on the table 2; Sa = the actual remaining rejection equal to the significance of the pile immersion by one hammer shock; m_1 = the mass of hammer; m_2 = the mass of the pile and its head; m_3 = the mass of the shock part of a hammer; ε^2 is the coefficient of shock restoration, equal to 0,2.

Table 2. Calculated energy of the vibroimmerser shock Ed

Power of the vibroimmerser, kN	100	200	300	400	500	600	700	800
Equivalent calculated energy of the vibroimmerser shock, Ed, kDt	45	90	130	175	220	265	310	350

Indicated formula (8) allows to make exact calculation of carrying capacity of the piles receiving rejections of more than 2 mm by one hammer shock. If rejections are less than 2 mm, it is recommended to carry out dynamic tests of the pile with measuring of elastic rejections, and to make calculations on special formulas:

$$F_u=\frac{1}{2\theta}\cdot\frac{2S_a+S_{el}}{S_a+S_{el}}\cdot\left[\sqrt{1+\frac{8E_d(S_a+S_{el})}{(2S_a+S_{el})^2}\cdot\frac{m_4}{m_2+m_4}\theta}-1\right] \quad (9)$$

$$\theta=\frac{1}{4}\left(\frac{n_p}{A}+\frac{n_f}{A_f}\right)\frac{m_4}{m_4+m_2}\sqrt{2gH} \quad (10)$$

where the following marks are introduced: S_{el} = the elastic rejection of a pile (elastic shift of the soil and pile as a result of one shock); Af = the area of the pile sides contacting with soil; g = the acceleration due to gravity; H = the height of the hammer shock part fall.

Choice of some method of calculation of the piles carrying capacity is usually made (in Russia) by a project organization. But if the results of dynamic testing of the piles are used, the number of tests must be no less than 6, except the static piles tests (no less than 2) with the condition that they are held on the building ground with the most unfavourable soil layers.

2 PROJECTING OF THE PILE FOUNDATIONS ON PERMISSIBLE SETTLEMENTS OF BUILDINGS AND CONSTRUCTIONS

Many experimental researches prove that concentrating deformations take place under the pile foundations. Therefore, in projection of the pile foundations on the permissible states, it is very important to calculate settlements and their accumulation with time. It is explained by the fact that if the distance between the piles is 3-4 d, foundations work as an indivisible massif. Threfore, in order to achieve the limiting state of the soil in the concentrated area under the piles of 5-15 m length and development of the plastic deformations area, much more load is required than for achievement of permissible settling. In this case, the pile and foundations loads are defined proceeding from the limiting permitted loads for the buildings.

On the basis of solution of the plane and area problems, theory of linear-deformed environment, filtration consolidation, phenomenological theory of hereditary creep and non-linear mechanics, the estimation methods of the deformed state of the pile foundations active areas, of settlements of different pile foundations with account of the main principles of their interaction with the surrounding ground, were elaborated (4).

In virtue of these researches, methods of the pile foundations projection proceeding from the limiting settlements of buildings and constructions, were worked out. It was noted that , according to the Russian Building Standards & Rules 2.02.03-85 (1), special significance of limiting resistance of a pile was equal to the load under which the influence of settling is equal:

$$S=\xi Su \tag{11}$$

where: Su = the limiting significance of the average settling of the projected building foundation; ξ = the coefficient of transition from the average limiting settling of the building foundation Su taken into account because of long lasting load, to the pile settling received after the static testing of the piles with the conventional dying-out of the settling.

According to the BS&R 2.02.03-85, the coefficient ξ is equal to 0,2. The standards permit more precise definition of the coefficient ξ on the basis of long lasting piles testing and on the results of observation of settlements of buildings erected on pile foundations in similar soil conditions. However, project organizations do not have nowadays sufficient data of long lasting testing of the piles or results of observation of the buildings settlements, that is why projection of pile foundations is usually held in accordance with the recomendations of Russian Building Standards and Rules.

Analysis of the numerous results of the pile testing proves that in calculated loads on the piles, according to the recomendations, the settlements constitute 2-8 mm. Under the influence of long lasting load, the settlements increase. Besides, in pile foundations there is the piles interaction and their settling will be bigger than the settling of the single piles in usual static tests.

We made researches of the actual settlements of the tape one-row pile foundations which carrying capacity was calculated according to recomendations of the BS&R. It was fixed that settlements were equal to 15-20 mm, i.e. 4-6 times less than the limiting significance of the average settling of the projected building or construction foundation. Therefore, adopted nowadays methods of calculation of carrying capacity of suspension piles give an enormous spare.

During the work of the piles as the parts of foundations, the character of transition of settling to the ground changes to a marked degree. The size of active area and settling are influenced by the distance between the piles, their number, leaning of raft on the ground or by lack of contact of raft with the ground and by other factors.

Settlements of single piles usually increase much in case of loading close to the limiting one. For the pile foundations, smooth increase of settling with the load increase is typical (Figure 5).

As a standard significance of limiting resistance of the piles we suggest taking the load ith which a total stabilized settling is close to the limiting permissible settling of buildings (4). Total stabilized settling is defined on the data of long lasting tests of piles or by calculations.

On the basis of experimental researches and analytic solutions and depending on the initial data, the following methods of the pile foundations projection are suggested.

1.The first method is based on calculation of settlements of the single piles and pile foundations in the course of time according to the methods described in (4). When using this method, it is necessary to have data of usual tests of the piles in analogous soil conditions, and graphs of settlements in time at least of 2 stages of loading. Settlements should be measured in 5, 10, 15, 30 minutes and 1, 2, 4, 8 hours after the loads applying.

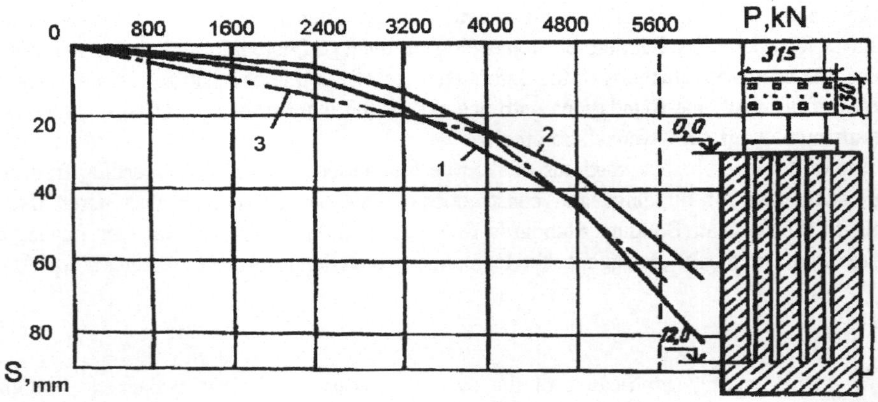

Figure 5.Comparison of calculated and real settlements of foundation from the piles of the section 30x30 cm, 12m long:
1 - in case of lack of raft and ground contact;
2 - in case of the raft leaning on the ground;
3- calculated data.

Formula for calculation of settlements in the course of time has the following form:
- when working in sandy soil:

$$S_t = \frac{1}{B}\operatorname{arctg}\frac{P}{b}[1 + D(1 - e^{-\eta t})] \tag{12}$$

- for the piles working in clayey soil:

$$S_t = \frac{1}{B}\operatorname{arctg}\frac{P}{b}[1 + A_t^{1-\lambda}] \tag{13}$$

- for the pile foundations there are the following formulas:

$$S_t = (1 + k_0)\frac{1}{B}\operatorname{arctg}\frac{P}{b}[1 + D(1 - e^{-\eta t})] \tag{14}$$

$$S_t = (1 + k_0)\frac{1}{B}\operatorname{arctg}\frac{P}{b}[1 + A_t^{1-\lambda}] \tag{15}$$

where: $B = \frac{1}{S_0} arctg \frac{P}{b}$, *So* = the settling measured in 15 minutes after the load *P* applying; *Ko* = the coefficient taking into account the piles interaction. It is defined on the Figure 6.; *b* - (kN) is defined according to the limiting meanings *tg, arctg* and limiting load on the pile P_{pr}.;

$$D = \frac{\frac{b}{P}\operatorname{tg}BS_n - 1}{1 - e^{-\eta t_n}} \qquad (16)$$

$$A = \frac{\frac{b}{P}\operatorname{tg}BS_n - 1}{t_n^{1-\lambda}} \qquad (17)$$

where: S_n = the settling in the moment t_n with the load P. Parameter λ is accepted equal to 0,7.

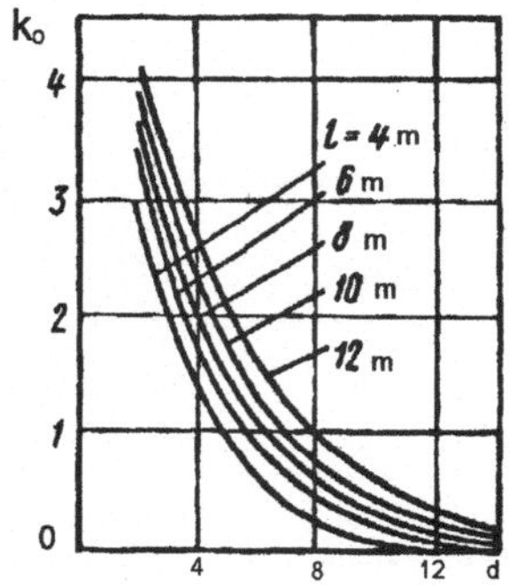

Figure 6.Dependence of coefficient K_o from distance between piles d and length of pile L .

Having defined (on the results of testing of 1 pile with the constant load) the coefficients A (1/time), B (1/cm), D and λ , it is possible to define (on the formulas (12) ,(13)) the settlements of the piles loaded by different loads P in any period of time.

The load when the total stabilized settlement is closed to the limiting settlement for buildings, is equal to the standard meaning of limiting resistance of a pile or a pile foundation. Bearing capacity of a pile is calculated by multiplying of the standard meaning of the pile limiting resistance on the coefficient of work conditions and by division into the safety coefficient on the ground. Standard meanings of the piles limiting resistance may be much higher than the ones received according to the existing now methods of the piles carrying capacity calculation.

On the figure 7 there are the results of research of the piles carrying capacity in case of settlements equal to the limiting ones. The pile with the section 30x30 cm and length of 7 m was tested on 1 of the building grounds of the block No.45 of Perm city. The load was equal 340 kN and stabilized settling was equal to 80 mm. This load may be accepted as a specific meaning of the pile limiting resistance. When using the traditional methods of the piles carrying capacity calculation, limiting resistance was equal to the load of 250 kN. In this case calculated load on the pile is equal to 272 and 200 kN. With the load of 270 kN, according to the tests results, the single pile settling constitutes about 3 cm. In the course of time the settling increases. On the figure 7(b) the graph of the settlement in the course of time, calculated on the formula (12), is given. Total stabilized settling is 4,5 cm. In case of the work of the piles as the parts of pile foundations, the piles interaction takes place. With the piles step 4d and the load on the pile of 270 kN, the settling in the course of time is calculated on the formula (15). Total settling does not exceed the ones permitted for the buildings.

2.When using the second method, the graph "load-settling" for the pile foundations on the results of analytic calculations of settlements on the worked out methods , is made and the pile foundations carrying capacity is defined proceeding from the buildings limiting settlements. When using this method, it is necessary to have results of the pile tests or the results of static sounding for calculation of the modulus of the ground deformation in the pressed area of pile foundations.

To make the graph "settling-load" with account of linearity, one may use approximate methods suggested by V.G. Berezantsev (1970): 1) substitution of the curve-linear part of the graph of settling for the broken line; 2) approximate account of influence of the limiting tense state areas on the size of settlement.

Taking into consideration these suggestions, it is possible to make the following formulas for calculation of settlements of the piles and tape pile foundations in case of their work in sandy soil:

$$S_K = \frac{P_{pr}}{El} W_0 + \frac{(P - P_{pr})}{E_n l} W_0 \tag{18}$$

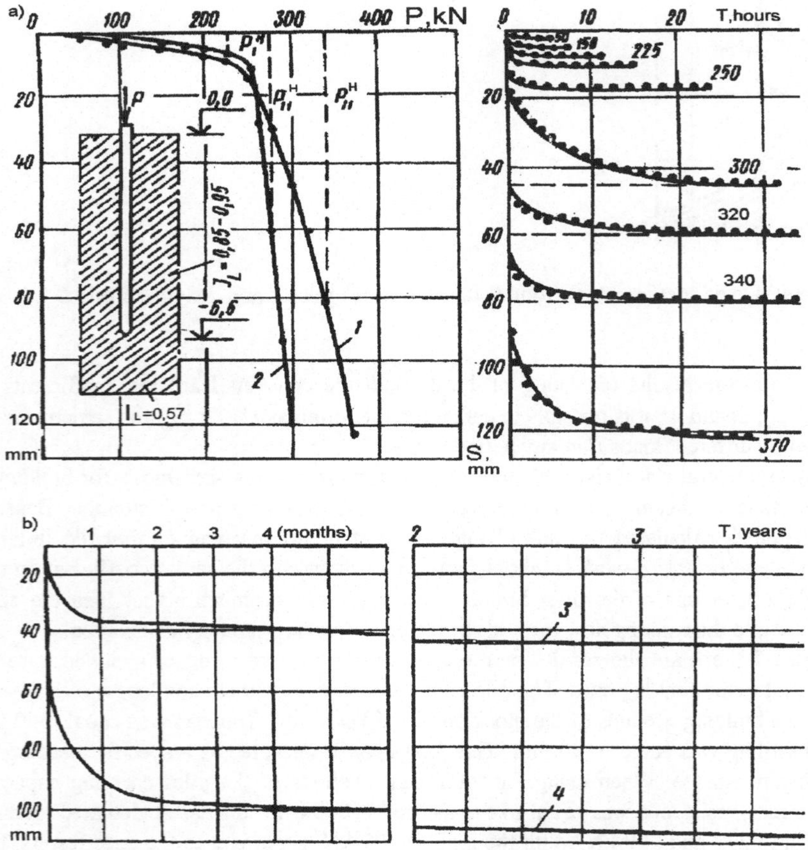

Figure 7. Results of the piles testing in case of the settlements equal to the limiting ones for the buildings (a), and dependence of settlements on the time with the load on the pile of 270 kN (b):
- 1 and 2 - for the pile with the section 30x30 cm, immersed on the depth of 7 and 6,5 m;
- 3 - for the single pile with the load of 270 kN;
- 4 - for the tape foundation with the piles step of 4d.

$$S_l = \frac{P_{pr}}{\pi E_1}\delta_0 + \frac{(P - P_{pr})}{\pi E_n}\delta_0 \qquad (19)$$

With the help of these formulas (18, 19) one can calculate the carrying capacity of the pile foundations proceeding from the permissible settlements of buildings and constructions.

Formulas for description of dependence "settling-load" for the piles and the tape pile foundations with account of influence of the limiting tense state areas on the size of settling and for the cases of foundations work in clayey soil may be presented in the following forms:

$$S_K = \frac{P_{pr}}{El}W_0 + \bar{S}\frac{c_p b}{2E_n} \qquad (20)$$

$$S_l = \frac{P_{pr}}{\pi E_1}\delta_0 + \bar{S}\frac{c_p d_0}{2E_n} \qquad (21)$$

where: S_k and S_l are the settlements of the pile cluster and the tape pile foundation; Ppr = the bearing capacity of the pile foundation within the limits of proportional dependence of settling on the load defined on the following formulas:

- for the piles (N)

$$P_{pr} = \frac{S_{pr.sh}El}{W_0} \qquad (22)$$

- for the tape pile foundations (N/cm)

$$P_{pr} = \frac{S_{pr}\pi E_1}{\delta_0} \qquad (23)$$

where: P = the load on the pile foundation when the settling is defined; Wo and δo are the nondimensional components of transition which are defined on the table (4); l = the length of the pile;

b and do is the width of the row of piles and pile foundation; E = the modulus of the ground deformation within the limits of linear dependence of settling on the load (4); E_n = the modulus of the soil deformation calculated on the formula (4) and on the graph of testing of the pile-stamp in case of different settlements out of the limit of proportionality; $Spr.sh.$ = the settlements when frictions on the sides of a pile are totally mobilized ;for the single piles the limiting transition of settlement with dependence on density of sandy soil and consistence of clayey soil is equal to 2-15 mm, for the pile foundations - 25-40 mm (4) ; S = the nondimensional value defined on the figure 8 with dependence on $\overline{P}$.

$$\bar{P} = \frac{P - P_{pr} - q}{c_p} \qquad (24)$$

where: q = the load on the level of the piles edge; Cp = the calculated significance of the ground cohesion.

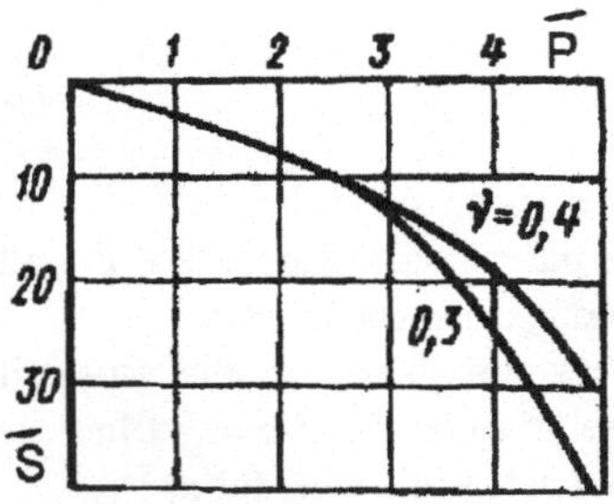

Figure 8. Graph for calculation of the settlements on the data of V.G. Berezantsev

$$c_p = c\frac{ctg\varphi}{ctg\varphi + \varphi - \frac{\pi}{2}} \qquad (25)$$

where: C = the ground cohesion.

For the tape pile foundations the meanings of q and Cp are taken on the unit of length of tape foundation.

Using the formulas (20, 21), one can find the carrying capacity of the pile foundations proceeding from the limiting settling in case of the foundations operation in clayey soil.Calculation is carried out by the following way. One calculates the nondimensional value $\overline{S}$ in case of settling equal to the permitted .For the pile cluster:

$$\bar{S} = \frac{2E_n}{c_p b}(S_u - S_{pr.sh.}) \qquad (26)$$

for the tape pile foundations:

$$\bar{S} = \frac{2E_n}{c_p d_0}(S_u - S_{pr.sh}) \qquad (27)$$

where : Su = the limit deformation of foundation and construction.

Having calculated $\overline{S}$, we can find in the Figure 8. the meaning of P. Load on the pile foundation when the limiting settling will be achieved, may be defined according to the condition:

$$P = \bar{P}c_p + P_{pr} + q \qquad (28)$$

Calculated load permitted on the pile foundation, is equal to

$$N = P / \gamma_K \qquad (29)$$

where: γ_k = the safety coefficient accepted according to the BS&R depending on the way of the piles carrying capacity calculation.

For example: it is required to calculate the carrying capacity of 1 meter of 2-row tape pile foundation, the section of the pile is 30x30 cm, the piles length is 12 m, the distance between the piles is 3d. Physical-mechanical features of the soil are as follows: modulus of the ground deformation in the pressed area $E_1 = 32$; $En = 8{,}7$ MPa, $Spr.sh. = 3$ cm, the angle of internal friction $\varphi = 16°$.

1. First we should find out the width of the pile foundation : $\beta = do/l = 1{,}3 / 12 = 0{,}108$;
2. Then we calculate the depth of active area. If $\beta = 0{,}108$, $\nu = 0{,}3$ g, according to the table (4) we find out that $\delta_2 = 1$ with the depth of $20/1 = 2{,}2$;
3. Now we define the component of transition δo on the monogram on the Figure.9. When $20/1 = 2{,}2$, $\beta = 0{,}108$; $\nu = 0{,}35$, $\delta = 2{,}54$;.
4. Now we should define the bearing capacity of 1-m tape pile foundation within the limits of linear dependence "settling-load":

$$P_{pr} = \frac{S_{pr.sh} \pi E_1}{\delta_0} = \frac{3 \cdot 3{,}14 \cdot 32 \cdot 10^2}{2{,}54} = 11800\ N / cm = 1180\ kN / cm$$

5. Then we should calculate the calculated meaning of cohesion for a unit of the foundation length with account of internal friction influence.

$$c_p = c \frac{\mathrm{ctg}\varphi}{\mathrm{ctg}\varphi + \varphi - \frac{\pi}{2}} = 0{,}023 \cdot 10^2 \frac{2{,}747}{2{,}747 + 0{,}349 - 1{,}571} = 4{,}15\,N / cm$$

6. Now we define the nondimensional value $\overline{S}$ when the settling is equal to the limiting one for the buildings 8 cm.;

$$\overline{S} = \frac{2}{c_p d}\left(S_u E_n - \frac{E_n P_{pr} \delta_0}{\pi E_1}\right) = \frac{2}{4{,}15 \cdot 120}\left(8 \cdot 8{,}7 \cdot 10^2 - \frac{8{,}7 \cdot 10^2 \cdot 11800 \cdot 2{,}54}{3{,}14 \cdot 32 \cdot 10^2}\right) = 17{,}49$$

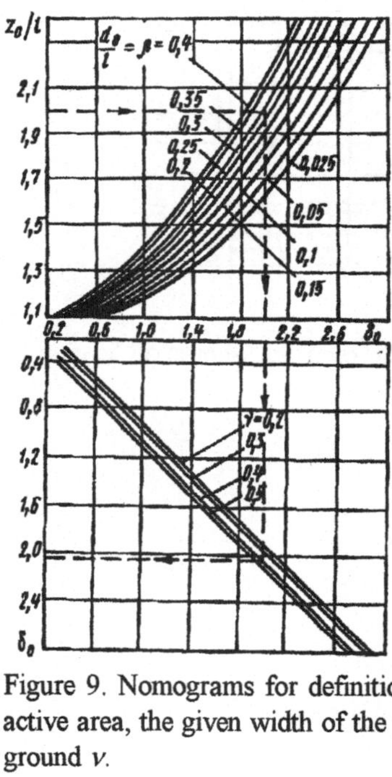

Figure 9. Nomograms for definition of the meanings δo and dependence on the given depth of active area, the given width of the pile foundation and the coefficient of the side widening of the ground ν.

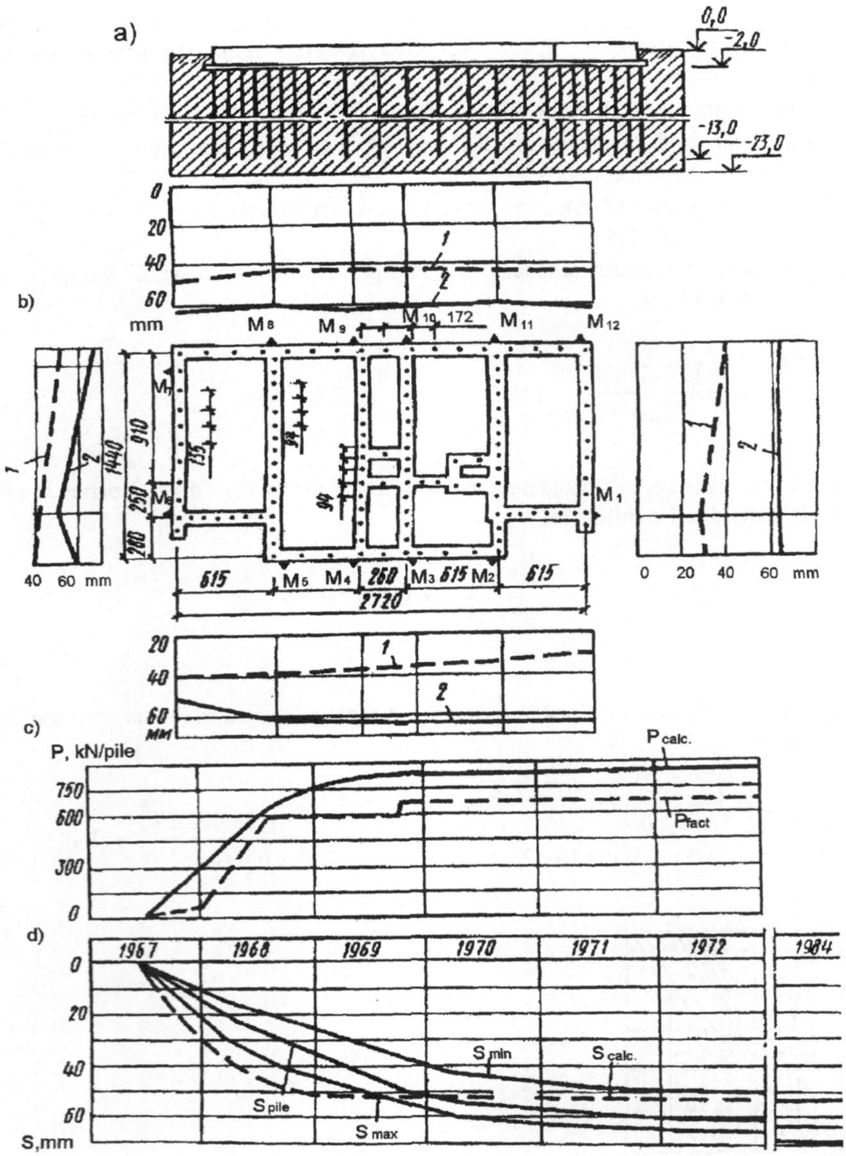

Figure 10. Results of observation of the settlements of the 9-storeyed block-house of the type 1-P-447 on Krupskaya Str., 79:

a) vertical section of the ground;

b) plan of the foundation;

c) diagram of the average speed of the load increase;

d) diagram of the marks settlements with dependence on time;

1 - settlements during the period of building;

2 - settlements during the period of observation.

7. Now we find P on the Figure 8. When $\overline{S}$ = 17,49, $\overline{P}$ = 3,25;

8. We calculate the load on the tape pile foundation when the limiting settling is achieved:

$$P=\overline{P}\,c_p+P_{pr}+q=3{,}75\cdot 4{,}15+11800+2310=1413\ kN/m$$

9. The load on the pile (as the part of foundation) will be equal to:

$$P_{cb}=\frac{PL}{n_p}=\frac{1413\cdot 0.9}{2}=636\ kN$$

10. Now we shall define the calculated carrying capacity of the pile with account of the work conditions coefficient and safety coefficient of the soil:

$$F_d=\frac{\gamma_c P_{cb}}{\gamma_q}=\frac{1\cdot 636}{1}=636\,kN$$

11. According to the results of static testing of the 2-row pile foundation, the limiting resistance of the piles in the structure of foundation is equal to 700 kN.

When projecting pile foundations proceeding from the limiting settlements of buildings and constructions, it is necessary to control the stress on the surface of the pile foundation edge, the voltage should not exceed the resistance of the foundation ground with account of its concentration when driving-in of the piles or the structure of the bore-driven piles (1).

Nowadays on the Department of Foundations and Bridges of the Perm State Technical University the programme "PLAST" (5,6) is elaborated. This programme allows to calculate the settlements of pile foundations with account of changing of the grounds features as the result of the piles driving, layering of the grounds of active area, of variable hardness of the erected building and preparation of the grounds loading.

It was noted, that divergence of calculated and nature settlements constitute 8-12% (Figure 5).The results of research of carrying capacity of the pile foundations in case of the settlements equal to the limiting ones, and description of the methods of projecting were used in building of more than 30 buildings and constructions in the Perm and Tyumen regions. The settlements of the most of the buildings and constructions were carefully observed (during 10-15 years). The experience of projection of the foundations on the limiting deformations proved that pile foundations may be projected more economically by means of increase of the loads or by cutting down of the piles number by 20-30%, and in case of raft and ground contact - by 50%.

For example, in the block No. 1405 on Krupskaya Street of Perm city, for the building of 9-storeyed brick houses of the type "1-R-447C-25/65", according to the RBS&R, 2-row pile foundations (with the pile section of 30x30 cm, 11 m length and with the load 400 kN) were projected. Building grounds were covered by a thick (23 m) layer of clay of the soft-plastic consistence. Experimental researches of the elements of pile foundations with the high and low raft and pile foundation projection proceeding from the limiting settlements of buildings showed the possibility of substitution of the 2-row pile foundations for the 1-row ones (Figure 10).

Tensometric rings for measuring of actual loads were put on some piles. Calculated load on the pile was equal to 800 kN, standard load - 666 kN, actual load - 640 kN.

In order to get full information about the settlings of this building, calculation of the pile foundations settlements with account of the piles internal rows interaction, was made. We also organized precise geodetic observation of deformations.

During the period of building, the average settling constituted 42 mm, during the whole period of observation - 64 mm (Figure 10).

Due to these examples, it is clear that definition of the piles carrying capacity proceeding from the limiting settling , is possible. Absolute dimensions, the difference and relative bends of real settlements of the buildings do not exceed those of the limiting settlements (4).

REFERENCES

1. Russian Building Standards and Rules 2.02.03-85. Pile Foundations. Official edition. State Committee of the USSR on the Building Affairs. Moscow, 1996.
2. Russian Building Standards and Rules 2.02.01-83. Foundations of Buildings and Constructions. Official edition. State Committee of the USSR on the Building Affairs. Moscow, 1985.
3. Bakholdin B.V. Calculation of the Piles Settlings with Account of Development of the Areas of Limiting Balance. Vol. No.84, Scientific Research Institute of Foundations. Moscow, 1985.
4. Bartolomey A.A., Omeltchak I.M., Yushkov B.S. Principles of Prediction of the Pile Foundations Settlings. Stroyizdat, Moscow, 1994.
5. Bartolomey A.A. Principles of Prediction of the Pile Foundations settlings. Magazine "Grounds, Foundations and Ground Mechanics", Stroyizdat, Moscow, 1995, No.3.
6. Bartolomey L.A. Solution of Geomechanical problems with the Help of the Programme "PLAST"// Works of the Ivth International Conference on the Problems of Pile Foundation Building. Part I - Perm: Perm State Technical University, 1994.

Design of Axially Loaded Piles – European Practice, De Cock & Legrand (eds)
© 1997 Balkema, Rotterdam, ISBN 90 5410 873 8

Design of axially loaded piles – Swedish practice

Lennart Svensson
Jacobson & Widmark AB
Connie Olsson
Swedish Geotechnical Institute
Carl-John Grävare
Pålanalys AB

1. Regional geology

The regional geology of Sweden is characterised by landforms caused by continental ice sheets that withdrew about 13.000 to 8.000 years ago. Large areas were isostatic subsided and later flooded by the sea when the ice melted. These areas are now farmland with fine-grained sediments deposited in water. The bedrock in Sweden is mostly of crystalline types. Sedimentary bedrock is only found in the most southern part (Scania), in some places in the middle parts and in the mountains along the border to Norway. Because of these conditions the dominated soil deposited by the ice is till. A layer of till covers most bedrock. In areas of or near to sedimentary rock the deposit is a clay till.
Fine-grained soils consisting of clays can be found along the coasts and along the shores of the large lakes. On the east coast the thickness of the clay layers is seldom more than 15 metres while thicknesses of 50 to 100 metres can be found on the west coast in the Gothenburg area. The clay is mostly soft and of glacial origin. The upper part of the layers is very soft and postglacial. The most upper part often has a high organic content. The clay till in the south-western part of Scania often has a thickness of 15 to 40 metres.
In the river valleys, especially in the northern half of Sweden, thick deposits of river sediments of sand and silt can be found. Large delta deposits can also be found near the highest prehistoric coastlines. The thicknesses of these river and delta sediments can be up to 50 metres. Also washed deposits of fine-grained soils with thicknesses of 20 metres can be found along hillsides.
In areas of former lakes organic soils of peat can be found. The thickness of theses deposits is seldom more than 5 metres.

2. Common practice for soil investigation

The objective of soil investigations for piling projects is to establish a reliable picture of soil strata to decide upon pile type and pile length. With the results also the risk of buckling and dragdown should be handled.
With regard to the geological conditions in Sweden the soil investigations mainly consist of penetration testing to determine depth to till layers or bedrock beneath soft fine-grained soils. The most common method is dynamic probing type HfA standardised by the Swedish Geotechnical Society. The method reminds of SPT but without preboring and soil sampling. In many cases penetration tests are carried out with hydraulic hammer or rock drilling equipment to determine the bedrock surface.
The soft soils are investigated nowadays mainly by cone penetration testing (CPT). The strength parameters are determined by field vane tests and the consolidation conditions are determined by tests in a CRS oedometer on undisturbed soil samples. For end-bearing piles the soft soil is investigated to get design parameters to calculate lateral support and negative skin friction or downdrag. For friction piles results from CPT and field vane tests are the basis for design of bearing capacity.
The Swedish Geotechnical Society has recently (1993) published reports in English of recommended standards for Cone Penetration Testing and the Field Vane Shear Test.

3. Piling technology

The Commission on Pile Research at The Royal Swedish Academy of Engineering Sciences has collected and published statistics of piled foundations in Sweden since 1962. In the last decade approximately 1 á 1.5 million metres of piles have been installed per year. Nearly 80 % are driven jointed precast concrete piles. For nearly 40 % of the end-bearing piles the geotechnical bearing capacity is verified by stress wave measurements. For friction piles in soft clay a composite pile of timber spliced to concrete is used. Steel piles are mainly used in special cases for example underpinning.

Piles are normally installed using hydraulic drop hammers or wire rope driven hammers with 3 to 5 tons weight.

4. National relevant documents

The basis for general design is the National Building Code (BKR 94), the Handbook for Concrete Structures (BBK 94) and the Handbook for Steel Structures (BSK 94) released by Boverket (Swedish Board of Housing, Building and Planning).

For highway bridges there is the National Code BRO 94 released by Vägverket (Swedish National Road Administration) and for railway bridges the National Code BV Bro released by Banverket (Swedish National Rail Administration).

Many handbooks and manuals for piling are released by IVA Pålkommissionen (The Royal Swedish Academy of Engineering Sciences. Commission on Pile Research).

All publications are written in Swedish.

Chapter No 7 in Eurocode 7 (EC 7) contain general rules for pile design. Rules are given for how to design piles on the basis of partly load tests and partly geotechnical soil data. The rules in EC 7 about safety factors are changed in the Swedish National Application Document (NAD). The rules given in EC 7 are good for static load tests when the number of tested piles are few. However, when dynamic tests (stress wave measurements) are used, sometimes 10 piles or more, are tested. For this situation which is rather common in Sweden special rules for chose of safety factors are given in the Swedish NAD.

5. Detailed description of the national design methods.

5.1 Definitions

Structural capacity of the pile: The pile material ultimate capacity

Geotechnical bearing capacity: The ultimate capacity of pile- soil interaction.

The pile bearing capacity: The lowest of structural capacity of the pile and geotechnical bearing capacity.

5.2 General philosophy

The piles, joints and shoes are to be designed structurally to withstand actions during handling, installation and during the final stage. This is called the structural capacity of the pile.

The pile- soil interaction shall withstand the actions at the final stage. This is called the geotechnical bearing capacity.

The design compressive force during the installation can either be derived from wave equation analysis or in a simplified way.

The maximum force F_{ck} from wave equation analysis gives the design axial force

$F_{cd} = \gamma_f F_{ck}$

where

γ_f = 1,0 in design for fatigue

γ_f = 1,3 in design for ultimate limit state

In the simplified way the design axial force can be reached from
$F_{cd} = 1,2\ \gamma_f R_k$
where
R_k = characteristic geotechnical bearing capacity

The design of the structural capacity of a pile shall take into account crushing and buckling. If the pile is designed for an axial force and moment the structural capacity for these actions shall be calculated. When designing for axial compression force the following factor is of importance:

- the pile cross section, bending stiffness and moment capacity,
- the strength of the pile material
- the initial curvature (after driving)
- driving work, and residual stresses caused by installation
- the lateral subgrade reaction

5.3 Design on basis of static load tests

Static load testing is seldom done in Sweden. Today dynamic load testing has taken over.Calibration of dynamic test method has been done with a number of static load tests during the last 20 years.

5.4 Design on basis of ground test results

5.4.1 Friction piles in soft clay

The ultimate bearing capacity of friction piles in soft clay can be geostatically analysed in general form on the basis of the undrained shear strength by the equation:

$$Q_f = \alpha\, c_{us} A_s + c_{ut} N_{c0} A_t$$

where

α = empirical factor. Value depends on soil characteristics, shape of pile, pile diameter, type of pile, installation method, time to failure, time after pile installation etc.
c_{us} = undrained shear strength along the shaft.
c_{ut} = undrained shear strength at the pile tip.
N_{c0} = bearing capacity factor.
A_s = shaft area.
A_t = tip area.

If the piles are installed in normally consolidated or slightly overconsolidated cohesive soils, the tip resistance is usually negligible in comparison with the shaft resistance.
For driven timber piles in soft clay α can be put equal to 1 for short-term loading and 0,7 for long-term loading (α = 0,7 can be considered representative of the creep failure of the pile, i. e. the load that leads to excessive creep settlements). For piles with constant cross-section driven into soft clay (such as concrete piles), α can be put somewhat lower (0,8 - 0,9 of the values given above).
In layered soil with varying characteristics, the total shaft resistance is obtained by summation of the contributions given by the various layers.
In the case of a pile whose diameter varies linearly with depth (timber piles) and which is installed in clay with an undrained shear strength increasing linearly with depth ($c_u = c_{u0} + kz$), the ultimate bearing capacity Q_f can be determined by the relation:

$$Q_f = \alpha\pi\, l_p \left[(1/2)\,(D_{ph} + D_{pt})\, c_{u0} + (1/6)\,(D_{ph} + 2\, D_{pt})\, k\, l_p \right]$$

where

D_{ph} = pile head diameter.
D_{tp} = pile tip diameter.

l_p = pile length in clay

The working load is determined considering acceptable settlements. Calculations of the working load is performed by dividing the shear strength values by partial safety factors γ_n for safety class according to the National Building Code and γ_m for uncertainties in the material properties. In the actual case this means that the ultimate bearing capacity along the shaft is only mobilised along the upper part of the pile due to the flexibility of the pile. If the strength value is the undrained shear strength determined by the field vane test values of γ_m according to the National Building Code can be used. If the piles are designed as creep piles beneath a raft then the ultimate creep bearing capacity should be used as working load. In such a case the bearing capacity is mobilised along the whole length of the pile.

5.4.2 Friction piles in sand, silt and heavily overconsolidated clay

The ultimate bearing capacity Q_f(the soil resistance to pile driving) of driven friction piles in sand, silt and heavily overconsolidated clay can be predicted by the following relation (Meyerhof, 1976) based on the results of SPT (Q_f in kN and A_t and A_s in m^2):

$$Q_f = 400\ N_{30t}\ A_t + 2\ N_{30s}\ A_s$$

where N_{30t} represents N_{30} count at the pile tip level and N_{30s} the average N_{30} count along the shaft of the pile. If the soil consists of layers with different characteristics the total bearing capacity is obtained by summation of the contributions given by each separate layer. In Sweden N_{20} obtained from the dynamic probe HfA is often put equal to N_{30} from SPT dependent on the lower driving energy.
The ultimate bearing capacity obtained by this relation is considered to correspond to characteristic values of pile driving resistance of precast concrete piles according to Swedish Standard.
The ultimate bearing capacity can also be predicted by a relation recommended by Bustamente & Gianeselli (1982) based on the results of CPT investigations.
The ultimate bearing capacity is often verified by dynamic load testing.

5.5 Design on basis of driving formulae

In 1960 a version of Hiley's formula was used for end bearing piles. Today design on basis of wave equation is used instead.

5.6 Design on basis of wave equation analysis

A driving criteria is often derived from a wave equation analysis. For standardised piles and driving equipment, a standard driving criteria for precast concrete piles derived from wave equation analysis is available in table 1. This is an example of standard driving criteria for precast concrete square pile with a crossectional area of 0.075 m^2.

Table 1: Design ultimate geotechnical bearing capacity in kN for end bearing piles in safety class 1 (SC1). For SC2 the values are divided by 1.1 and in SC3 with 1.2 (Reference table 1.5 in IVA RAPPORT 94)

		Hydraulic hammers			Wire rope driven hammers		
mm/ 10 blows	Drop height (m)	3 ton	4 ton	5 ton	3 ton	4 ton	5 ton
10	0.20	410	475	500			
	0.30	525	620	670	410	475	500
	0.40	630	730	800	460	550	585
	0.45			825 *	525		
	0.50	725	825		525	620	670
	0.60				605	705	775

* if pile length < 8m drop height 0.40 m

Dynamic load tests are done on around 40% of all piling site. Tests are performed at start of work as test piling operation. The number of piles tested is approximately 5 % but not less then 4 piles. If construction control is performed 10-25 % of the piles are tested.

5.7 Safety factors.

In Sweden the partial factoring is used but also a probability study can be used.
There are three partial factors used γ_{Rd}, γ_m and γ_n where γ_{Rd} is for the model uncertainties , γ_m is for variations in the material parameters and γ_n is depending on safety class.
Sometimes this factors are given as total safety factor: $\gamma_{tot} = \gamma_{Rd}\gamma_m\gamma_n$
The values for the product $\gamma_{Rd}\gamma_m\gamma_n$ in the use of wave equation should not be less than 2.3 in safety class 2 in accordance with BRO 94.
If dynamic load testing is performed and 4 piles are tested in one soil type the characteristic value of the measurements (average) shall be divided by the product $\gamma_{Rd}\gamma_m\gamma_n$. This value shall in safety class 2 not be less then 1.6 if the piles are driven to bedrock and not less then 1.85 if driven into soil in accordance with BRO 94.

5.8 Rules for serviceability

Rules can be found in relevant national standards.

6. Quality control and monitoring

The monitoring of the installation of piles follows the European pre-standard PR EN 288005 "Displacement piles". The position of the piles must be measured after installation to see that both structural capacity and geotechnical bearing capacity is higher than the action on the pile in its real position.

7. Particular national experiences

Extensive research on friction piles in soft clay has been performed at The Chalmers University of Technology in Gothenburg and published in theses by Torstensson (1973), Beigler (1976) and Jendeby (1986).

Extensive research of general piling technology has also been performed by The Swedish Academy of Engineering Sciences, Commission on Pile Research. The results are published in several reports (most of them in Swedish, recent reports have a summary in English). Some of them is listed in the reference list.

8. References

Beigler, S-E: Soil-structure interaction under static loading. Thesis. Chalmers University of Technology. (1976).

BKR 94, BFS 1993:58 med ändringar BFS 1995:18.

BRO 94, Publ 1994:1-9 med Supplement 2, Publ 1996:1.

BV BRO BVH 583.10 utgåva 3.

Hansbo, S. Foundation Engineering, (Elsevier, 1994)

IVA Pålkommissionen Rapport 58. Grävpåleanvisningar. (1979).

IVA Pålkommissionen Rapport 75. Förtillverkade betongpålar. Förslag till standard med dimensioneringsunderlag. (1984).

IVA Pålkommissionen Rapport 77. Grävpålar i friktionsjord. Anvisningar för dynamisk förbelastning. (1985).

IVA Pålkommissionen Rapport 79. Expanderkroppar. Anvisningar för dimensionering, utförande och kontroll. (1988).

IVA Pålkommissionen Rapport 81. Systempålar. Stödpålar av höghållfasta, korrosionsskyddade stålrör, slagna med lätta höghastighetshejare. Anvisningar för beräkning av dimensionerande bärförmåga. (1989).
IVA Pålkommissionen Rapport 84a. Beräkning av dimensionerande lastkapacitet för slagna pålar med hänsyn till pålmaterial och omgivande jord. (1995).
IVA Pålkommissionen Rapport 90. Grova stålrörspålar - anvisningar för dimensionering, utförande och kontroll. (1993).
IVA Pålkommissionen Rapport 94. Standardpålar av betong - lastkapacitet och geoteknisk bärförmåga. (1996).
Jendeby, L.: Friction piled foundations in soft clay A study of load transfer and settlements. Thesis. Chalmers University of Technology. (1986).
Olsson, C.; Holm, G.: Pålgrundläggning. (Svensk Byggtjänst, 1993).
SGF 1:93E. Recommended Standard for Cone Penetration Tests. (1993)
SGF 2:93E. Recommended Standard for Field Vane Shear Test. (1993)
SGF 1:96. Geoteknisk fälthandbok. (1996).
Torstensson, B-A: Kohesionspålar i lös lera - En fältstudie i modellskala. Thesis. Chalmers University of Technology. (1973).

Design of Axially Loaded Piles – European Practice, De Cock & Legrand (eds)
© 1997 Balkema, Rotterdam, ISBN 90 5410 873 8

Design of axially loaded piles – Swiss practice

F. Bucher
Swiss Federal Institute of Technology, Zürich

ABSTRACT: In Switzerland piles are frequently used in foundation engineering. The main reasons are the following: the general scarcity of building land and therefore the need to construct new buildings often on soils with unfavourable conditions, the increase in building loads, the application of the popular top-down construction method and the availability of efficient and competitive piling systems. Thus, already twenty years ago a national code for pile foundations appeared. It has been revised recently and is the most important basis for this national report.

1 REGIONAL GEOLOGY

With respect to the geology, Switzerland is commonly divided into three main areas: the Alps, the Central Plain and the Jura. An overview of the geological conditions of these three areas is given below following mainly Schindler/Nievergelt (1987), von Moos (1953) and Jäckli (1989).

The Alps exhibit rocks of most varied nature, sometimes even over short distances. The core of the Alps is formed by crystalline rocks as gneiss, mica schists and granites. North of it limestones, calcareous and argillaceous schists, sandstones and conglomerates are predominant.

In the Jura, limestones, marls, clay-shales, sandstones and gypsum are the most frequently found rocks.

With a few exceptions of tectonically stressed or highly weathered zones, the physical properties of the rocks found in the Alps and the Jura are not of great interest in connection with pile foundations. This is in contrast to the soils found in the Central Plain and in the Alpine and Jura valleys.

The soils of Switzerland may very generally be divided into the deposits of the glacial period and the deposits of the post-glacial period.

During the last glaciation, the major part of the Central Plain was covered by glaciers. Their main deposits are ground moraines consisting typically of grains of all sizes from the clay and silt fractions up to gravels and even boulders. They are usually very compact and often used in connection with end bearing piles. The lateral and terminal moraines are less homogeneous and coarser.

The glacial streams separated the morainic material into different fractions. Coarse material was deposited as outwash gravels. Their bearing capacity is very high. As they are further very permeable, sometimes tension piles are required in cases of deep excavations and high ground water levels.

Fine materials of the silt and clay fractions were deposited in water-filled basins during the melting of the glaciers. These glacial clays may be normally consolidated or overconsolidated. Their bearing capacities are very often not sufficient for flat foundations and may require pile foundations.

During the post-glacial time, the filling up of the wide valley bottoms and the lakes continued by the rivers. The deposits are generally coarse grained although intermixed with finer materials and occasionally organic materials. At the lower ends and along the shores of the many lakes in the Central Plain of Switzerland, problematic soils are often encountered. A typical soil profile may include (from top to bottom): peat, lake marls, lacustrine clays. Below theses soft and usually sensitive soils, the ground moraine is found in depths of up to 30 m and more. Building loads have frequently to be transferred down to the depth of the ground moraine by end-bearing piles. Friction piles may only be considered if the ground moraine is deep seated or if the pile loads are rather small.

2 COMMON PRACTICE FOR SOIL INVESTIGATION

Pile foundations require a detailed investigation of the soil profile and of the geotechnical properties of the subsoil over the entire depth which is influenced by the piles.

The investigation has to cope with all aspects which could be of importance for the pile project as

- the geological profile and the geotechnical properties of the soil and rock layers to a sufficient depth
- cleavage and cavities in the bedrock
- the ground water with respect to its hydrostatic, hydrodynamic and chemical conditions
- geological, hydrogeological and atmospheric conditions related to the durability of pile system and material
- stray currents.

For the investigation generally borings are required which allow to carry out the necessary in-situ tests and to collect soil samples for complimentary tests in the laboratory.

As far as in-situ tests are concerned, the Standard Penetration Test (SPT) is quite common in deposits of glacial and fluvio-glacial deposits. If these deposits are fine-grained, the static cone penetration test (CPT) could be considered as well and has recently been described by Amann (1992). But as the glacial deposits are very often coarse grained and heterogeneous in Switzerland, the application of the CPT is not common in soils of the glacial period.

In fine grained cohesive soils of the post-glacial time, the in-situ vane shear test is often applied for the determination of the undrained shear strength. In case of stiff fissured clays, the Standard Penetration Test is carried out instead of the vane test. The cone penetration test is increasingly considered for the investigation mainly of the post-glacial lacustrine deposits.

It is general practice to install open tube micropiezometers in the boreholes. Depending on the soil profile this is done in one or several levels. Emphasis is placed on an efficient sealing of all the boreholes, with or without micropiezometers.

The laboratory soil programme consists besides the classification tests mainly of triaxial, direct or vane shear tests, of fall cone tests and of compression tests.

3 PILE TECHNOLOGY

As outlined in chapter 2, the geological conditions are quite mixed in Switzerland. This results in the fact, that in Switzerland many different pile types are in use. The selection is done mainly on the basis of the actual geological situation and the loading.

For large depths of the bearing stratum and for high loads, bored non-displacement piles are selected. On the overall, they are used in the majority of the applications. Bored piles with diameters of 0.5 to 1.2 m have usually a temporary casing. Bored piles with diameters of 1.8 to 3.5 m are excavated using a bentonite suspension. For diameters from 1.2 m to 1.8 m, they may be carried out in connection with one or the other method (Merz, 1992). An example of large bored piles with a diameter of 3.2 m and a length of up to 60 m is described by Faes (1989).

If the total pile length is less than about 30 m and the working load is less than 2000-2500 kN, displacement piles may be considered as well. The main types are driven prefabricated concrete, steel and timber piles and driven or screwed cast in place piles. In terms of the total length, displacement piles are applied in Switzerland about half as often as non-displacement piles.

Some average production figures of the different types of piles used in Switzerland have been published by Arz (1992).

4 NATIONAL RELEVANT DOCUMENTS WITH REGARD TO PILE DESIGN

There is one national relevant document with regard to pile design which has to be mentioned. It is the Recommendation SIA V 192 "Pfähle" published in German and French in the year 1996. It has replaced the Swiss Code SIA 192 "Pfahlfundationen" of the year 1975.

The Swiss Recommendation for piles SIA V 192 comprises the following main chapters:

- Definitions
- Principles of design and execution
- Calculation, design and proofs
- Construction and materials
- Execution
- Testing and quality control.

An outline of the Recommendation SIA V 192 gives a publication of Vuilleumier (1995).

5 DESCRIPTION OF THE NATIONAL DESIGN METHODS

5.1 *General philosophy*

According to the Swiss Standard SIA 160 "Einwirkungen auf Tragwerke" published in 1989 two limit states have to be proved for all civil engineering structures:

- the ultimate limit state
- the serviceability limit state.

The Swiss Recommendation for Piles SIA V 192 follows this design principle in all aspects and is therefore essentially also in agreement with EC7, Part 1.

Both, the ultimate limit state of the soil and the ultimate limit state of the pile have to be proved. In this publication, however, the evaluation of the ultimate limit state of the soil is mainly discussed.

Generally, design values are compared. The design values of the actions and of the resistance are obtained from the characteristic values with partial safety factors.

The change from the traditional design methods using global safety factors to the design methods applying partial safety factors is not unproblematic. If constant values for the global, respectively partial safety factors are considered, different results may be obtained depending on the shear strength parameters and the design method as shown by Lang (1992) and Bucher (1990, 1995).

Swiss standards and recommendations do not contain specific methods or formulas to be used in design. Thus, the engineer is free to select the design method which is considered to be the best for a given application with the following restriction: design methods based on empirical or analytical methods must be validated by static load tests for comparable conditions as far as the geological profile, the pile type and the loads are concerned.

5.2 *Definitions*

According to the Swiss Recommendation for Piles SIA V 192, each property (actions, shaft or base resistance, soil properties etc.) may be considered at three different levels.

Values from experimental determinations are at one level. For example, R is the bearing capacity of a pile obtained from a pile load test, or φ' is a friction angle obtained from a triaxial shear test.

Characteristic values are at a second level. They are carefully determined or selected values and denoted by the subscript k. For example, R_k is the characteristics value of the bearing capacity of a pile obtained from two pile load tests.

Design values are at a third level. They are calculated from the characteristic values with partial safety factors and denoted by the subscript d. For example, R_d is the design bearing capacity of a pile.

5.3 *Design on basis of static load tests*

Static load tests are considered to be the most reliable basis for the design of piles according to the Swiss Recommendation for Piles SIA V 192.

Therefore, the procedure for static load tests is outlined in details in SIA V 192. The sequence of the loading includes loading, unloading and reloading up to the working load. Failure of the soil should be reached by end of the load test. It is emphasised to the duration of each loading stage should be sufficient to reach a stable settlement. The criteria for this condition is that the time-settlement curve is over a period of 30 minutes or 0.09 Δt_1(whatever is larger; Δt_1 being the time of the load increase until the observation to fulfil this criteria starts) at least linear in the semi-logarithmic plot and the increase in the settlement is equal or less than 0.1 mm.

SIA V 192 recommends that a pile load test is carefully planned in terms of the loading and the measuring programme. The instrumentation has to be selected according to the load distribution. In many cases, measurements which allow to separate the friction and the end bearing pressure of the pile shall be considered. A successful load test in which the load cell has been placed near the foot of the pile has been reported by Vollenweider (1988).

In the evaluation of static load tests, SIA V 192 follows essentially EC7.

5.4 *Design on the basis of ground test results and of driving formulae*

The Swiss Recommendation for Piles SIA V 192 recommends that design methods on the basis of ground test results and of driving formulae may only be used if they are validated by static load tests for similar conditions. As the soil conditions in Switzerland are very variable the applicability of these design methods are quite limited. Thus, these methods are rather suited for estimating than for calculating the axial pile capacity.

The most common methods on the basis of in situ test results are based on static or dynamic penetration tests (CPT or SPT), pressuremeter tests or shear vane tests. With regard to laboratory tests, generally the triaxial and the direct shear test are considered for determining the effective shear strength parameters. Based on them the axial capacity is obtained with the bearing capacity equation.

In the textbook of Lang, Huder and Amann (1996), some of the most frequently used design methods are given. A compilation and description of the different design procedures is included also in Stuckrath (1992).

There is a considerable variety of pile driving formulae in use. Their results may differ by a factor of 10 or more [Locher 1992] and are useful only if they are validated by load tests. SIA V 192 stresses mainly also the importance of the influence of time on the bearing capacity.

The main application of driving formulas is to evaluate the behaviour of driven piles on a particular site.

5.5 *Design on the basis of wave equation analysis*

The development of the design method based on the wave equation has resulted in an increasing application of this method in Switzerland. Detailed comparisons of the results obtained from static load tests and from the CAPWAP simulations have been published by Stuckrath (1992) and by Stuckrath and Vuillet (1992). Bored and driven piles have been examined in model-scale tests. The results showed that the ultimate load values obtained from the two methods give in general a remarkably good correspondence. However, the load distribution with depth may be quite different. For the bored pile, the CAPWAP simulation resulted in an overestimation of the shaft resistance and an underestimation of the toe resistance. For the prefabricated piles, it resulted in an underestimation of the shaft resistance and an overestimation of the toe resistance. The CAPWAP displacement predictions are for all piles higher than the results from the static load tests.

5.6 *Prescribed factors of safety*

In the old Swiss Standard on Piles SIA 192, a global safety factor with respect to the calculated ultimate end bearing resistance of at least 2.0 and to the calculated ultimate shaft resistance of at least 3.0 was required. With respect to static load tests, a minimum global safety factor of 2.0 was required if failure was reached. If soil rupture was not reached during the load testing and if under the maximum test load a stabilisation of the settlement of the pile conforming given specifications was reached a global safety factor of at least 1.50 with respect to the maximum test load was required.

In the new Swiss Recommendation for Piles SIA V 192, partial safety factors and load factors have been introduced. This is in principal accordance with the Swiss Standard SIA 160. The individual partial safety factors and load factors recommended in SIA V 192 agree essentially with those given in EC1 and EC7, part. 1.

5.7 *Rules for serviceability*

The Swiss Recommendation for Piles SIA V 192 states that the serviceability of the piles must correspond to the serviceability of the supported structure. Specific recommendations are given with respect to the durability of piles, mainly timber and steel piles.

6 QUALITY CONTROL AND MONITORING

The control of the quality of the piles has to cover the following aspects according to SIA V 192

- the quality of the building materials
- the integrity of the pile over its full length
- the bearing behaviour
- the observance of the tolerances in the execution of the piles.

For the control of the integrity of the piles, acoustic tests are mainly carried out. Two or more holes are preformed with plastic tubes over the entire length of each pile. Also the low strain test is sometimes applied for the integrity testing.

For the measurements of deformations and strains in axially loaded piles, mainly extensometers, sliding micrometers and strain gauges are selected. Examples of instrumented piles are given by Amstad and Kovari (1992). In this publication, also the results from measurements obtained in load tests are given.

7 PARTICULAR NATIONAL EXPERIENCES

The major research project on piles which has been carried out in Switzerland in the last few years is that of Stuckrath (1992).

Full-scale tests on bored and jacked piles, in an instrumented test pit in homogeneous deposits, were undertaken in order to clarify load transfer mechanisms as a function of both soil type and installation method.

The pile load tests revealed that installation by jacking brings about compaction in sand, remoulding in clayey silt, and soil tightening in both soil types. Pile surface effects were found to be determinant in load transfer through shaft resistance for both bored and jacked piles.

Two major conclusions of this study are the following:

- The experimental pile behaviour revealed in the instrumented tests points to the impossibility for a general equation to predict load-head movement behaviour adequately for different soils and installation methods in homogeneous deposits, let alone real site conditions.
- More instrumented pile tests should be carried out, most preferably in collaboration with research centres, in order to clarify load transfer mechanisms in natural deposits.

REFERENCES

Amann, P. 1992: *Baugrunduntersuchungen mit der Drucksonde im Hinblick auf Pfahlgründungen*. Zürich: ETH, Institut für Geotechnik.

Amstad, Ch. & K. Kovari 1992: *Deformationsmessungen bei Pfahlfundationen*. Zürich: ETH, Institut für Geotechnik.

Arz, P. 1992: Keynote Paper: Germany, Austria and Switzerland. *Proc. of Piling Conference, London*. London: Thomas Telford.

Bucher, F. 1990. *Anwendung der neuen Tragwerksnormen des SIA im Grundbau*. Zürich: SIA Dokumentation D 064.

Bucher, F. 1995. *Pfahlbemessung: Vergleichsberechnungen Alte-Neue Norm SIA 192*. Zürich: SIA Dokumentation D 0130.

Faes, M. 1989. *Grossbohrpfähle mit Durchmesser 3.2 m bis 60 m Länge*. Zürich: SIA Dokumentation D 038.

Jäckli, H. 1989. *Geologie von Zürich*. Zürich: Orell Füssli.

Lang, H.J. 1992. *Pfahldimensionierung und neuere Tragswerksnormen*. Zürich: ETH, Institut für Geotechnik.

Lang, H.J., Huder, J. & P. Amann 1996. *Bodenmechanik und Grundbau, 6. Auflage*.Berlin: Springer.

Locher, P. 1992. *Verdrängungspfähle - neue Herstellungsmethoden*. Zürich, ETH, Institut für Geotechnik.

Merz, H. 1992. *Stand und Entwicklung bei der Herstellung von Bohrpfählen*. Zürich: ETH, Institut für Geotechnik.

Schindler, C. & P. Nievergelt 1987. *Einführung in Geologie und Petrographie*. Zürich: Verlag der Fachvereine.

SIA 1975. *Code SIA 192: Pfahlfundationen*. Zürich: SIA.

SIA 1989. *Norm SIA 160: Einwirkungen auf Tragwerke*. Zürich: SIA.

SIA 1996. *Empfehlung SIA V 192*. Zürich: SIA.

Stuckrath, L. & L. Vuillet 1992. Large-scale static and dynamic tests under laboratory-controlled conditions. *4th Int.Conf. on the Application of Stress-Wave Theory on Piles, The Hague*. Rotterdam: Balkema.

Stuckrath, L. 1992. *A study of pile behaviour in different soil types including installation effects*. Lausanne: EPFL, Thesis No. 1074.

Vollenweider, U. 1988. *Prüfung nach dem Pfahlfusspressverfahren, Zürcher S-Bahn, Viadukt Neugut*. Zürich: Spezialtiefbau AG.

von Moos, A. 1953. The subsoil of Switzerland. *Proc. of the 3rd Int.Conf. on Soil Mechanics and Foundation Engineering*. Zürich.

Vuilleumier, F. 1995. *Nouvelle Norm SIA 192: Foundation sur pieux*. Zürich: SIA Dokumentation D 0130.

Design of Axially Loaded Piles – European Practice, De Cock & Legrand (eds)
© 1997 Balkema, Rotterdam, ISBN 90 5410 873 8

Design of axially loaded piles – United Kingdom practice

J.D. Findlay, N.J. Brooks, J.N. Mure & W. Heron
Stent Foundations Ltd, Hampshire

1 INTRODUCTION

This is a view on the current state of design practice for commercial piling in the United Kingdom. It was prepared at the request of the ISSMFE European Regional Technical Committee 3 "Piles". It excludes offshore piling.

Time did not allow for any questionnaire to be sent to main contractors, specialist contractors, consultants and client organisations who are known to be actively involved in piling but is based on the wide knowledge of the authors who work for one of the UK's largest piling organisations.

The following geological description is taken from Powell, Clarke and Shields.

2 GEOLOGICAL AND GEOTECHNICAL DESCRIPTION OF THE UK

2.1 Geology

Naturally occurring deposits formed prior to the Pleistocene glaciation are generally referred to by geologists as rocks and comprise the 'solid' formations shown on geological maps. The strength of these 'solid' formations varies such that some of them are classified for engineering purposes as soils. These include, for example, many of the deposits in the south east of the UK such as London Clay and Gault Clay. Naturally occurring deposits formed after the Pleistocene glaciation are generally referred to as 'drift' deposits and these are predominantly soils.

The geology of the UK and hence the distribution of soil and rock types is complex. The UK can be divided approximately into two areas by the diagonal line shown on Figure 1. To the north and west of that line the dominant character is upland dissected by plains and valleys of limited extent. This area is underlain by rocks which are mainly older than the Carboniferous Coal Measures formed during the Palaeozoic period. The land to the south and east is undulating lowland underlain by rocks younger than the Carboniferous Coal Measures.

Most of the UK has been affected by glaciation which has resulted in extensive drift deposits of till (boulder clay), laminated clays and other glacial materials as far south as London. In the upland area these deposits are confined to the plains and valleys which are where most development has taken place. The rocks forming the upland are either exposed or covered by a thin layer of soil. Periglacial and lacustrine deposits are also found in the plains and valleys together with estuarine and coastal muds and silts. Peat and other organic soils are found throughout the upland region. The majority of site investigation in the upland area is confined to testing drift deposits and establishing rock head.

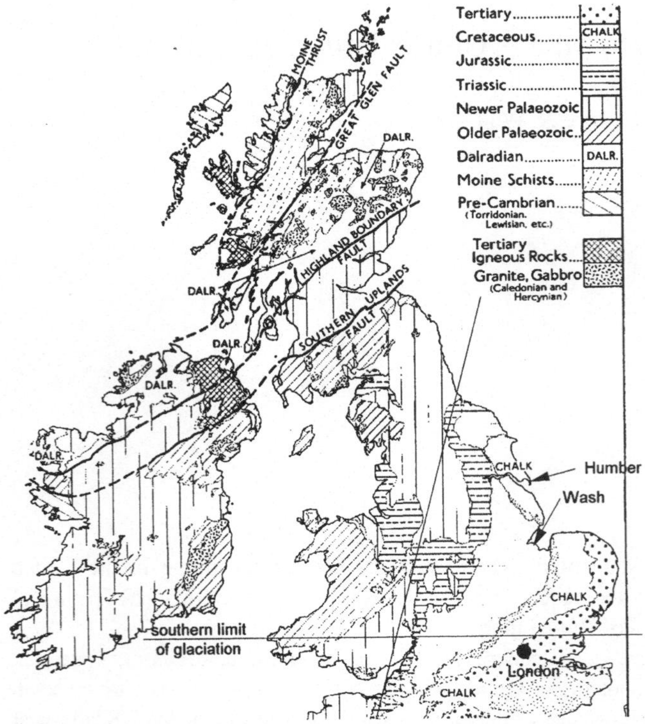

Figure 1 Main geological features of the UK (after Blyth and de Freitas, 1984)

Most of the rocks in the lowland area are sedimentary deposits laid down in marine or estuarine environments. Many of these strata, formed during the Mesozoic and Tertiary periods, have a relatively low strength and so are classed for engineering purposes as 'soils' (e.g. London Clay) although, at depth, they may become sufficiently strong to be classed as 'rock'. It is also possible to have a profile consisting of alternating layers of 'soil' and 'rock' due to varying degrees of lithification of the sedimentary deposits within the profile.

The strata are generally named after the principal locality in which the formation is well displayed (e.g. London Clay, Oxford Clay). The actual lithology of a deposit varies and maybe different from that inferred by the name (e.g. Lower Greensand includes clay layers).

The exposures tend to run in the north-east south-west direction and have an influence on the topography since the harder, more resistant layers give rise to scarps. Unlike the upland areas, much of the rock is overlain by drift comprising glacial, periglacial and lacustrine deposits. Major deposits of estuarine and coastal muds and silts are found in areas around the Humber and the Wash.

The Midlands is essentially a drift-covered Triassic plain which includes Mercia mudstones (Keuper marls), limestones, Lias Clay and Bunter sandstone. To the south east of this plain lie the scarplands which are formed of outcrops of Jurassic and Cretaceous rocks which comprise alternate layers of weak and strong rocks. The sequence of rocks is very varied and includes, in the Jurassic sequence, massive limestones and sandstones, Oxford Clay and Kimmeridge Clay. The Cretaceous sequence includes the Lower Greensand, Gault Clay and Chalk. The south east area of the UK consists of the London and Hampshire basins which lie to the north and west of the Weald. The basins, which are bounded by chalk uplands, comprise Oligocene and Eocene deposits including London Clay, Thanet Sand and Bagshot Beds. Terrace gravels and extensive alluvial deposits are found in the London Basin. The Weald, in the south east corner of the UK, is an area bounded by a chalk rim which varies in colour and strength enclosing exposures of Upper and Lower Greensand, Gault Clay, Weald Clay and Purbeck clays, shales and limestones.

2.2 Geotechnical Properties

The variety of materials ranging from organic deposits to strong rocks means that the soils and rocks in the UK encompass the complete range of strengths and stiffnesses. Drift deposits in the upland region are predominantly firm to stiff to hard gravelly sandy clay containing boulders, lenses of sand and gravel and lenses of laminated clay. The more recent deposits comprise softer organic and alluvial clays and loose sands.

Pre glacial soils in the lowland area are generally either stiff to hard clays or dense to very dense sands. These are particularly prevalent in the south east and include, for example, London Clay and Thanet Sand. Glacial soils of similar properties to those in the upland region are found as far south as London. More recent drift deposits include medium dense terrace gravels, soft organic clays and peats and head deposits.

3 SITE INVESTIGATION PRACTICE IN THE UK

3.1 General

Site investigation in the United Kingdom is a subject of much debate. Over more than a quarter of a century those close to the site investigation industry have bemoaned their lot. It is generally considered that clients are not prepared to pay sufficient money for quality site investigations which should save many times their cost in the avoidance of the results of encountering unexpected ground conditions. There is unfortunately less rational argument over why the industry is not able to command sensible reward for a vital ingredient for a high proportion of the construction industry's output.

There have been a number of efforts to improve the situation by demonstrating the cost savings which could accrue from properly executed site investigations and providing guidance on good practice. The latest documents are a series of four publications under the general title Site Investigation in Construction. The main documents which guide good practice in the UK are BS5930:1981 Code of Practice for Site Investigations, currently undergoing revision, and BS1377:1990 & 1995: Methods of tests for soil for civil engineering purposes. In addition there are many publications by the Geological Society, the Association of Geotechnical and Geoenvironmental Specialists, the British Drilling Association as well as textbooks used in university teaching which provide a rich source of how to investigate the very varied geology of the UK. Despite the extensive knowledge of good practice in the UK the general standard of site investigation presented to piling contractors is not of a high standard.

For medium and large projects, a client will appoint a professional team which may be drawn from many of the disciplines recognised as part of the construction industry. A member of the team, probably an Engineer, but not necessarily one with direct access to a specialist in geotechics, will draw up a specification and scope of works for which a competitive tender will be sought. For small projects, as well as some more substantial undertakings, the client or his advisors may rely on the expertise of the site investigation contractor to suggest a suitable scope of works albeit in competitive tender. For complex projects a consulting engineer will be appointed and they will usually have a geotechnical department or will appoint a consultant or firm which specialises in geotechnical engineering. Specialist practices and departments will have a mixture of engineering geologists and civil engineers, a large proportion of whom will have a post-graduate degree. Thus there is a wide variety of expertise behind the selection of the scope of work for any site investigation. Well conducted site investigations with desk studies and two phase ground investigations will usually be practised by the knowledgeable organisations but across the whole industry, all too often site investigation is reduced to a ground investigation of very limited scope.

Across the whole industry execution of the physical investigation work is most frequently undertaken by self employed drillers working for an umbrella organisation or small firms with low overheads and little formal training in geotechnics. There have, however, been efforts over recent years to provide an accreditation scheme for drillers but as many inadequacies arise from the scope of works as from the skill employed in undertaking the work.

3.2 Boreholes

By far the most common method of undertaking ground investigations is by a 'Shell and Auger Rig' which comprises a set of tripod legs and a diesel engine together with temporary casing and sinking tools. The hole is normally advanced with a hollow shell fitted with a internal retaining ring or a gravity one way clack valve, the latter being necessary to retain granular soils. Only disturbed samples can be obtained and these are the principal source for borehole log soil descriptions. Where rock is encountered, the hole will either be advanced using a chisel or a rotary drilling rig. The latter will either be an attachment to the tripod or a purpose built machine will be used with either double or triple tube core barrels. Most core barrels using water or foam flush are now thin walled and capable of reasonably high core recovery although triple tube barrels will need to be specifically requested in most circumstances. Especially in coal mining areas (current and former) thick wall air flush barrels may be used unless specifically forbidden by the unwary.

The soft ground technique although being very old and fairly crude has the advantage that it can deal with all the geological formations encountered in the UK and with adequate supervision can certainly give good stratigraphic identification. Disturbed samples from boreholes are taken at frequent intervals normally supplemented with U100 samples and Standard Penetration Tests (SPT) for cohesive and granular soils respectively. Vane testing, piston samples, large diameter undisturbed sampling, pressuremeter testing and piezometer installation will also take place less frequently and where the investigation is not routine.

Dynamic probing and sampling are increasingly being used to obtain soil profiles when designing foundations for straightforward developments and to assess areas of instability or rapidly changing profiles.

Although most of the larger site investigation companies attempt to present borehole logs in accordance with the guidelines set out in BS5930:1981, across the whole industry there is usually some relevant data missing from borehole logs. This can include lack of timing to complete the borehole, omission of temporary casing data, lack of water observations and poor recording of chiselling. BS5930 illustrates two logs with data presented in slightly different ways. Table 1 indicates the data which should be recorded on a typical borehole log.

Rock cores are adequately logged by reputable companies to give comprehensive descriptions. Mechanical logging is not recorded in an ideal fashion for the piling designer or contractor as only Total Core Recovery and RQD are recorded on the log illustrated by BS5930:1981. It is hoped that there will be an improvement in the recording of data with the publication of 'Piling in Weak Rocks' by the Construction Industry Research and Information Association, CIRIA in 1997.

3.3 Laboratory and In-situ Testing

Piling practice in the UK tends to be based on indirect design methods rather than direct design methods. Design parameters are most commonly obtained from classification, quick undrained

Table 1 Items to be noted on a borehole log undertaken in accordance with BS5930:1981

Company name
Borehole number and sheet number
Equipment and methods
Location
For whom carried out
Ground level
Co-ordinates
Dates for execution
Description, Reduced Level, Legend, Depth & Thickness for each stratum
Depth, Sample type and number and test type for each sample or test
Field records which would include blow counts, recovery remarks and standpipe/piezometer details
Water level observations during boring noting date, time, depth of hole, depth of casing, depth to water and remarks
Notes, Sample/test key and Remarks
By whom logged
Scale of borehole log

triaxial and oedometer tests on samples of cohesive soil and SPT tests in granular deposits.

The requirements for laboratory testing are set out in BS5930:1981 which makes reference to the appropriate part of BS1377:1990 or 1995 for the details of the test procedure to be adopted. The aims of laboratory testing are stated as being:

(a) to identify and classify the samples with a view to making use of past experience with materials of similar geological age, origin and condition

(b) to obtain soil and rock parameters relevant to the technical objectives of the investigation

The codes do not address the analysis and use of the test results and practitioners are expected to refer to appropriate design manuals and textbooks. BS5930:1981 warns that it is essential to consider whether any proposed test simulates probable conditions in the field sufficiently well to be relevant to the problem in hand. It also notes that tests may be done because a design is based on a body of experience or form of analysis which requires a particular test to obtain the relevant parameters. Too often laboratory testing is carried out without bearing this advice in mind.

By far the most widely used in-situ test in the UK is the SPT. Its main use is as an indicator of the density and compressibility of granular soils. It is also commonly used to check the consistency of stiff or stony cohesive soils and weak rocks. BS1377:1990 part 9 sets out the requirements to adhere to the standard but because of the wide spread use of the test, the provisions of the standard are not universally adopted. Indeed the requirements of the previous version of the standard dating from 1975 had not been fully implemented. The current standard requires a 63.5 kg drive weight with 760 mm free fall, drive head to weigh between 15 and 20 kg, total drive assembly weight not greater than 115 kg, rods must be AW or stiffer attached to a split spoon sampler of 51mm diameter of 684 mm length, internal diameter 35mm and fitted with a shaped cutting shoe at its open end and an internal ball check valve. A 60 degree solid cone is often screwed on the end of the split spoon in place of the cutting shoe when tests are made in gravel or gravely sand or in stony material such as glacial till or chalk.

The SPT tool penetrates under the dead weight of the rods and trip hammer, before being driven the 150mm of the seating drive. The blows for two 75mm increments of penetration are recorded. The sampler is then driven a further 300mm generally recording the blows for each 75mm of penetration. In hard soils a maximum of 25 blows for the seating drive and 50 blows for the test drive in soils or 100 blows in weak rocks is recommended by BS1377:1990.

Many things can go wrong with such an apparently straight forward test and experienced designers will consider whether the blow values quoted in an investigation report are credible. Additional comprehensive advice can be found in the CIRIA report 'The Standard Penetration Test (SPT): Methods and Use' written by Professor C R I Clayton.

On larger projects, effective stress testing of soils and weak rocks will be carried out although cost limitations are likely to require that careful selection for a limited number of samples will prevail. Static cone penetrometer testing (CPT/CPTU), prebored pressuremeter testing in weak to strong rocks and self boring pressuremeter testing in clays, sand and weak rocks are likely to be included as appropriate. More information about their place in UK practice is included in Sections 3.4 and 3.5. In competent rock, unconfined compressive testing will be the normal laboratory test but other tests will be used on the more complex projects. Point load tests are frequently used to supplement triaxial testing.

Standpipes and piezometers are frequently installed but apparently less frequently read judging by the evidence of site investigation reports. A wide variety of instruments are available but where piling is likely to be the foundation solution only the simplest devices are normally used. Permeability testing using falling or rising tests, constant head tests and pumping tests are used in the more comprehensive investigations. However, it is a constant complaint from piling contractors that the observation of the water regime associated with a site is often the most poorly observed and reported aspect of UK investigations.

Geophysical testing including radar, seismic and resistivity techniques are becoming more common although many designers remain concerned over the engineering accuracy which can be obtained from such techniques.

3.4 CPT/CPTU Testing

Static cone testing (CPT/CPTU) is not infrequently used but in many areas of the country the use is restricted by obstructions or density of materials with the result that it is rather less used in suitable soils than would be the case in many European countries. However, there has been a steady growth since the 1970s and a range of trucks, crawler mounted rigs and demountables are now available. Nearly all the CPT/CPTU testing in the UK is undertaken by specialist contractors most often working as sub-contractors to the main site investigation contractor, but also directly for a client.

The decision to use and specify CPT/CPTU is most often taken by consultants who are predominately geotechnical and environmental consultants. CPT/CPTU testing in the UK is now being carried out in all types of soil and occasionally weak rock. Table 2 gives an approximate breakdown of testing by soil type. Fill is also tested but usually only to determine its extent.

Around 90% of the UK testing is in England with over 60% being concentrated in the southern part of the country, illustrating the geological influences.

The main use of the CPT/CPTU in design is in indirect design methods based on derived soil properties although there is an increasing use directly in empirical design formulae in pile design.

Table 2 Types of soils in which CPT/CPTU are used

Soil Type	% of CPT/CPTU Testing
alluvium	25
stiff overconsolidated clay	10
till (boulder clay)	10
sands and gravels	20
chalk	15
fill	20

Both 10 cm^2 and 15 cm^2 CPT and CPTU equipment is currently in use in the UK. Most cones have inclinometers fitted and increasing use is made of seismic and pressuremeter cones to determine stiffness profiles. Over recent years there has been a contractor led move towards 15 cm^2 cones which now account for about 70% of testing. CPT testing using a 15 cm^2 cone on 10 cm^2 rods is preferred for the robustness and friction reducing effects in the variable UK soils.

The basic equipment in use for CPT/CPTU satisfies either BS1377:1990 or the International Reference Test Procedure (IRTP). Where the documents do not deal with items directly, similar tolerances to those cited are used.

3.5 Pressuremeter Testing

Early use of pressuremeters in the late 1960s in the UK drew heavily on the French empirical approach. In the subsequent quarter century the development of pressuremeter testing in the UK has pursued the general aim of attempting to make direct measurements of the particular fundamental soil parameter required. Thus it is hardly ever used as an index tester of properties in the UK. It is normally undertaken by a specialist sub-contractor, which has the high degree of operator skill required in the pressuremeter's UK applications, to a main site investigation contract.

A device such as a pressuremeter is attractive where in theory the boundary conditions are controlled and well defined, as are the stress and strain conditions in the surrounding soil mass. The self-boring type of pressuremeter has gained most academic interest as it potentially offers the closest approach to undisturbed soil testing of any in-situ test by its ability to tunnel its way into the ground. Since methods of interpretation are not so simple to standardise and there is active academic interest compared with the rest of site investigation, widespread usage has not been enhanced.

In the UK, the pressuremeter seems to offer great potential with the list of parameters which may be deduced:

1 deformation modulus
2 strength
  (a) undrained strength for clays or weak rocks, S_u
  (b) angle of shearing resistance for sands, ϕ'
  (c) angle of dilation for sands, Ψ
3 in-situ total horizontal stress, σ_{ho}
4 additional parameters from more specialist tests i.e. coefficient of horizontal consolidation, c_h, and angle of shearing resistance for clays, ϕ'

The degree of success in obtaining any of these parameters is critically dependent upon the details of the type of test and on the interpretation of the data and this has driven the test into a very specialist market. The two broad categories of test can be distinguished in terms of the installation method:

1 Menard-type pressuremeter (MPM) test - in which the device is placed into a pre-formed hole (which is usually slightly oversized)
2 Self-boring pressuremeter (SBP) test - in which the device bores its own way into the ground.

A third category, the push-in pressuremeter (PIP) test, developed principally for offshore use has gained relatively little experience.

There are two major types of pressuremeter, either pressurised liquid expands the membrane and cavity volume changes are measured or pressurised gas expands the membrane and cavity radius changes are measured by means of displacement transducers. In some cases pressurised oil is used to expand the membrane in devices designed for use in weak rocks. In all types great emphasis is placed on calibration as without proper controls, pressuremeter tests can be rendered meaningless.

Table 3: Summary of bearing pile types+

REPLACEMENT			SMALL DISPLACEMENT		LARGE DISPLACEMENT	
Mini diameter	Small diameter	Large diameter	Mini diameter	Small and large diameter	Small and large diameter	
Rotary and percussion bored cast-in-place			Driven and jacked,	Driven, hollow section, open	Driven, cast-in-place	Timber
Continuous flight auger grout and concrete injected			preformed steel and	ended steel*	Steel tube method:	Steel, hollow section,
	Prebored grouted precast units (partially preformed)	Excavated cast-in-place	concrete Driven, cast-in-place concrete shell, steel cased	Driven, solid section steel (H-piles) Driven, hollow cylinder, open ended concrete*	Franki type, Vibroform, Closed ended (constant and taper section) Concrete: West shell partially pre-formed	closed ended Concrete, solid section precast segmental precast

There is much debate around the value of the particular parameters obtained from any test method and its interpretation. Such concerns are outside the scope of this paper but it is fair to note that the results from pressuremeter tests are treated with caution in the UK. There are, however, several publications which can aid the designer in assessing the value of results obtained.

4 PILING TECHNOLOGY

4.1 General

The pile types in use in the UK have, of course, been influenced by the basic factors which affect the installation of piled foundations. Load transference to the soil or rock is achieved by a combination of both end bearing and shaft resistance and is dependent not only on the behaviour of the pile, but also on the characteristics of the chosen installation procedure. The selection of appropriate pile types is generally dependent upon the following:

1. underlying soil conditions, including groundwater levels
2. the nature and size of the loads to be supported by the foundations
3. properties of the pile materials and characteristics of installation equipment
4. effects of environmental and cost constraints

4.2 Pile Type Categorisation

For subsequent discussion it is useful to refer to a table which categorises pile types into a recognised pattern. An acceptable division is shown on Table 3

. + For all piles of non circular cross section, diameter refers to characteristic width

* If hollow section, open ended steel and concrete piles are driven and a soil plug forms at the base, the displacement approaches that of closed-ended section piles.

4.3 Ground Condition Influences

The following is a snapshot of some of the conditions frequently met in the UK. It aims to give a very brief insight into the considerations that would be brought to bear in the UK.

1 Weak soil overlying competent strata: Displacement piles and replacement piles are both used and have many advantages and disadvantages which need to be considered along with the

influences other than ground conditions. In output terms CFA, Driven cast-in-place (DCIP), Precast segmental and rotary bored are the main types used.

2 Piling in stiff clay: Rotary bored piles are likely to be the most economic because the piled length can be bored without temporary casing. Often small diameter piles will be carried out by CFA depending on rig availability. Enlarged based (under-reamed) piles may often be used in clay, without any lens of water bearing layers soils, for piles carrying significant load. Such piles are used for the highest loaded piles in the UK as these conditions apply to London. Displacement piles are unlikely to be economic although technically possible provided they can resist the heavy driving.

In glacial tills, boulders and bore stability may require the use of oscillators or bentonite for rotary bored piles.

3 Weak rock overlying stronger rock: Displacement piles may in some rocks be driven to considerable depths without apparent increase in driving resistance. Care is required in establishing acceptance criteria. Small diameter bored piles may be uneconomic if the groundwater level is high and tremie placement of concrete is required. CFA piles and DCIP are frequently considered.

4 Ground variability: Significant lateral and vertical variability is a frequent feature of UK geology. Pile selection has, therefore, to be able to adapt to variable length without undue cost in material wastage or difficulty in adapting technique. This tends to count against preformed piling types.

5 Negative skin friction: Reducing the number of piles by increasing diameter is an effective economic solution for the pile foundation but may not suit the superstructure arrangements. Friction reducing coatings are used but may not be effective for driven piles because of damage. They are expensive for rotary bored piles because of the need for a permanent casing on which to place the low friction material. It is often cheaper to increase the capacity of the pile to account for the potential negative skin friction.

6 Ground disturbance: The sensitivity of a site to ground disturbance is likely to be a consideration as the majority of sites are not greenfield in the UK because of the industrial history and the population density. Preboring may be required for displacement piles, if necessary, supplemented with a support fluid for temporary support in sensitive cases. Preboring may also be used to avoid heave and displacement of piles already installed in large groups, although it is common to use redriving for modest groups.

7 Obstructions: Glacial soils, indurated layers within otherwise uncemented formations and natural and man-made obstructions within fill are all situations which can interrupt pile installation. The approaches to overcome them include excavation prior to piling where at a practical depth, large diameter rotary bored equipment using heavy chisels, core barrels, down the hole hammers, mini diameter piles on a flexible basis to avoid obstructions, robust driven piles capable of penetrating the obstructions and occasionally rock roller equipment and blasting.

8 Aggressive ground: Corrosion of piles is a common problem in disturbed soils with industrial contaminants, in zones of fluctuating groundwater and where there are naturally elevated values of certain substances such as sulphates and chlorides.

Guidance is given in BRE Digest 363 and BS5328 for reinforced concrete, Romanoff, Morley and Bruce for corrosion and protection of steel and in BS5268: Part 5 for the treatment of timber.

9 Environmental: Normally this refers to noise, vibration, traffic movements and water pollution. Again there are references to assist the designer and constructor. Starting to take a high profile and can be a prime driver in the choice of techniques.

4.4 Piling Equipment

There are no statistics in the UK on the distribution of the different types of piling equipment used. Equally there is no point in describing the common piling techniques which are sensibly

Table 4 Common equipment types used in the UK

EQUIPMENT TYPE	REMARKS
Percussion boring	Known as tripod piling, 300 to 600mm dia usually limited to 20m depth, used for difficult access and limited headroom. Reasonable availability.
Rotary boring	Crawler crane mounted, increasingly on dedicated hydraulic machines, a few on lorry chassis. 300 to 3000mm dia depending on equipment. Under-reams up to 6.3m dia on crawler crane mounted machines. Dominated by less than 10 companies.
Continuous flight auger	Mainly dedicated hydraulic carriers, 300 to 1000mm dia although up to 600mm dominates. Extensively available with reputable companies having instrumentation systems.
Mini piles	Both replacement and small displacement used. Many small companies, with a few bigger operators who undertake a large percentage of the work which is about 20% of total piling market by value.
Tube or H driven steel	Often used adjacent to railways and for marine work and may be installed by non specialists.
Driven cast-in-place	Wet shaft type mainly. Very efficient where limited thickness of dense granular soil exists. In UK normal max. depth is 18m and usually less than 500mm dia. Greater depths and diameters exceptional. Less than about 10% of market.
Driven pre-cast concrete	Commonly up to 350mm square and usually segmental reinforced concrete. Individual lengths up to about 13m because of transport. About 25-30% of total market. Has lost market share to CFA piling.

described in several publications, eg, Tomlinson and CIRIA Report PG1. In Table 4 below is listed the most common techniques with a few remarks about usage.
It is emphasised that these are only the most frequently used types of piling, but that a much larger range of piling, of about 25 types of piling, is available in the UK.

5 PILE DESIGN DOCUMENTS

There is no national design code for piles within the UK. However there is a wide body of knowledge from which a designer can draw in order to guide him towards a sensible pile design. Much of the published data recognises piling is a process in which the calculations for the pile design only form a small part. Significant emphasis is put on local experience, principally because of the great geological variability which can often be encountered.

The most commonly used sources of information are briefly noted below and listed in the references. Specialists will also refer to conference papers as well as the specialist journals of various learned societies throughout the world although the emphasis is very much on English language papers. The list does not claim to be exhaustive.

Site Investigation: BS5930, BS1377, QJEG Working Party paper, technique specific CIRIA guides, Site Investigation in Construction series. In addition many very useful articles, papers and case histories appear in the Quarterly Journal of Engineering Geology, Geotechnique, Journal of Geotechnical Engineering and Ground Engineering.

Pile foundation design: BS8004, Reports prepared by CIRIA/DoE Piling Development Group, geological stratum specific CIRIA guides, Guidance notes of the ICE Specification for Piling and Embedded Walls, Books by Tomlinson and by Fleming et al. As well as the journals listed above, the Proceedings of the Institution of Civil Engineers and the Insitituion of Structural Engineers contain many useful case histories and papers.
Structural design: BS8110 and BS5400 may be specified. There are many standards dealing with concrete and steel which fall outside the scope of this paper.

6 DETAILED DESCRIPTION OF NATIONAL DESIGN METHODS

6.1 General Philosophy

The British Standard Code of Practice for Foundations, BS8004, states that "the ultimate bearing capacity can be estimated by customary soil mechanics procedures following tests on undisturbed samples of cohesive soil or from in situ tests in boreholes or test pits.". Elsewhere the same Code says "the ultimate bearing capacity may be assessed by applying a dynamic formula, or by using stress wave analysis, or on the basis of soil tests....... or from a loading test,.." Hence, for the majority of soils and rocks national design methods do not exist. The exception being particular geological units, such as Chalk and Mercia Mudstone (Keuper Marl), for which CIRIA, has produced detailed reports including advice on the design of load bearing piles. In place of national methods, common practice prevails with the opportunity to introduce new methods should a worthy proposal meet with general approval. However, books such as, "Piling Engineering" (Fleming et al 1992) and "Pile Design and Construction Practice" (Tomlinson 1995) have assumed *code like* status.

Pile designs are generally carried out using a lumped safety factor (see Section 6.7) which is applied to the estimated working load for a pile and the ultimate capacity of the pile assessed using soil and rock parameters. Eurocode EC7 should not, in itself, affect the general philosophy, although the partial factoring system is a fundamental change to safety factoring.

6.2 Definitions

Bearing Pile - A pile driven or formed in the ground for transmitting the weight of a structure to the soil by the resistance developed at the pile point or base and by friction along its surface.

Ultimate Pile Capacity - The maximum resistance offered by the pile when the strength of the resisting soil is fully mobilised.

In the UK this is usually determined as the point at which the vertical settlement is equivalent to 10% of the pile base diameter, or, the point at which the pile continues to penetrate the ground at a constant rate for no further increase in applied loading.

Allowable Pile Capacity - A capacity which takes into account the piles bearing capacity, the materials from which the pile is made, the required load factor, allowable settlement, pile spacing, negative skin friction, the overall bearing capacity of ground beneath the piles and any other relevant factor.

Preliminary Test Pile - A pile installed before the commencement of the main piling works for the purpose of establishing the suitability of the chosen type of pile and for confirming the design, dimensions and bearing capacity.

Pile failure - In the UK a pile can be deemed to have failed when either

- the measured ultimate capacity is less than that required, or
- the recorded settlement(s) is in excess of predetermined settlement limits for the structure.

6.3 Design on basis of static load tests

BS8004 asserts that a preliminary pile should be tested except where there is extensive local experience or high factors of safety; "the piles should be tested to determine the load settlement characteristics and ultimate bearing capacity.". For infrastructure projects preliminary proof tests have traditionally been carried out, these have checked the adequacy of an existing design without attempting to determine the ultimate capacity. For commercial developments the general approach in the UK is only to have preliminary test piles where there is a large number of piles such that the cost of a test can be offset against a reduction in factor of safety from 2.5 to 2.0. Even with commercial developments, the preliminary test is often only used to test the adequacy of an existing design.
Analysis of incomplete pile tests, in order to predict ultimate end bearing and shaft resistance, is carried out, but the settlement recorded is not always sufficient to permit a meaningful analysis.

Test piles of smaller diameter are frequently used; in such cases it is recommended that the diameter is not less than 50% of that of the contract piles.

6.4 Design on basis of ground tests

6.4.1 Bored piles: Capacity in cohesive soils

It is almost universal to calculate base resistance of piles in clay as the product of the undrained shear strength and a bearing capacity factor, *Nc,* where for piles *Nc* is 9.
qb=Nc*Su
For shaft resistance the undrained total stress approach is again generally employed.
$\tau s = \alpha * Su$
The value of α selected will usually depend on shear strength, size of test specimen and plasticity, although other factors will also have an influence on actual shaft resistance. Typically, α will be between 0.4 and 0.6 for overconsolidated marine clays.

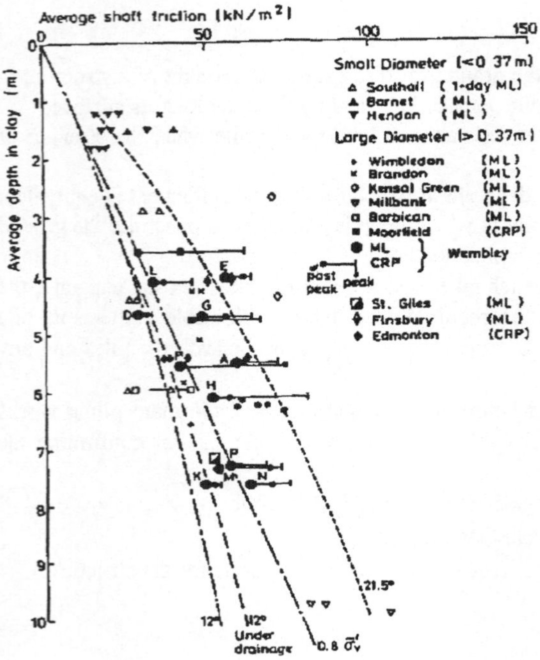

For Boulder Clay αSu is normally taken as a constant of approximately 70kN/m2 for undrained shear strengths between 80 and 200kN/m2.

Undrained shear strength is obtained from quick undrained triaxial tests, or indirectly from standard penetration tests. Experience indicates that a SPT profile will generally display less scatter than a triaxial Su verses depth profile.

The alternative to the undrained approach is an empirical effective stress method, based on that proposed for London Clay (Burland and Twine 1988), see Figure 2. In this method ultimate shaft resistance is related solely to depth into London Clay; such that using the formula: $\tau sf = \sigma'v*k*\tan\phi'r$, $\sigma'v$ ignores all material above the clay surface and can be considered as a minimum historic value, and $k*\tan\phi'r$ can be given a β value of 0.7 for lower bound results or 0.8 for a moderately conservative ultimate capacity. This method is simple and does not require any soils testing, but does assume that high horizontal stresses exist, thus is not suitable for sloping sites. When using this method for shaft resistance, the end bearing is still normally determined using undrained strengths.

6.4.2 Bored piles: Capacity in granular soils

A semi-empirical effective stress method is generally employed for calculating shaft resistance in granular soils: $\tau = p'v*k*\tan\delta'$
k is usually taken as being between 0.7 and 0.9
$\delta = \phi$ for cast in place piles.

A limiting value in often placed on τ based on a critical depth related to pile diameter or the relationship between peak ϕ' and $\phi'crit$ may be used to restrict τ at depth.

End bearing is determined using drained bearing capacity analysis, $qb = Nq*\sigma'v$. Nq is most widely taken from the graph produced by Berezantzev et al, with ϕ' reduced by 3° for bored piles. Again a limiting value is usually adopted, of between 10 and 15 MPa.
Occasionally the average insitu stress is considered, $Po = P'v*(2ko+1)/3$, for consistency the Nq derivation would be modified.

The effective angle of friction, $\phi°$, is selected by reference to SPT "N" values, grading and angularity.

6.4.3 Bored piles: Capacity in Rocks

Presumed end bearing values are sometimes invoked, particularly when end bearing on the surface of strong rocks. Rock socket design is generally based on overseas experience (eg. Williams Johnson and Donald); $\tau s = \alpha\beta Co$.
α an adhesion value related to the unconfined compressive strength, Co
β a reduction factor to account for a jointed rock mass.
Ultimate end bearing will range between Co and 4.5Co depending on the reliability placed an the recorded or estimated Co. Concrete stress will frequently control pile loading, such that it is not necessary to adopt a high end bearing factor.

6.4.4 Bored piles: Other materials

Chalk has been the subject of two CIRIA reports, PG6 (1979) and PR11 (1994). The design of piles in Mercia Mudstone (Keuper Marl) is often based on advice given in CIRIA Report 47. Table 5 summarises the recommendations given in these reports.

Table 5 Bored Piles in Chalk and Mercia Mudstone

Bored Piles in Chalk		
	CIRIA report PG6	CIRIA report PR11
Shaft resistance related to SPT N value	Yes	
Shaft resistance related to overburden pressure		Yes
Base resitance related to SPT N value	Yes	Yes
Bored Piles in Keuper Marl		
Empirical relationship between weathered grade and ultimate shaft resistance		
Base resistance as for granular material, phi usually indirectly derived from moisture content and plasticity index		

6.4.5 Driven piles: Capacity in cohesive soils

The design approach for driven piles in clay is the same as that for bored piles except that the value of α will typically lie between 0.6 and 0.8. The increase in α is considered to be due to the different construction process in driving displacement piles. For Boulder Clays where Su lies in the range 80 and 200 kN/m^2, αSu is normally taken to be 80 kN/m^2.

6.4.6 Driven piles: Capacity in granular soils

The same semi-empirical approach as for bored piles is generally used for calculation of shaft resistance but K is often taken as unity and δ as 0.7ϕ. A limit value of 100kN/m^2 would normally be applied. End bearing resistance is generally calculated in the same way as for bored piles using Berezantzev et al. However this approach for end resistance of displacement piles can be conservative because the construction process can densify the soil and enhances its bearing properties. An alternative wholly empirical approach proposed by Meyerhof is also often employed which is based upon direct conversion factors of SPT N values to unit shaft and end bearing values. An upper limit end bearing stress is normally applied in the range 10 - 15 MPa The lower value - 10kN/m^2, is nowadays considered conservative.

6.4.7 Driven piles: Capacity in rocks

Where strong relatively unweathered rock can be reached by the pile base then a 'structural' approach is often used i.e. pile capacity is based on pile properties rather than rock properties. As regards weak and weathered rocks very little guidance is available except in the case of chalks (see below). A rock socket philosophy may be pursued using bored pile parameters; however pile refusal will often practically limit socket length. The tendency is to select conservative bearing parameters and either check by load testing after which design modification may be adopted or to select reasonably conservative factors of safety.

6.4.8 Driven piles: Other Materials

The two CIRIA chalk reports mentioned previously suggest design parameters for driven piling.

However CIRIA Report 47 ducks the issue and recommends design be based on piling contractor experience.

6.5 Design on Basis of Driving Formulae

BS 8004 makes the following recommendations regarding dynamic pile formulae (driving formulae)

> ''..7.5.2 Calculation by dynamic pile formulae
>
> 7.5.2.1 *General* In non-cohesive soils an approximate value of the ultimate bearing capacity may be determined by a dynamic pile formula. All dynamic formulae are based on the following two assumptions.
>
> (a) The resistance to driving of a pile can be determined from the kinetic energy of the driving hammer and the movement of the pile under a blow.
>
> (b) The resistance to driving is equal to the ultimate bearing capacity for static loads.''

It goes on to explain that the various assumptions made in the calculations of these losses give rise to a variety of dynamic formulae as noted in Whitaker.

Cautionary notes are given about using driving formulae in deposits such as saturated silt, mud, marl, clay and chalk and where on redriving after a period of rest the resistance has decreased. It notes that where capacities are generated predominantly at the toe in gravels, sands and other non-cohesive soils of this type, then one of the more reliable formulae should give a calculated result within the range of 40% to 130% of the ultimate bearing capacity that would be determined by a test load. Having added further cautionary words about the reliability of such formula, it states ''The Hiley formula is one of the more reliable and is probably the most commonly used in Britain (Cornfield 1963; Whitaker 1976). ....''

A pile driving formula, therefore, is based on the premise that hammer energy input can be equated to work done by the pile on the soil, and in its simplest form can be represented as:-

$$W.h = R.s$$

where W = hammer weight
h = hammer fall height
R = resistance of ground
s = penetration of pile, or 'set'

The above formula is not in practical use as it is a gross over-simplification of the actual pile driving process, but it serves the purpose of demonstrating the principle. As stated in BS 8004 the Hiley Formula is the formula in most common use in U.K. It is normally applied in its so-called simplified form, which is:-

$$W.h.\eta = R\,(s+c/2)$$

where η = a factor taking account of efficiency and cushioning system
c = temporary compression of soil, pile, and driving system

It can readily be observed that the particular inaccuracies of this model are:-

a) Hammer transfer energy

Pile driving hammers normally consist of a driving assembly involving the hammer weight itself, a cushioning system, and a driving helmet. The potential energy (represented by W h) is greater than that energy which is transferred by the hammer to do work on the pile. The Hiley formula recognizes this in the reduction factor η; however the problem is in the accurate evaluation of η.

This problem has, in recent years, been overcome by pile driving contractors utilising dynamic pile testing methods for the direct measurement of actual hammer transfer energy at the pile top. The value of transfer energy is then used as input quantity to the pile driving formula, and, as long as the contractor is consistent in his use of hammers and cushioning systems the same transfer energy value can be repeated. Regular checks should then be made to ensure the measured value of transfer energy remains consistent.

b) Temporary compressions or quakes of soil, pile and drive system

The evaluation of the drive system temporary compression is eliminated by the measurement of transfer energy directly at the pile top, which is of course, at a location following the transfer of the blow through the drive system. The combined pile and soil temporary compressions can then be readily measured on the pile during the driving operations on site, for input into the driving formula.

The driving formula is normally applied in the form:-

$$s = \frac{Wh\eta}{R} - \frac{c}{2}$$

so that the input of transfer energy, resistance force (R), and measured temporary compression gives the set to which the pile should be driven for the input resistance value. The input resistance value (R) would normally be the pile working load plus any other applied loads e.g. downdrag load, multiplied by a factor of safety. This factor of safety would normally be 2 or greater.

It would be incorrect to consider that pile design by driving formula alone is carried out in U.K. However, driving formulae are quite routinely used and applied in driven pile installation as a guide as to when to stop driving a pile into the soil, and as stated above, the Hiley formula, normally in a modified form, is in most common usage.

Having identified the stratum to which the pile should be driven in order to provide adequate bearing from a site investigation report, it may be the case that preliminary pile testing is considered desirable in order to prove the pile bearing characteristics and to 'calibrate' the driving formula and 'set' to which the pile has been driven. This would normally be the case where there was little or no local experience of the pile type in the prevailing soil conditions. Where there was adequate local experience, pile installation on the basis of the driving formula alone would be acceptable, as long as the piles were installed consistently to the recognized bearing stratum.

Most U.K. pile driving contractors will tend to control (or 'design') their pile installation using a pile driving formula of which they have experience, whatever the soil type. Nevertheless, it is generally recognized that some soil types are more suited to this approach than others. As a general rule cohesive soil types (clays etc..) are not suited, similarly chalks. However, local experience does sometimes dictate otherwise, and it is reasonable to state that where piling equipment, pile type, and the local soil type has been calibrated by some form of pile testing, normally static testing, then even so-called unsuitable soil types may employ driving formula techniques in the design and construction control process.

The measurement of 'set' at the end of pile driving is normally routinely carried out whatever the soil type whether or not a pile driving formula is employed. Recognising the great disturbance to the soil caused by the process of pile installation remeasurement of the pile 'set' with time after installation is often undertaken.. This is normally only practical for preformed pile types and not practical for driven cast-in-place pile types or marine piles requiring manouvering of floating plant.

The qualitative check is:-

(i) Set the same as at pile installation - little change of bearing capacity
(ii) Set decreased since pile installation - increase of bearing capacity
(iii)Set increased since pile installation - decrease of bearing capacity

It is most important that the hammer transfer energy at time of restrike(s) is the same, in practice, as that which was used for pile installation, otherwise the results can be misleading. Practical examples of some soil types in UK in which such phenomenon could be observed in practice would be:

condition (i) : piles in gravel soils,
condition (ii) : piles in clay soils,
condition (iii): piles in some weak rocks e.g. Coal Measure mudrocks.

In summary, pile driving formulae are used in the design process of driven piles but only in conjunction with local experience, site investigation information, in many instances in conjunction with dynamic pile testing, and in some instances in conjunction with load testing. Pile driving formulae are normally used where the bearing stratum consists of a cohesionless soil type, weathered or hard rock type, or glacial till. In addition it is routine practice to measure the 'set' at end of drive of preformed piles whatever the soil type, and then to restrike a sample of the piles, where practicable, to measure change of 'set' with time.

6.6 Design on Basis of Wave Equation Analysis

BS8004 states the following with regard to wave equation analysis:

> 7.5.2.2 *Wave equation analysis.* The ultimate bearing capacity of a pile may be predicted from analysis of the stress wave resulting from the hammer blow. ..... estimates of the dynamic resistance can be made which can be correlated with the static bearing capacity....

The BS8004 definition applies to dynamic pile testing, involving the measurement of force and acceleration in a pile subjected to a hammer blow and the subsequent analysis of the wave form so measured, usually by a wave matching program such as CAPWAP. It is not an accurate representation of the current use of wave equation analysis.

Wave equation analysis is commonly carried out via a computer program that simulates a pile under the action of an impact pile driving hammer. A number of programs are available within the UK. GRLWEAP is probably the program in most common use. The program will, given a set of dynamic soil resistance parameters and a hammer and driving system, compute the following pile parameters:

- Blow count of a pile for one or more *assumed* values of ultimate resistance
- The *axial* stresses in a pile corresponding to the calculated blow count
- The energy transferred to the pile

Thus the program is commonly used to predict:

- The bearing capacity at time of driving (or at restrike) given a particular blow count (or *set*)
- The set (or blow count) required to achieve a particular bearing capacity.
- The stresses in the pile during pile driving

Wave equation analysis is in widespread use in offshore piling. In this environment, it is used primarily to predict driving stresses in large diameter steel tubular piles and also for hammer selection. In on-shore piling, wave equation analysis is less widespread. Notwithstanding the better representation of the pile, soil and pile driving system that is offered by wave equation analysis, it has not yet superseded the Modified Hiley Formula in determining pile sets. Its use is, however, gaining ground. In addition to the ability to predict blow count, pile stresses and the ability to consider a number of piling hammers, wave equation analysis allows the prediction of pile driveability.

With good knowledge of the soil profile leading to a model of resistance with depth, wave equation analysis is used by some contractors to simulate the action of pile driving. This analytical option is used to assess:

- the ability to drive a pile to a given depth
- the depths at which pile stresses (compressive or tensile) are most critical
- the time to drive the pile (and therefore predict pile production)

Wave equation analysis is not restricted by soil type and can be applied in a mixed soil profile. Succesful use of wave equation analysis, of course, requires good soils information, reliable static analysis, accurate information on the driving system, dynamic soil parameters with estimated rate changes preferably based on existing results. Like other design methods, its reliability is improved by local experience.

6.7 Global Factors of Safety - Axial Compression

The current method used in the UK is to use lumped or global Factors of Safety (FoS), applied to either the applied working load or to the calculated ultimate capacities. The global FoS is used to cater for all the uncertainties with the design.

The value of the FoS depends on a number of criteria, namely;

- Sensitivity of the structure
- Nature/variability of loading
- Adequacy of site/ground information
- Variability of ground conditions
- Design method
- Local experience

For piles founded on superficial and solid deposits, a global FoS of between 2.0 to 3.0 is applied to the total ultimate capacity. The total ultimate capacity includes the total contribution from both skin friction and end bearing effects.

The nature of the proposed pile load testing is often used to dictate the required minimum FoS.

Factor of Safety	Testing Proposals
2.0	Preliminary test piles + Working test piles
2.5	Working test piles
3.0	No load testing Previous local experience is however essential

The above concept is related only to 'standard' loading and ground conditions. The following information gives brief guidance on dealing with some other common conditions.

(a) Wind Loading
BS 8004 Clause 2.3.2.4.3, states " Where the foundation loading beneath a structure due to wind is a relatively small proportion of the total loading, it may be permissible to ignore the wind loading in the assessment of the allowable bearing pressure, provided the overall factor of safety against shear failure is adequate. For example, where individual foundation loads due to wind are less than 25% of the loadings due to dead and live loads, the wind loads may be ignored in this assessment. Where the ratio exceeds 25%, foundations may be so proportioned that the pressure due to combined dead, live and wind loads does not exceed the allowable bearing pressures by more that 25%".

(b) Negative Skin Friction
If negative skin friction (or downdrag) forces are considered to be applied to a pile shaft, unless a proprietary slip coating/ sleeving is utilised on the affected pile shaft area, the additional loading must be taken into account in assessing the pile allowable working load.

As an example, where the pile settlement under working conditions is sufficient to reduce the effects of negative skin friction the following approach is used.

$Q_a = (Q_u - NSF)/FoS$

Different approaches are used when the negative skin friction forces are large in relation to the working load, or where the settlement of the pile is not sufficient to reduce the negative skin friction effects.

(c) Material Stresses
The induced working compressive stress on a pile cross sectional area should not exceed the allowable stress value of the pile material.

Pile Material	Allowable Material Stress
Cast in Place Concrete	25% of 28 day concrete cube strength
Precast Concrete	25% of 28 day concrete cube strength
Steel Sections	30 - 50 % of the characteristic yield strength

(d) Nature of Loading
When the induced loadings are caused by significant impacts, cyclic effects, vibrations or where the properties of the ground are expected to deteriorate significantly with time, the previously quoted factors of safety should be increased.

(e) Tension Loads
From published test results, it is evident that the skin friction developed during uplift forces is generally less than that developed during compressive loading. It is normal practice, therefore, to increase the global FoS to a minimum of 3.0.

(f) Weak Rocks
For piles installed into weak rock a different approach to the use of a global FoS is necessary as serviceability constraints usually govern the pile design. It is normal practice to separate the shaft friction and end bearing components of the ultimate capacity and to apply different FoS to each component.
CIRIA Report R509 - Piled Foundations in Weak Rock, recommends the following FoS.

Condition	Allowable Working Load	Factors of Safety
$Q_b >> Q_s$	$Q_a = Q_b / F_1 + Q_s / F_2$	$F_1 = 2$ to 3
		$F_2 = 1.5$ to 2
$Q_b <= Q_s$	$Q_a = Q_s / F_3$	$F_3 = 1.2$ to 1.5

For piling in Chalk material, the guidelines given in CIRIA Report PG6 are utilised. In order to control settlements a nominal FoS of 1.5 is applied to the skin friction component with a higher FoS applied to the end bearing, depending on the method of piling.

(g) Group Effects
All the previous methods relate to the design of isolated axially load piles. For groups of piles the calculated working capacity should be adjusted to allow for group action, pile spacing and any other factors which influence the overall efficiency of the pile group. Traditional empirical group analysis formula are no longer used in the UK.

6.8 Serviceability

The analysis of a piled foundation needs to be checked to ensure that the serviceability performance is acceptable. BS 8004 Clause 2.1.2, define that;

Foundation design should ensure that foundation movements are within limits that can be tolerated by the proposed structure without impairing its function. Since the capacity of the structures to accommodate movement varies, the design of the structure should be interrelated to its foundation'.

Serviceability is a highly subjective issue, with dependency on both the function of the proposed structure and the possible reaction of the end users to the effects of settlement. As structures are variable in size and sensitivity there is difficult in setting general guidelines as to what is an allowable movement. Of more importance to a structure is the possibility of inducing excessive differential settlements which will result in unacceptable structural deformations. This form of settlement is more likely to cause perceived 'structural damage'. The publication 'Soil - Structure Interaction', gives guidance on the classification of visible damage to walls and relates this degree of damage to typical groups of structures.

In general structural design engineers do not have a full understanding/appreciation of what constitutes an allowable level of movement for the structure being designed. Although there is practical guidance available in 'Soil-Structure Interaction', most design engineers tends to be over conservative and specify allowable settlements, at working load conditions, in the range 5 to 15mm.

In the design of piles the general situation is that for piles which derive their greater resistance from skin friction, and are provided with a minimum factor of safety of 2.5 applied to the working load, then the deflections will generally be restricted to less than 10mm (Tomlinson). In this case it is normally not necessary to carry out any settlement predictions.

For piles of large diameter or piles which derive their greater resistance from end bearing, it is normal practice to apply higher FoS to the calculated capacities, or to carry out settlement calculations to ensure that the predicted settlements are acceptable.

7 APPLICATION OF DESIGN METHODS

It will be observed from Section 6.4 and 6.5 that the calculation methods used require simple substitution of arithmetic values for the various coefficients and variables. By far the most difficult task is to select a realistic design line for the appropriate parameter required by the calculation formula.

Space does not permit a fully documented design example to be presented. In general terms, UK design line selection follows the philosophy of a 'moderately conservative' evaluation of soil parameters from the set of derived data. The term 'moderately conservative' is a conservative best estimate and may be considered to be similar but probably a little higher than the 'characteristic value' used in Eurocode 7.

The designer has to consider whether the test results give a measure of the parameter which will actually govern the situation being considered. There could, for example, be a systematic error caused by a scale effect. Once a correction, if required, has been applied then a 'moderately conservative' design line can be drawn. It is essential to consider the resultant in the light of experience. A review will indicate how the values compare with the typical range for the parameter being considered. If there is no comparable experience, a more cautious view would be adopted. More cautious or optimistic views may be applied as a result of this review and the chosen design line may also need to be compared with design values derived from a different set of measurements. The process is likely to require the assessment of a body of data which is not entirely consistent but this is not unusual in geotechnical design.

In the UK, it is not considered that any formal statistical procedures are applicable and indeed most designers would consider that statistical analysis would lead to errors. Experience in the selection of appropriate design parameters comes from the results of actual prototype performance together with peer group discussion and review of any published data. Training takes the approach of passing selection philosophy onto younger and less experienced members of a design group by mentors.

8 QUALITY CONTROL AND MONITORING

Traditionally in the UK, an Engineer requested piling to be undertaken in accordance with a specification and scope of works provided by him. Tests on materials would be undertaken in accordance with relevant British Standards. Inspection of the workmanship would take place throughout the piling process and a few piles, usually 1-2% of the total number of piles would be subjected to proof load testing to 1.5 times the working load. Where ground conditions or piling practice was unfamiliar, preliminary pile testing of one or two piles to 2.5 times working load would be included.

Against a background of conservative design and conscientious inspection this system of quality control worked well and there have been very few piling failures in the UK.

Over recent years competitive pressures on both consulting engineers and specialist piling sub-contractors have resulted in a changed market place where regular inspection of workmanship is not normal, apart from some public works. Piling sub-contractors have become more expert and undertake most of their work in accordance with their own designs, except for most road projects. Such designs will be submitted to the Client's Engineer for approval. In addition, the rise of the use of CFA piling has diminished the value of traditional inspection.

The current approach in the UK, where most substantial firms belong to the Federation of Piling Specialists (FPS), is to operate a self regulated quality control system which will deliver piles within a Quality Assurance (QA) scheme. The FPS requires members to operate a Quality Assurance scheme and this has undoubtedly spread the use of such schemes throughout the industry. BS 5750 Quality Systems which covers QA for design/development, production, installation, servicing, final inspection and testing, first published in 1987, is the bible of all schemes. It is now referred to under its ISO numbers BS EN ISO 9001, 9002 and 9003, the latest version having been published in 1994. Company schemes operated in accordance with these standards will include several hold and inspection points within the piling process and these will involve representatives from other organisations, where they are available, to witness specific items in the piling process.

Within the normal quality control systems operated within the UK, pile testing continues to be the most frequently used test of the efficacy of a design. Only in well known ground with a substantial history of successful piling will the use of high factors of safety allow tests be avoided. Integrity testing as an assurance of structural integrity has become a normal part of a QA scheme.

9 PARTICULAR NATIONAL EXPERIENCES

It is not appropriate to try to examine in any detail the steps in the evolution of piling and pile design in the UK. The pressure for heavier buildings in urban areas especially London from the 1950s onwards certainly advanced the cause of large diameter bored piles. Initially American machinery was imported but the ground conditions to be faced were very different from those normally associated with skyscrapers. The work of BRE especially from the 1950s to the 1970s certainly helped to validate the design methods and allowed the use of large diameter underream piles to become a common and trusted technique.

In the last 25 years piling advances have been as commonly brought about by specialist contractors as academics or consulting engineers but generally in an evolutionary way. Burland and Twine advocated the use of effective strength for shaft adhesion in 1988 but the method has not been widely adopted. The performance of large driven piles in normally and overconsolidated clays was the subject of some specific research work by an oil company. Jardine and co-researchers have used their offshore work as a basis of a large onshore model study of pile behaviour which is now entering the public domain.

New techniques have been suggested such as the wedge pile (Burland) but adopted in a slightly different form by a specialist contractor. Some techniques have been adopted by co-operation with continental firms such as the Atlas and Omega piles.

REFERENCES

Babtie Group Ltd ‘Piled Foundations in Weak Rock’ Report R509, CIRIA, London, 1997

Berezantsev,V.G. et al. ‘Load Bearing Capacity and Deformation of Piled Foundations’ Vol 2 P11-12, Proc. 5th Int. Conf., ISSMFE, Paris, 1961

Blyth, F.G.H. and de Freitas, M.H. ‘A Geology for Engineers’ Edward Arnold, London, 1984

British Standards Institution ‘Code of Practice for Foundations’ BS8004, BSI, London, 1986

British Standards Institution ‘Code of Practice for Site Investigations’ BS5930, BSI, London, 1981

British Standards Institution ‘Methods of tests for soil for civil engineering purposes, Parts 1-9’ BS1377, BSI, London, 1990,1994 & 1995

British Standards Institution ‘Quality Systems’ BS5750, BSI, London, 1990 and 1994

British Standards Institution ‘Steel, Concrete and Composite Bridges’ BS5400, BSI, London, 1978-1990

British Standards Institution ‘Structural Use of Concrete’ BS8110, BSI, London, 1985

Burland, J.B. and Twine, D. ‘The Shaft Friction of Bored Piles in terms of Effective Strength’ Deep Foundations on Bored and Asugered Piles, Van Impe (Ed.), Bakema, Rotterdam, 1988

CAPWAP, Case Pile Wave Analysis Program, CAPWAP Users Manual, Goble Rausche Likins and Associates Inc., 1996

Chandler, R.J. and Davis, A.G. ‘Further Work on the Engineering Properties of Keuper Marl’ Report 47, CIRIA, London, 1973

Clayton, C.R.I., 'The Standard Penetration Test (SPT): Methods and Use' Report R143 CIRIA, London, 1995
Cornfield, G.M. 'Hiley Formula Simplified without Graphs' Vol 196, P781, Engineering, London, 1963
Fleming, W.G.K. and Sliwinski, Z.J. 'The Use and Influence of Bentonite in Bored Pile Construction' Peport PG3, DoE/CIRIA Piling Development Group, 1977 reprinted 1991
Fleming, W.G.K., Weltman, A.J., Randolph, M.F. and Elson, W.K. 'Piling Engineering' (Second Edition), Blackie, Glasgow & London, 1992
Geological Society Engineering Group Working Party Report 'The Description and Classification of Weathered Rocks for Engineering Purposes' Q. J. Eng. Geol., Vol 28, Part 3, P207-242, 1995
GRLWEAP, Wave Equation Analysis of Pile Driving, General Users Manual for GRLWEAP, Goble Rausche Likins and Associates Inc., 1996
Head, J.M. and Jardine, F.M. 'Ground-bourne Vibrations Arising from Piling' Report TN142, CIRIA, 1992
Healy, P.R. and Weltman, A.J. 'Survey of Problems Associated with the Installation of Displacement Piles' Report PG8, DoE/CIRIA Piling Development Group, 1980
Hobbs, N.B. and Healy, P.R. 'Piling in Chalk' Report PG6, DoE/CIRIA Piling Development Group, 1979 reprinted 1990
Institution of Civil Engineers 'Specification for Piling and Embedded Retaining Walls' Thomas Telford Publishing, London, 1996
Institution of Structural Engineers, ICE, Int. Assoc. for Bridge & Sructural Engineering 'Soil-Structure Interaction' Institution of Structural Engineers, London, 1989
IRTP. Report of the ISSMFE Technical Committee on Penetration Testing of Soils - TC16 with Reference Test Procedure: CPT-SPT-DP-WST. 1989
Lord, J.A., Twine, D. and Yeow, H. 'Foundations in Chalk Report PR11, CIRIA, 1994
Mair, R.J. and Wood, D.M. 'Pressuremeter Testing: Methods and Interpretation' Report B3, CIRIA, London, 1987
Meigh, A.C., 'Cone Penetration Testing: Methods and Interpretation' Report B2, CIRIA, London, 1987
Meyerhoff, G.G. 'Bearing Capacity and Settlement of Pile Foundations' Vol GT3, P197-228, Proc. Amer. Soc. Civ. Engineers, ASCE, 1976
Morley, J. 'The Corrosion and Protection of Steel Piling' Britsh Steel Corporation Report T/CS/1115/1/79C, November 1979
Morley, J. and Bruce, D.W. 'Corrosion and Surface Protection. Survey of Steel Piling Performance' British Steel Corporation, 1983
Powell, J. J. M., Clarke, B. G., Shields, C. H. 'CPT95 - National Report on UK Practice' Pressuremeters. Proceedings of the Third International Symposium on Pressuremeters. BGS, Thomas Telford, 1990
Romanoff, M. 'Performance of Steel Pilings in Soil' 25th Conf. National Assn. of Corrosion Engineers, Houston Texas, March 1969
Site Investigation Steering Group. 'Site Investigation in Construction 1 Without Site Investigation Ground is a Hazard' Thomas Telford, 1993
Site Investigation Steering Group. 'Site Investigation in Construction 2 Planning, Procurement and Quality Management' Thomas Telford, 1993
Site Investigation Steering Group. 'Site Investigation in Construction 3 Specification for Ground Investigation' Thomas Telford, 1993
Site Investigation Steering Group. 'Site Investigation in Construction 4 Guidelines for the Safe Investigation by Drilling of landfills and Contaminated Land' Thomas Telford, 1993
Thorburn, S. and Thorburn, J.Q. 'Review of Problems Associated with the Construction of Cast-in-place Concrete Piles' Report PG2, DoE/CIRIA Piling Development Group, 1985

Tomlinson, M..J. 'Pile Design and Construction Practice' 4th Edition, E.F.N. Spon, London, 1995
Weltman, A.J. 'Integrity Testing of Piles: A Review' Report PG4, DoE/CIRIA Piling Development Group, 1977
Weltman, A.J. 'Pile Load Testing Procedures' Report PG7, DoE/CIRIA Piling Development Group, 1980
Weltman, A.J. and Head, J.M. 'Site Investigation Manual' Report SP25, CIRIA, 1983
Weltman, A.J. and Healy, P.R. 'Piling in 'Boulder Clay' and other Glacial Tills' Report PG5, DoE/CIRIA Piling Development Group, 1978 reprinted 1990
Weltman, A.J.(Ed.) 'Noise and Vibration from Piling Operations: Proceedings of the DoE and CIRIA Informal Meeting on Piling Noise and Vibration, 24 April 1979' Report PG9, DoE/CIRIA Piling Development Group, 1980
Whitaker, T. 'The Design of Pile Foundations' Pergamon Press (Oxford), 2nd Edition, 1974
Williams, A.F., Johnston, I.W. and Donald, I.B. 'The Design of Socketed Piles in Weak Rock' Vol 1 P327-347, Proc. Int. Conf. Structural Foundations on Rock, Sydney, 1980
Wynne, C.P., 'A Review of Bearing Pile Types' Report PG1, DoE/CIRIA Piling Development Group, Second Edition, 1988.
Jardine, R.J. and chow fc. 'New Design Methods for offshore Piles'. MTD publication 96/103.
British Standard Institution. 'Structural use of Timber': Part 5; Preservative Treatments for Constructional Timber. BS 5268, BSI, London 1989.
British Standard Institution. Methods for specifying concrete, including ready mixed concrete, BS 5328, 4 Parts, BSI, London 1990-1991.
Building Research Establishment. 'Sulphate and Acid Resistance of Concrete in the Ground'. Digest 363, 1991.

Design of Axially Loaded Piles – European Practice, De Cock & Legrand (eds)
© 1997 Balkema, Rotterdam, ISBN 90 5410 873 8

Author index